NEW DIRECTIONS
IN LEMUR STUDIES

NEW DIRECTIONS IN LEMUR STUDIES

Edited by

Berthe Rakotosamimanana
Hanta Rasamimanana
Antananarivo University
Antananarivo, Madagascar

Jörg U. Ganzhorn
University of Hamburg
Hamburg, Germany

and

Steven M. Goodman
Field Museum of Natural History
Chicago, Illinois

Kluwer Academic / Plenum Publishers
New York, Boston, Dordrecht, London, Moscow

Library of Congress Cataloging-in-Publication Data

New directions in lemur studies / edited by Berthe Rakotosamimanana
... [et al.].
     p.   cm.
  Includes bibliographical references and index.
  ISBN 0-306-46187-0
  1. Lemurs Congresses.   I. Rakotosamimanana, Berthe.
  II. International Primatological Society. Congress (17th : 1998 :
Antananarivo, Madagascar)
  QL737.P95N48   1999
  599.8'3--dc21   99-37306
                                                                   CIP

Proceedings of the XVIIth Congress of the International Primatological Society, held August 9–14, 1998, in Antananarivo, Madagascar

ISBN: 0-306-46187-0

© 1999 Kluwer Academic / Plenum Publishers
233 Spring Street, New York, N.Y. 10013

10 9 8 7 6 5 4 3 2 1

A C.I.P. record for this book is available from the Library of Congress

All rights reserved

No part of this book may be reproduced, stored in a retrieval system, or transmitted in any form or by any means, electronic, mechanical, photocopying, microfilming, recording, or otherwise, without written permission from the Publisher

Printed in the United States of America

PREFACE

Over the course of the past decade, there has been an enormous augmentation in the amount of information available on the lemurs of Madagascar. These advances are closely coupled with an increase in the number of national and international researchers working on these animals. As a result, Madagascar has emerged as one of the principal sites of primatological studies in the world. Furthermore, the conservation community has a massive interest in the preservation of the natural habitats of the island, and lemurs serve as one of the symbols of this cause.

Between 10 and 14 August 1998, the XVIIth International Primatology Society (IPS) Congress was held in Antananarivo, Madagascar. For a country that about a decade ago was largely closed to foreign visitors, this Congress constituted a massive event for the Malagasy scientific community and was assisted by about 550 primatologists from 35 different countries. Naturally, given the venue and context of the Congress, many of the presentations dealt with lemurs and covered a very wide breadth of subjects.

As the field of primatology has grown, particularly the study of lemurs, many new ideas and directions have emerged. Congresses, where colleagues and students exchange ideas and present new working hypotheses, are often the events where interactions give rise to major advances in practical applications of techniques, as well as theoretical formation of new avenues of research. What we have tried to assemble in this book is a series of chapters based on presentations made at the IPS Congress in Antananarivo which comprise many of these new directions in lemur studies, covering the areas of phylogeny, paleontology, physiology, behavioral ecology, education, and conservation of lemurs. Perhaps some of these new directions will prove to be incorrect or flawed, while others will certainly lead to major advances in our understanding of lemurs, primates in general, and the natural world.

> Berthe Rakotosamimanana
> Hanta Rasamimanana
> Jörg U. Ganzhorn
> Steven M. Goodman

Antananarivo, 13 December 1998

ACKNOWLEDGMENTS

The production of a collection of papers such as this in a timely fashion is not a simple matter. We apologize to the contributors, reviewers of the various chapters, and other colleagues that we jostled, pushed, or harassed in various forms to complete this book on schedule. Adam Cohen, Joanna Lawrence, and Robert Wheeler of Kluwer Academic/Plenum Publishers have been extremely helpful and efficient in seeing this book to fruition within a short period of time.

A considerable amount of time and effort was devoted to this volume by numerous reviewers, several of whom read and commented on more than one chapter on very short notice. We are grateful to Raimund Apfelbach, Simon Bearder, David Burney, Ian Colquhoun, Brigitte Demes, Todd Disotell, Joanna Durbin, Joanna Fietz, Dieter Glaser, Claire Hemingway, Jodi Irwin, Alison Jolly, William Jungers, Peter Kappeler, Christoph Knogge, Andreas Koenig, John Oates, Deborah Overdorff, Alexandra Müller, Thomas Mutschler, Leanne Nash, Richard Prum, Joel Ratsirarson, Jean-Pierre Sorg, Robert Sussman, Chia Tan, Ian Tattersall, Urs Thalmann, Ruth Warren, Gwen Wehbe, Larry Wolf, Patricia Wright, and Anne Yoder. Lucienne Wilmé was responsible for translating, editing, and correcting the vast majority of résumés. We are grateful for her aid with this task.

The IPS held in Antananarivo could not have been possible without the gracious financial help of numerous organizations. We would like to thank the following organizations, companies, and foundations: Association Nationale pour la Gestion des Aires Protégées, American Society of Primatology, BIOCULTURE, Brasserie STAR, Caisse d'Epargne de Madagascar, Conservation International, Cortez Travel Agency, Duke University, Ecole Normale Supérieure, Institute for the Conservation of Tropical Environments, Jersey Wildlife Preservation Trust, Jiro sy Rano Malagasy, Maison de Tourisme, Margot Marsh Biodiversity Foundation, MATERA Ocean Indien, Ministère de l'Enseignement Supérieur, Organisation Nationale pour l'Environment, PACT Madagascar, Parc Botanique et Zoologique de Tsimbazaza, QIT Madagascar Minerals, Ltd., Société Moritani, The John D. and Catherine T. MacArthur Foundation, United Kingdom Embassy, Université d'Antananarivo, USAID Madagascar, Wenner Gren Foundation, and World Wide Fund for Nature. The aid of these institutions and all of the students and other volunteers who helped before and during the Congress is gratefully acknowledged.

CONTENTS

1. Ancient DNA in Subfossil Lemurs: Methodological Challenges and Their Solutions 1
   Anne D. Yoder, Berthe Rakotosamimanana, and Thomas J. Parsons

2. Past and Present Distributions of Lemurs in Madagascar 19
   Laurie R. Godfrey, William L. Jungers, Elwyn L. Simons, Prithijit S. Chatrath, and Berthe Rakotosamimanana

3. Skeletal Morphology and the Phylogeny of the Lemuridae: A Cladistic Analysis 55
   Gisèle Francine Noro Randria

4. Support Utilization by Two Sympatric Lemur Species: *Propithecus verreauxi verreauxi* and *Eulemur fulvus rufus* 69
   Léonard Razafimanantsoa

5. Field Metabolic Rate and the Cost of Ranging of the Red-tailed Sportive Lemur (*Lepilemur ruficaudatus*) 83
   Sonja Drack, Sylvia Ortmann, Nathalie Bührmann, Jutta Schmid, Ruth D. Warren, Gerhard Heldmaier, and Jörg U. Ganzhorn

6. Metabolic Strategy and Social Behavior in Lemuridae 93
   Michael E. Pereira, Russ A. Strohecker, Sonia A. Cavigelli, Claude L. Hughes, and David D. Pearson

7. Cathemeral Activity of Red-fronted Brown Lemurs (*Eulemur fulvus rufus*) in the Kirindy Forest/CFPF 119
   Giuseppe Donati, Antonella Lunardini, and Peter M. Kappeler

8. Social Organization of the Fat-tailed Dwarf Lemur (*Cheirogaleus medius*) in Northwestern Madagascar 139
   Alexandra E. Müller

9. Demography and Floating Males in a Population of *Cheirogaleus medius* 159
   Joanna Fietz

10. Influence of Social Organization Patterns on Food Intake of
    Lemur catta in the Berenty Reserve 173
    Hantanirina Rasamimanana

11. The Importance of the Black Lemur (*Eulemur macaco*) for Seed
    Dispersal in Lokobe Forest, Nosy Be 189
    Christopher R. Birkinshaw

12. Taste Discrimination in Lemurs and Other Primates, and the
    Relationships to Distribution of Plant Allelochemicals in
    Different Habitats of Madagascar 201
    Bruno Simmen, Annette Hladik, Pierrette L. Ramasiarisoa,
    Sandra Iaconelli, and Claude M. Hladik

13. Folivory in a Small-bodied Lemur: The Nutrition of the Alaotran
    Gentle Lemur (*Hapalemur griseus alaotrensis*) 221
    Thomas Mutschler

14. Conservation of the Alaotran Gentle Lemur: A Multidisciplinary
    Approach 241
    Anna T. C. Feistner

15. Teaching Primatology at the Université de Mahajanga (NW Madagascar):
    Experiences, Results, and Evaluation of a Pilot Project 249
    Urs Thalmann and Alphonse Zaramody

16. Lemurs as Flagships for Conservation in Madagascar 269
    Joanna C. Durbin

Index 285

ANCIENT DNA IN SUBFOSSIL LEMURS

Methodological Challenges and Their Solutions

Anne D. Yoder,[1,2] Berthe Rakotosamimanana,[3] and Thomas J. Parsons[4]

[1] Department of Cell and Molecular Biology
  Northwestern University Medical School
  303 East Chicago Ave., Ward Building
  Chicago, Illinois, 60611-3008, USA
  ayoder@nwu.edu (corresponding author)
[2] Department of Zoology, Field Museum of Natural History
  Roosevelt Road at Lake Shore Drive
  Chicago, Illinois, 60605, USA
[3] Département de Paléontologie et d'Anthropologie Biologique
  BP 906, Faculté des Sciences, Université d'Antananarivo
  Antananarivo (101), Madagascar
  brakoto@syfed.refer.mg
[4] Armed Forces DNA Identification Laboratory
  The Armed Forces Institute of Pathology
  Rockville, Maryland, USA

ABSTRACT

We present preliminary results from an ongoing study of ancient DNA in Madagascar's subfossil lemurs. These animals, though extinct, are the evolutionary contemporaries of the living lemurs. Any phylogenetic study that focuses only on the extant Malagasy primates is therefore incomplete. Here, we present a cytochrome *b* phylogeny that includes sequences for *Palaeopropithecus* and *Megaladapis*. We also discuss the various methodological challenges that we have faced in acquiring these sequences along with our solutions for overcoming them. In addition to serving as a report of primary data, we hope that this contribution will serve as a guide for other projects facing similar challenges. Our analyses suggest that *Palaeopropithecus* is sister to the living indrids, as predicted by morphological studies. Contrary to morphological data, however, *Megaladapis* appears to belong to an independent lemuriform lineage rather than form a clade with *Lepilemur*. Our results are subject to further testing as longer sequences and additional subfossil taxa are added to the study.

RÉSUMÉ

Nous présentons des résultats préliminaires d'une étude en cours sur l'ADN ancien des lémuriens sub-fossiles. Ces animaux, bien qu'éteints, sont les contemporains dans l'évolution des lémuriens actuels et toute étude phylogénétique qui se concentrerait sur les seuls primates malgaches actuels serait ainsi incomplète. Nous présentons ici un cytochrome b phylogénie? qui comprend des séquences pour *Palaeopropithecus* et *Megaladapis*. Nous présentons également les divers défis que nous avons eu à surmonter dans la méthodologie d'acquisition de ces séquences et les solutions que nous y avons apportées. Si des données de base sont présentées, nous espérons que cette contribution servira de guide à d'autres projets qui auront à surmonter des défis similaires. Nos analyses suggèrent que *Palaeopropithecus* et l'actuel Indri sont issus d'un même parent comme le suggèrent d'ailleurs les études morphologiques. Par contre et contrairement aux données morphologiques, *Megaladapis* appartient à une lignée lémuriforme indépendante et n'appartiendrait pas à un clade avec *Lepilemur*. Nos résultats sont soumis à des tests complémentaires au fur et à mesure que des séquences plus longues et que de nouveaux taxons subfossiles sont disponibles pour l'étude.

INTRODUCTION

Within the past 2000 years, approximately 25 species of large-bodied vertebrates went extinct on the island of Madagascar (Burney & MacPhee, 1988); the vast majority of these were primates. As many as 16 species of primates from at least seven genera are known from subfossil remains, many of which have been discovered only recently (Godfrey et al., 1990; Jungers et al., 1991; Simons et al., 1992; Simons et al., 1995b). Although extinct, the fact that at least some species existed as recently as 500 years b.p. (Godfrey et al., 1997) indicates that in an evolutionary sense, subfossil lemurs should be considered contemporaries of the living lemurs. When both the extinct and extant forms are considered, lemuriforms span more than four orders of magnitude in body size, exhibit every possible activity cycle, display at least three different types of social system, consume a large range of food types, and show a remarkable array of locomotor patterns (Richard & Dewar, 1991; Godfrey et al., 1997; Simons, 1997). Of course, these ecological and behavioral categories must be inferred from the skeletal and dental remains of the subfossils rather than observed directly in nature. Even so, numerous functional-morphological studies have revealed that the subfossil taxa were even more remarkable in their diversity than the living lemurs. For instance, *Megaladapis* has been likened as an ecological analog more to the koala than to any other primate (Walker, 1974; Jungers, 1980). *Palaeopropithecus* has been likened to an arboreal sloth (MacPhee et al., 1984) whereas *Archaeoindris* has been described as a ground sloth analog (Jungers, 1980; Vuillaume-Randriamanantena, 1988). Early classifications based on craniodental evidence of *Archaeolemur* and *Hadropithecus* placed them as monkeys (Major, 1896; Standing, 1908), and subsequent postcranial studies have found them to be remarkably terrestrial for strepsirrhine primates (Walker, 1974), thus strengthening the monkey analogy.

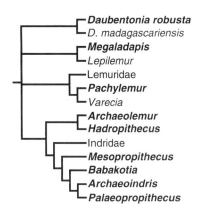

Figure 1. Summary estimate of subfossil lemur phylogenetic relationships based on published morphological analyses (subfossils are depicted in bold italics).

Although morphological data have been employed to identify the phylogenetic relationships of the subfossil lemurs, these studies have been handicapped by the homoplasy that is typical of endemic Malagasy primates (Eaglen, 1980). Even so, morphological data have provided a number of discrete phylogenetic hypotheses. Current understanding of subfossil lemur phylogeny holds that a number of the subfossil genera (*Palaeopropithecus*, *Mesopropithecus*, *Babakotia*, and *Archaeoindris*) comprise a monophyletic family Palaeopropithecidae that is in turn the sister group of the extant Indridae with the Archaeolemuridae as sister of the combined palaeopropithecid-indrid clade (Jungers et al., 1991; Simons et al., 1992). The genus *Megaladapis* is proposed to be the sister taxon to the extant genus *Lepilemur* (Schwartz & Tattersall, 1985), *Daubentonia robusta* is assumed to be sister to the smaller, living *D. madagascariensis* (Simons, 1994), and *Pachylemur* has been determined to be most-closely-related to the extant genus *Varecia* within the Lemuridae (Seligsohn & Szalay, 1974; Crovella et al., 1994). In Figure 1 we summarize these hypotheses as a best estimate of subfossil, relative to extant, lemuriform relationships. Ideally, we would like to employ genetic data to test this morphology-based phylogeny.

Recent molecular-phylogenetic investigation of the extant lemurs indicates that they are a monophyletic assemblage and are thus the product of a single colonization of Madagascar (Adkins & Honeycutt, 1994; Yoder, 1994; Porter et al., 1995; Yoder et al., 1996a; Porter et al., 1997; Yoder, 1997). These phylogenetic results are compatible with geological data that show that Madagascar has existed as an isolated refugium for the duration of eutherian mammal evolutionary history (Krause et al., 1997). At present, Madagascar lies approximately 480 km to the east of Africa at the narrowest point of the Mozambique Channel and is otherwise completely isolated from other significant landmasses. Moreover, there has been a deep oceanic rift separating Madagascar from Africa for at least the past 150 million years (Rabinowitz et al., 1983) for which changing sea levels would have had little effect. Thus, Madagascar's separation significantly predates the first appearance of any of the modern mammalian lineages in the fossil record. Phylogenetic analysis of the subfossil lemurs is therefore critical, not only for understanding the array of morphological and ecological adaptations that they display, but also for further testing the hypothesis that primates colonized Madagascar only once.

Within the past decade or so, the Polymerase Chain Reaction (PCR) technique has been employed to access DNA from extinct organisms, with variable success. PCR is a biochemical technique whereby free nucleotides, DNA primer sequences, and a polymerase (*Taq*) are employed to amplify DNA in vitro. A solution containing these ingredients, along with target DNA, is taken through a series of thermal cycles whereby copies of the target DNA are manufactured exponentially. Theoretically, one can begin with a single target DNA molecule and end with millions of copies that can then easily be sequenced with routine methods. Just as there have been spectacular successes using this technique (Krings et al., 1997), there also have been equally spectacular failures (Woodward et al., 1994). The failures relate directly to the universal challenges of ancient DNA (aDNA) studies: aDNA is always fragmented, usually in sizes from 100–500 bp (Hoss et al., 1996), and in low copy number. In both empirical efforts, and in tests of amino acid racimization (which occurs at rates nearly equivalent to DNA degradation), workers have found that age is not the fundamental determinant of DNA preservation (Beraud-Colomb et al., 1995; Bailey et al., 1996; Hoss et al., 1996; Poinar et al., 1996), thus suggesting that environmental factors play a key role. Of critical importance to DNA preservation is that the specimen be protected from the ravages of ultraviolet irradiation, hydrolysis, and oxidation (Lindahl, 1993; Lindahl, 1995). Empirical results also indicate that cool environments are more likely to yield specimens for which DNA is amplifiable (Poinar et al., 1996; Yang et al., 1996) and that the degree of histological preservation can be an indicator of DNA preservation (e.g., Herrmann & Hummel, 1994). Nonetheless, even in ideal preservation conditions (such as cool, dry caves), there are theoretical limits to DNA preservation (Lindahl, 1993). Thus far, there have been no repeatable studies to demonstrate that aDNA older than 100,000 years can be recovered.

Fortunately for the subfossil lemur problem, most of the known specimens are less than 25,000 years old, with the majority being considerably younger (Simons et al., 1995a). Furthermore, a significant percentage of subfossil lemur specimens were (and continue to be) collected from cave sites in Madagascar which, in addition to being cool, are also dark and often dry, thereby affording at least some protection against hydrolysis and ultraviolet irradiation. Certainly, the majority of subfossil lemur teeth appear to be well-preserved at the macroscopic level, though detailed histological analysis has not been performed. Given the advantages of recent extinction, favorable environmental conditions, and good histological preservation, aDNA studies offer real hope for the reconstruction of subfossil lemur evolutionary history. Even so, the challenges are significant and cannot be underestimated.

Ancient DNA Challenges

The two greatest challenges for the recovery and characterization of DNA from extinct organisms is that aDNA is fragmented, whether from 4-, 100-, or 13,000-year-old samples, and it is typically in very low copy number (Pääbo, 1993). Both qualities relate to the fact that DNA is chemically fragile. Once the mechanism of DNA repair in the living organism is turned off with death, DNA quickly degenerates and is modified by breakage due to loss of bases and/or is rendered inaccessible to enzymatic amplification due to crosslinkage. These characteristics conspire to introduce contamination as a continual nightmare for aDNA researchers (Cimino et al., 1990; Handt et al., 1994; Hardy et al., 1994; Schmidt et al., 1995; Zischler et al., 1995; Austin et al., 1997). Although there have been published reports of aDNA of astonishing antiquity (Poinar et al., 1994; Woodward et al., 1994), all of the sequences reported from samples more than a million years old are thought to derive from contamination (Zischler et al., 1995; Austin et al., 1997; Gutierrez & Marin, 1998). In any PCR amplification, the DNA polymerase will always favor intact contaminating DNA to the desired aDNA. Contamination usually stems from two possible sources: 1) human genomic DNA from handling and preparation of ancient samples and 2) PCR amplicons from one's own lab. One particularly problematic and unexpected source of contamination has been human mitochondrial DNA (mtDNA)-like nuclear pseudogenes. A notorious example of the misleading effects of these contaminants concerns the report of DNA sequences from Cretaceous bone fragments (Woodward et al., 1994; Allard et al., 1995; Hedges & Schweitzer, 1995; Henikoff, 1995; Zischler et al., 1995). The most abundant and insidious source of contaminants comes from homologous PCR amplicons, however. One aerosol droplet of a robust PCR reaction has been estimated to contain as many as 10,000 copies of the target DNA sequence (Sykes, 1993) and thus PCR products can quickly cover lab surfaces, clothes, reagents, etc.

Damaged DNA can also result in amplification of artifact sequences. If aDNA is poorly represented in the PCR reaction, modern DNA of any abundance can compete for primer annealing, even if the PCR primers are well-designed and specific to the target DNA. Mis-priming of modern DNA, combined with weak priming of target DNA, can result in the generation of artifact sequences early in the reaction that can then be exponentially amplified in subsequent PCR cycles (Huang & Jeang, 1994). If such is the case, at best, one ends up with a heterogeneous population of amplicons—some representing the desired target and some representing contaminants or artifacts.

The problems described above can be compounded or (even worse) amplification made impossible by the co-purification of PCR inhibitors. One of the many consequences of the organic decay process is that soil-derived degradation products (collectively known as humus) can become associated with the subfossil specimen. These products often act as strong inhibitors of *Taq* polymerase (Tuross, 1994). Another cause of inhibition results when cross-links between reducing sugars and amino groups occur as a consequence of the Maillard reaction (Poinar et al., 1998). Maillard products can usually be identified by a brownish tint in the DNA extract and by their blue fluorescence under ultraviolet light when the DNA extract is run through an agarose gel (Pääbo, 1990). Although lack of PCR amplification can relate to numerous characteristics of aDNA, inhibition can be unambiguously identified. If adding an aliquot of aDNA extract to a positive control (i.e., "spiking") prevents amplification, then polymerase inhibition is immediately confirmed.

Finally, primer design is critical to the success of an aDNA project. Whereas modern DNA can be amplified in long sections, with primers conveniently designed to anneal to conserved regions of the genome, aDNA must be amplified in short segments due to its fragmentary nature. Not only does this substantially increase the cost per base pair sequenced, it introduces numerous challenges for primer design. Because one must design primers more frequently, perhaps every 50–100 bases versus every 500–1000 bases with modern DNA, one must always confront the possibility that the aDNA primers will either be too specific (if they are designed from the sequences of a single organism) or that they will be too general (amplifying numerous organisms, but poorly). Moreover, due to the demand to design many primers, and often in nonideal regions of the genome, it is unavoidable that relative efficiency of individual primer pairs will vary. Whereas certain primer pairs will be capable of amplifying few (or even single) molecules, others will be far less reliable (Krings et al., 1997). The issue of primer design is a particular challenge for the subfossil lemur problem. There are very few primers that can be designed to recognize the full array of subfossil lemur diversity because the lemuriform radiation is quite old, dating to at least the middle

Eocene (Yoder et al., 1996a; Porter et al., 1997), and because many of the primary lemuriform lineages have been evolving independently of one another for most of this time period (Yoder, 1997).

MATERIAL AND METHODS

As this project's ultimate goal, individuals for at least one representative, for at least one species, for each of the nine genera (*Archaeolemur, Archaeoindris, Babakotia, Daubentonia, Hadropithecus, Megaladapis, Mesopropithecus, Pachylemur,* and *Palaeopropithecus*) will be sampled to yield cytochrome *b* and cytochrome oxidase subunit II (COII) sequences. Subfossil teeth and bone fragments are being sampled, with teeth preferred for reasons described in the following section. Teeth are surface decontaminated with a 5% household bleach solution for 10 minutes. Bone fragments, when used, are ultraviolet irradiated for fear that bleach can penetrate the bone, thereby destroying endogenous DNA. After surface decontamination, specimens are reduced to powder in either a counter-top coffee grinder or in a Spex freezer mill.

DNA is extracted using a phenol/chloroform (PCI) protocol, modified from several published protocols (Hagelberg & Clegg, 1991; Richards et al., 1995; Bailey et al., 1996; Parr et al., 1996; Lalueza et al., 1997). Negative extraction controls (i.e., water is substituted for bone or tooth material) are always included with each set of extractions. After PCI extraction, DNA is collected via filtration with Centricon-50's (Amicon) rather than via ethanol precipitation. DNA is then amplified with PCR that employs *Taq* Gold (PE Applied Biosystems) in order to achieve a "hot start" (i.e., DNA is completely denatured before primer annealing and extension can occur) which improves specificity and sensitivity (Pääbo, 1990). Sequencing of the PCR products is accomplished with cycle sequencing using a dye terminator sequencing kit (PE Applied Biosystems). Sequences are then edited and compiled with AutoAssembler 1.3.0 (PE Applied Biosystems). A sequence is not accepted as accurate until both strands have been sequenced from at least two independent PCR reactions.

For this report, sequences were analyzed with maximum parsimony using the program PAUP* (Swofford, 1998). Heuristic searches were conducted with 100 replicates of the random addition option and all other options set by default. Relative support for internal nodes was estimated using the bootstrap (Felsenstein, 1985). For all bootstrap tests, 100 replicates were run with the random addition option selected from the heuristic search menu. For all analyses, a step matrix was imposed in which transversions received a weight of ten and transitions received a weight of one (10X weighting). Analyses of two different character sets were conducted: one in which all sites were included and one in which only third position sites were considered. For both types of analysis, three different samples of taxa were considered: one in which all taxa were included, one that excluded *Megaladapis* and one that excluded *Palaeopropithecus*. The rationale for character weighting and taxon sample is discussed in the Phylogenetic Results and Discussion section.

Ancient DNA Solutions

Like all ancient DNA studies, this study faces the challenges of maximizing DNA yield while also minimizing contamination and inhibition, of designing primers that are both specific to and universal for the organisms under investigation, and of designing

Table 1. Methods employed for verifying aDNA authenticity

All preparation of subfossils performed in "clean" facilities.
Contamination control measures (bleach; ultraviolet irradiation; protective clothing).
Analyze multiple individuals per taxon when possible.
Perform multiple DNA extractions per individual.
Perform multiple PCR amplifications per extraction.
Amplification and sequencing strategy that allows for overlapping sequences.
Obtain sequences for two genes (cytochrome *b* and COII).
Always include negative extraction and PCR controls.
Replicate results in two independent laboratories.

primers that are efficient and can be placed at intervals such that target sequences are within the range expected for DNA that is severely fragmented. Fortunately, subfossil lemur specimens are large enough, abundant enough, and recent enough so that these challenges are not insurmountable. The methods by which we have overcome many of the obstacles are reviewed below with the hope that this contribution will serve as a guide for other projects facing similar challenges.

Minimizing Contamination

This study has taken extreme measures (summarized in Table 1) to control for contamination and to authenticate the sequenced DNA as intrinsic to the subfossil samples. Two modes of sequence authentication have been completed. First, different specimens of two taxa (*Palaeopropithecus* and *Megaladapis*) have been independently processed (decontaminated, extracted, amplified, and sequenced) in two labs (ADY's at NU and TJP's at AFDIL) and have yielded either identical or highly similar sequences. This strategy follows the idea that if specimens of the same species, processed in different labs, show closely-related sequences, it will strengthen the argument that they are of authentically-ancient origin (Handt et al., 1994). Second, some of the specimens have been divided and processed in parallel between the two labs. This strategy follows the idea that material taken from the same specimen and processed in two labs should yield identical sequence. In all cases, subfossil samples and extracts are never exposed to post-PCR facilities. For both labs, primers, and all other reagents are shipped directly from the respective manufacturers. This circumvents the possibility that contamination can be transmitted from one lab to the other via reagents. Final verification of sequences is determined by phylogenetic analysis (DeSalle & Grimaldi, 1994; Handt et al., 1994; Richards et al., 1995; Young et al., 1995). Such analysis, for example, could have readily identified the Cretaceous bone-fragment sequences as spurious in that the reported sequences would have grouped closely with humans rather than with birds, reptiles, or non-human mammals.

Maximizing DNA Yields

We are focusing on subfossil lemur teeth as the primary source of DNA for this study. Several workers have found that DNA is less fragmented and can be more readily amplified in teeth than in bone (Ginther et al., 1992; DeGusta et al., 1994; Zierdt et al.,

1996), probably due to the higher concentration of hydroxyapatite which binds and thus preserves DNA (DeGusta et al., 1994). The fact that there are literally thousands of subfossil lemur teeth is therefore a distinct advantage of this project. mtDNA is the molecule of choice for this and other aDNA studies due to the fact that most cells possess multiple mitochondria but only a single nucleus. For every single-copy nuclear gene within a given cell, there will be approximately a 1000-fold excess of mitochondrial genes. Given the fact that only very small amounts of DNA can typically be recovered from ancient specimens, one has a far greater chance of recovering a mitochondrial gene than a nuclear gene via PCR. Two methods of DNA extraction are commonly employed in aDNA studies. The first is a modified silica protocol (Boom et al., 1990; Hoss & Pääbo, 1993; Hoss, 1994). The silica method is considered desirable for two reasons: 1) silica beads, in the presence of salt, act as "DNA magnets" and for that reason are highly efficient in binding even minute quantities of DNA and 2) because the silica beads bind only DNA, all other potential co-purifiers are washed away, leaving behind only purified DNA. Also, it has been recently determined that guanidine thiocyanate (a component of the silica protocol) provides greater yields relative to number of years post-mortem than does EDTA (Tuross, 1995). The second method is a modified PCI protocol (Hagelberg & Clegg, 1991; Richards et al., 1995; Bailey et al., 1996; Parr et al., 1996).

Early efforts in this project focused on the silica method, with zero success. Although subsequent experience indicates that the lack of success might relate more to primer issues than to low DNA yield, we soon moved on to the PCI method. Originally, we employed a PCI extraction method that did not include an EDTA decalcification stage. DNA yields were quite low, however (80 μl of extract from 1–3 g of tissue, of which 10 μl were necessary for a 50 μl reaction). In an effort to increase DNA yields, we modified the protocol to introduce an EDTA decalcification step as is commonly employed in other aDNA protocols. This we achieved with dialysis, rather than washing, as one of us (TJP) has found significant loss of DNA with the latter. We have also found that full agitation of the sample during tissue digestion and cell lysis is important in that it allows for nearly full digestion of tooth and/or bone powder, thus further increasing DNA yields. Finally, we employ Centricon 50's (Amicon) to concentrate the DNA rather than the more typically-employed Centricon 100's. This modification is due to the fact that the membrane pore size is smaller with the former and thus will retain shorter fragments of DNA than will the latter.

Minimizing PCR Inhibition

One immediate and unfortunate consequence of enhanced DNA yields is that the very same extraction modifications that are favorable to DNA retention are also favorable to the co-purification of PCR inhibitors. As we progressively enhanced DNA yields, we noted a trend for the resulting extractions to show the brownish tinge associated with the probable co-purification of Maillard products. Moreover, we observe a correspondence between the degree of brownish discoloration of the extract and the degree of blue "haze" under ultraviolet illumination at a position roughly corresponding to 500 bp reported by Pääbo (1990) to be indicative of the co-purification of Maillard products. When these characteristics are observed, PCR amplification is always diminished or prevented entirely. Also, by using the positive-control spike test in these cases, we have confirmed PCR inhibition.

Several methods for overcoming inhibition have been employed by us with variable success: 1) simple dilution of the sample, in some cases up to 1:50, 2) nested PCR, 3) the addition of 12.5 units of *Taq* Gold ("Max Taq"), and 4) isopropanol precipitation (Hanni et al., 1995). Of the four methods, we prefer the dilution technique. Nested PCR can be problematic in that it requires a starting PCR product long enough to permit a second PCR with at least one internal primer—something that is often not possible with highly fragmented DNA. "Max Taq" is not always effective but is always costly. Contrary to the Hanni et al. (1995) report, isopropanol precipitation did not eliminate inhibitors in the three samples attempted. In summary, inhibition has been problematic though not insurmountable and we will continue our efforts to enhance DNA yields and thus overcome inhibition through dilution.

Balancing Primer Sensitivity and Specificity

In the initial stages of this project, we attempted to amplify a 198 bp (plus primer sequences) target of the cytochrome *b* gene—a target size that is well within the expected limits of aDNA fragment lengths. Aside from a single bout of contamination, however, we had no success in amplifying this fragment. It was not until we reduced our target length to 98 bp (plus primer sequences) that we were first able to amplify and sequence authentic subfossil lemur DNA. These and subsequent experiments have confirmed that typical subfossil lemur DNA is even more fragmented than it is for many other extinct organisms of Holocene age (Stone & Stoneking, 1993; Cooper, 1994; Hagelberg et al., 1994; Yang et al., 1996). Thus, a significant emphasis on primer design has been introduced to the subfossil lemur project.

The combined challenges of working within a diverse evolutionary radiation and of targeting a rapidly-evolving region of the total genome constrain our ability to design efficient primers. For a study that focuses on extant organisms, these competing constraints can easily be overcome by designing primers such that they lie within conserved regions of the mitochondrion (e.g., the tRNA regions) but frame and amplify the more variable and informative regions (Fig. 2). Because of the extreme degree of fragmentation in the subfossil lemurs, however, we do not have this luxury. Our primers must be designed such that the target sequences are no more than 150 base pairs (bp) in length. Moreover, the primers must be placed such that the amplicons, when sequenced and aligned, overlap to form a continuous stretch of at least 500 bp for each gene (as an example, see Fig. 3). We have been forced to design primers in genomic regions that are too variable to allow for universal amplification of the lemuriform radiation. Thus, it is frequently the case that universality must be sacrificed for primer sensitivity.

In Figure 4 we illustrate this effect in extant lemuriforms for four different primer pairs designed to amplify portions of the cytochrome *b* gene. The figure demonstrates that whereas a primer pair combination might be quite sensitive in one taxon, it may be functional yet inefficient in another taxon, and completely non-functional in yet another. For example, primer pair L14839-H14954 (Fig. 4b) amplifies its target with great efficiency in *Propithecus* and *Cheirogaleus*, with only moderate efficiency in *Eulemur*, and with none in *Daubentonia* and *Lepilemur*. The results are even less encouraging for certain other primer pairs (e.g., Fig. 4c). Given that we don't know a priori within which, if any, of the living lemuriform lineages the extinct lineages belong, early efforts towards amplification must of necessity be experimental.

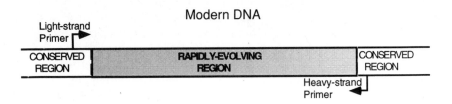

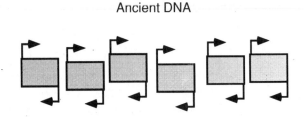

Figure 2. Comparison of PCR amplification and sequencing possibilities for modern DNA versus ancient DNA. Drawing on top represents long strand of modern DNA wherein only two primers are necessary to amplify and sequence entire region of interest (rapidly-evolving region). Drawing below indicates that because ancient DNA is so fragmented, twelve primers are required to amplify and sequence the equivalent target stretch of DNA sequence.

PHYLOGENETIC RESULTS AND DISCUSSION

Thus far, we have been successful in obtaining intrinsic DNA sequences for three subfossil genera: *Palaeopropithecus*, *Megaladapis*, and *Archaeolemur*. From *Palaeopropithecus* we have obtained cytochrome *b* sequence from two individuals, from *Megaladapis* three individuals, and from *Archaeolemur* one individual. We have also obtained COII sequence from one *Megaladapis* individual. In this report, we present the phylogenetic analysis of cytochrome *b* sequences for one *Palaeopropithecus* (AM

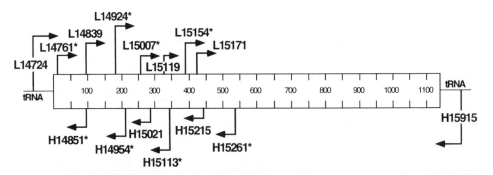

Figure 3. Example of PCR and sequencing strategy used in this study. Figure is map of cytochrome *b* gene with primer positions indicated by arrows. Primers are lettered to indicate light (L) and heavy (H) strands and numbered to match Anderson et al. (1981) sequence. In extant lemurs, entire cytochrome *b* gene can be amplified and sequenced with two primers (L14724 and H15915). In subfossil lemurs, up to thirteen primers are required to amplify and sequence only half of the gene (approximately 550 bp). Primers shown with asterisk (*) were designed to prevent amplification of human sequences.

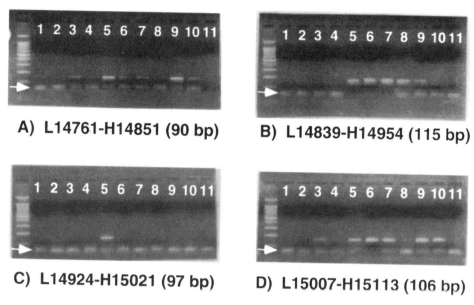

Figure 4. Comparison of agarose gels for which cytochrome *b* PCR amplicons have been run out for four different primer pairs. Primers are lettered and numbered as in Figure 3. Length of target sequence is shown in parentheses. Lanes contain amplicons from the following starting templates: 1) *Daubentonia* (100 pg/μl), 2) *Daubentonia* (10 pg/μl), 3) *Lepilemur* (100 pg/μl), 4) *Lepilemur* (10 pg/μl), 5) *Propithecus* (100 pg/μl), 6) *Propithecus* (10 pg/μl), 7) *Cheirogaleus* (100 pg/μl), 8) *Cheirogaleus* (10 pg/μl), 9) *Eulemur* (100 pg/μl), 10) *Eulemur* (10 pg/μl), 11) PCR negative control. Figure compares performance of primer pairs in different lemuriform lineages as well as sensitivity of primers (i.e., all taxa are tested at two different DNA concentrations). White arrows indicate primer dimer.

6184; Ankazoabo) and one *Megaladapis* (UA 4822—AM 6479; Bevoha) as these individuals have thus far yielded the most complete sequences.

The subfossil sequences were analyzed in a data matrix that contains numerous lemuriform taxa, representing all five lemuriform families, and three lorisiform outgroups. This matrix contains an alignment of the first 550 bp of the cytochrome *b* gene (Anderson numbers 14747–15296; Anderson et al., 1981). Of the two subfossils, *Palaeopropithecus* has the most complete sequence with 508 bp of the possible 550 sites. *Megaladapis* is less complete with only 350 bp. Regions of the gene that contain missing sequence for *Palaeopropithecus* do not overlap with regions where *Megaladapis* is missing sequence due to the fact that there was differential primer failure in the two taxa (much as is illustrated for extant taxa in Fig. 4).

In Figure 5 we illustrate the most parsimonious arrangement of the ingroup taxa wherein all sites are analyzed with a 10X weighting of transversions. This weighting scheme has been empirically determined to be appropriate and typically more informative for strepsirrhine cytochrome *b* sequences than is equal weighting, especially for small data sets (Yoder et al., 1996b). When all taxa are included in the analysis, a surprising arrangement of subfossil taxa is suggested; *Palaeopropithecus* and *Megaladapis* are shown to form a clade that is basal to all lemuriforms except *Daubentonia* (Fig. 5a). This result is surprising in that it has not been anticipated by any of the morphological data. One possible interpretation of this result is the unsettling possibility that the two taxa are joined because they are similar due to contamination. There is much evidence to prove that this is not the case. First, the bootstrap value in support of this

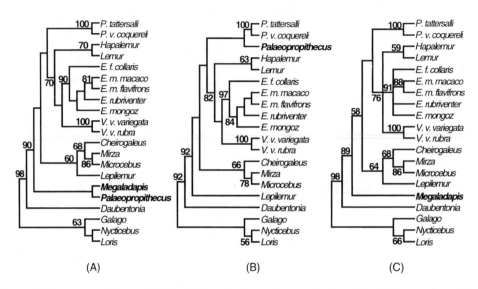

Figure 5. Parsimony analyses of 550 bp of the cytochrome *b* gene for two subfossil taxa, various extant lemuriforms, and three lorisiform outgroups. All sites were included in analysis and transversions were weighted ten times transitions. A) All study taxa included in analysis, B) *Megaladapis* excluded, and C) *Palaeopropithecus* excluded. Numbers on branches represent bootstrap support for node above that branch. Lack of a number indicates bootstrap support less than 50%. Subfossil taxa are in bold font.

clade is less than 50%, indicating that the grouping is very weak. Second, pairwise distance analysis shows that the two sequences are 5.5% divergent which is well within the range for other intergeneric comparisons (e.g., *Propithecus* and *Varecia* are 4.6% divergent). If shared contamination was the explanation for the grouping, we would expect the sequences to be identical. Third, as discussed above, multiple individuals have been sequenced for both subfossil taxa, and in two different labs, with the result that intra-genus sequences are either identical or nearly so.

Another explanation for the putative *Palaeopropithecus/Megaladapis* clade is that it is real. Perhaps, contrary to all predictions based on morphological data, these two taxa are more closely related to each other than either is to other lemuriform taxa. We suggest that this is unlikely. Certainly, the morphological data in support of a close evolutionary relationship between *Palaeopropithecus* and living indrids is persuasive (Jungers et al., 1991; Simons et al., 1992) A more likely explanation for the association of the two subfossil taxa is that they, like extant lemuriform family lineages (Yoder, 1997), began independent evolution many millions of years ago. In other words, it is likely that they are long branches (sensu Felsenstein, 1978), and thus, long-branch attraction my be playing a role. This problem is probably exacerbated by the fact that the data set is small and that both taxa are missing a subset of the data. To tease apart the potential effects of long-branch attraction and sampling error, we conducted two similar analyses in which one or the other subfossil taxon was removed from the analysis (Figs. 5b and c). It is interesting to note that in the absence of the other subfossil, both genera fall in very different parts of the lemuriform radiation. As predicted by the morphological data, *Palaeopropithecus* is placed as the basal member of an indrid clade

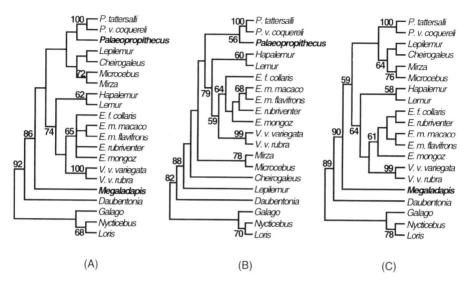

Figure 6. Parsimony analyses of same region of the cytochrome *b* gene analyzed in Figure 5 except that first and second position sites are excluded. As in Figure 5 analysis, transversions were weighted ten times transitions. A) All study taxa included in analysis, B) *Megaladapis* excluded, and C) *Palaeopropithecus* excluded. Numbers on branches represent bootstrap support for node above that branch. Lack of a number indicates bootstrap support less than 50%. Subfossil taxa are in bold font.

(Fig. 5b), but contrary to the morphological data, *Megaladapis* appears to be an independent lemuriform lineage (Fig. 5c) rather than sister to *Lepilemur*.

Another possible explanation for the seemingly spurious results in Figure 5a is that codon sites under selection could be biasing the results. Empirical exploration of protein-coding mitochondrial genes in the Strepsirrhini suggests that this may be a factor (Yoder et al., 1996b; Stanger-Hall & Cunningham, 1998). To investigate this possibility, we omitted first and second position sites (wherein 96% and 100% respectively of nucleotide substitutions will yield amino acid replacements; Li & Graur, 1991) and analyzed third position sites only (wherein only about 30% of nucleotide substitutions yield an amino acid replacement). Again, transversions were weighted 10X and analyses were run for all taxa, all taxa minus *Megaladapis*, and all taxa minus *Palaeopropithecus*. In Figure 6 we illustrate the results. In this case, there is perfect agreement between the complete analysis (Fig. 6a) and both subset analyses (Figs. 6b and c) with regard to the placement of the subfossil taxa. In all three analyses, as with the taxon-subset analyses of the complete data set (Figs. 5b and c), *Palaeopropithecus* joins the indrid clade and *Megaladapis* is basal to other lemuriforms except *Daubentonia*.

CONCLUSIONS

The most significant conclusion reached by this study is that intrinsic DNA, though highly fragmented, exists in sufficient quantities to be extracted, amplified and sequenced in many if not all subfossil lemur taxa. The phylogenetic analyses presented here should nonetheless be considered preliminary due to the small data set analyzed. In addition to the surprising placement of the two subfossil taxa in Figure 6a, it is notable that certain other lemuriform taxa (e.g., *Lepilemur*) tend to be variably placed in the analyses depending on taxon and character sample. In fact, with regard to the two subfossil taxa, it may be seen as significant that their relative placement can be said to be more stable than some of the extant taxa for this small data set. Thus, the conclusion that *Palaeopropithecus* joins the indrid clade and *Megaladapis* is another independent long branch within the lemuriform radiation, and not sister to *Lepilemur*, may ultimately be born out by future analyses. It is our intent that these future analyses will enjoy the benefits of longer DNA sequences and larger subfossil-taxon samples.

ACKNOWLEDGMENTS

We thank the editors of this volume for inviting the submission of this progress report. The manuscript was significantly improved thanks to comments from Todd Disotell, Jörg Ganzhorn, Steve Goodman, Jodi Irwin, and Bill Jungers. ADY is grateful to Alan Cooper for his instruction in aDNA techniques and to Kasia Witkos for her painstaking organizational efforts during the early phases of the project. Thanks go to Jodi Irwin for keeping ADY's lab operating at optimum levels. We look forward to future collaboration with David Burney and Elwyn Simons in achieving our long-term goals. We thank the American Registry of Pathology for support. The opinions and assertions expressed herein are solely those of the authors and are not to be construed as official or as the views of the United States Department of Defense or the United States Department of Army. This research has been supported by the Leakey Foundation and by NSF grants SBR-9729014 and DEB-9707559 to ADY. All sequencing in ADY's lab was performed on an ABI 377 automated DNA sequencer purchased with funds from NIH grant 1 S10 RR12996-01. This is DUPC publication number 669.

REFERENCES

ADKINS, R. M., AND R. L. HONEYCUTT. 1994. Evolution of the primate cytochrome *c* oxidase subunit II gene. Journal of Molecular Evolution, **38**: 215–231.

ALLARD, M. W., D. YOUNG, AND Y. HUYEN. 1995. Detecting dinosaur DNA. Science, **268**: 1192.

ANDERSON, S., A. T. BANKIER, B. G. BARRELL, M. H. L. DE BRUIJN, A. R. COULSON, J. DROUIN, I. C. EPERON, D. P. NIERLICH, B. A. ROE, F. SANGER, P. H. SCHREIER, A. J. H. SMITH, R. STADEN, AND I. G. YOUNG. 1981. Sequence and organization of the human mitochondrial genome. Nature, **290**: 457–465.

AUSTIN, J. J., A. B. SMITH, AND R. H. THOMAS. 1997. Palaeontology in a molecular world: the search for authentic ancient DNA. Trends in Ecology and Evolution, **12**: 303–306.

BAILEY, J. F., M. B. RICHARDS, V. A. MACAULAY, I. B. COLSON, I. T. JAMES, D. G. BRADLEY, R. E. M. HEDGES, AND B. C. SYKES. 1996. Ancient DNA suggests a recent expansion of European cattle from a diverse wild progenitor species. Proceedings of the Royal Society of London, Series B, **263**: 1467–1473.

BERAUD-COLOMB, E., R. ROUBIN, J. MARTIN, N. MAROC, A. GARDEISEN, G. TRABUCHET, AND M. GOOSSENS. 1995. Human β-globin gene polymorphisms characterized in DNA extracted from ancient bones 12,000 years old. American Journal of Human Genetics, **57**: 1267–1274.

BOOM, R., C. J. A. SOL, M. M. M. SALIMANS, C. L. JANSEN, P. M. E. WERTHEIM-VAN DILLEN, AND J. VAN DER NOORDA. 1990. Rapid and simple method for purification of nucleic acids. Journal of Clinical Microbiology, **28**: 495–503.

BURNEY, D. A., AND R. D. E. MACPHEE. 1988. Mysterious island: What killed Madagascar's large native animals? Natural History, **97**: 46–55.

CIMINO, G. D., K. METCHETTE, S. T. ISAACS, AND Y. S. ZHU. 1990. More false-positive problems. Nature, **345**: 773–774.
COOPER, A. 1994. Ancient DNA sequences reveal unsuspected phylogenetic relationships within New Zealand wrens (Acanthisittidae). Experientia, **50**: 558–563.
CROVELLA, S., D. MONTAGNON, B. RAKOTOSAMIMANANA, AND Y. RUMPLER. 1994. Molecular biology and systematics of an extinct lemur: *Pachylemur insignis*. Primates, **35**: 519–522.
DEGUSTA, D., C. COOK, AND G. SENSABAUGH. 1994. Dentin as a source of ancient DNA. Ancient DNA Newsletter, **2**: 13.
DESALLE, R., AND D. GRIMALDI. 1994. Very old DNA. Current Opinion in Genetics and Development, **4**: 810–815.
EAGLEN, R. H. 1980. The systematics of living Strepsirhini, with special reference to the Lemuridae. Unpublished Ph.D. thesis. Duke University, Durham, NC.
FELSENSTEIN, J. 1978. Cases in which parsimony or compatibility methods will be positively misleading. Systematic Zoology, **27**: 401–410.
———. 1985. Confidence limits on phylogenies: an approach using the bootstrap. Evolution, **39**: 783–791.
GINTHER, C., L. ISSEL-TARVER, AND M.-C. KING. 1992. Indentifying individuals by sequencing mitochondrial DNA from teeth. Nature Genetics, **2**: 135–138.
GODFREY, L. R., W. L. JUNGERS, K. E. REED, E. L. SIMONS, AND P. S. CHATRATH. 1997. Subfossil lemurs, pp. 218–256. *In* Goodman, S. M., and B. D. Patterson, eds., Natural Change and Human Impact in Madagascar. Smithsonian Institution Press, Washington, D.C.
GODFREY, L. R., E. L. SIMONS, P. CHATRATH, AND B. RAKOTOSAMIMANANA. 1990. A new fossil lemur (*Babakotia*, Primates) from northern Madagascar. Comptes Rendus de l'Académie des Sciences, Paris, **310**: 81–87.
GUTIERREZ, G., AND A. MARIN. 1998. The most ancient DNA recovered from an amber-preserved specimen may not be as ancient as it seems. Molecular Biology and Evolution, **15**: 926–929.
HAGELBERG, E., AND J. B. CLEGG. 1991. Isolation and characterization of DNA from archaeological bone. Proceedings of the Royal Society of London, Series B, **244**: 45–50.
HAGELBERG, E., S. QUEVEDO, D. TURBON, AND J. B. CLEGG. 1994. DNA from ancient Easter Islanders. Nature, **369**: 25–26.
HANDT, O., M. HOSS, M. KRINGS, AND S. PÄÄBO. 1994. Ancient DNA: methodological challenges. Experientia, **50**: 524–529.
HANNI, C., T. BROUSSEAU, V. LAUDET, AND D. STEHELIN. 1995. Isopropanol precipitation removes PCR inhibitors from ancient bone extracts. Nucleic Acids Research, **23**: 881–882.
HARDY, C., D. CASANE, J. D. VIGNE, C. CALLOU, N. DENNEBOUY, J.-C. MOUNOLOU, AND M. MONNEROT. 1994. Ancient DNA from Bronze Age bones of european rabbit (*Oryctolagus cuniculus*). Experientia, **50**: 564–570.
HEDGES, S. B., AND M. H. SCHWEITZER. 1995. Detecting dinosaur DNA. Science, **268**: 1191.
HENIKOFF, S. 1995. Detecting dinosaur DNA. Science, **268**: 1192.
HERRMANN, B., AND S. HUMMEL. 1994. Introduction, pp. 59–68. *In* Herrmann, B., and S. Hummel, eds., Ancient DNA. Springer-Verlag, New York.
HOSS, M. 1994. More about the silica method. Ancient DNA Newsletter, **2**: 10–12.
———, P. JARUGA, T. H. ZASTAWNY, M. DIZDAROGLU, AND S. PÄÄBO. 1996. DNA damage and DNA sequence retrieval from ancient tissues. Nucleic Acids Research, **24**: 1304–1307.
HOSS, M., AND S. PÄÄBO. 1993. DNA extraction from Pleistocene bones by silica-based purification method. Nucleic Acids Research, **21**: 3913–3914.
HUANG, L.-M., AND K.-T. JEANG. 1994. Long-range jumping of incompletely extended polymerase chain fragments generates unexpected products. BioTechniques, **16**: 242–246.
JUNGERS, W. L. 1980. Adaptive diversity in subfossil Malagasy prosimians. Zeitschrift für Morphologie und Anthropologie, **71**: 177–186.
———, L. R. GODFREY, E. L. SIMONS, P. S. CHATRATH, AND B. RAKOTOSAMIMANANA. 1991. Phylogenetic and functional affinities of *Babakotia* (Primates), a fossil lemur from northern Madagascar. Proceedings of the National Academy of Sciences, U.S.A., **88**: 9082–9086.
KRAUSE, D. W., J. H. HARTMAN, AND N. A. WELLS. 1997. Late Cretaceous vertebrates from Madagascar: implications for biotic change in deep time, pp. 3–43. *In* Goodman, S. M., and B. D. Patterson, eds., Natural Change and Human Impact in Madagascar. Smithsonian Institution Press, Washington, D.C.
KRINGS, M., A. STONE, R. W. SCHMITZ, H. KRAINITZKI, M. STONEKING, AND S. PÄÄBO. 1997. Neanderthal DNA sequences and the origin of modern humans. Cell, **90**: 19–30.
LALUEZA, C., A. PEREZ-PEREZ, E. PRATS, L. CORNUDELLA, AND D. TURBON. 1997. Lack of founding Amerindian mitochondrial DNA lineages in extinct Aborigines from Tierra del Fuego—Patagonia. Human Molecular Genetics, **6**: 41–46.

LI, W.-H., AND D. GRAUR. 1991. Fundamentals of Molecular Evolution. Sinauer, Sunderland, MA.
LINDAHL, T. 1993. Instability and decay of the primary structure of DNA. Nature, **362**: 709–715.
———. 1995. Time-dependent degradation of DNA. Ancient DNA III, Oxford, U.K.
MACPHEE, R. D. E., E. L. SIMONS, N. A. WELLS, AND M. VUILLAUME-RANDRIAMANANTENA. 1984. Team finds giant lemur skeleton. Geotimes, **29**: 10–11.
MAJOR, C. I. FORSYTH. 1896. Preliminary notice on fossil monkeys from Madagascar. Geological Magazine, **3**: 433–436.
PÄÄBO, S. 1990. Amplifying ancient DNA, pp. 159–166. *In* Innis, M., D. Gelfand, J. Sninsky, and T. White, eds., PCR protocols: a guide to methods and applications. Academic Press, San Diego.
———. Ancient DNA. Scientific American, **5**: 86–92.
PARR, R. L., S. W. CARLYLE, AND D. H. O'ROURKE. 1996. Ancient DNA analysis of Fremont Amerindians of the Great Salt Lake wetlands. American Journal of Physical Anthropology, **99**: 507–518.
POINAR, G. O., H. N. POINAR, AND R. J. CANO. 1994. DNA from amber inclusions, pp. 92–103. *In* Hermann, B., and S. Hummel, eds., Ancient DNA. Springer-Verlag, New York.
POINAR, H. N., M. HOFREITER, W. G. SPAULDING, P. S. MARTIN, B. A. STANKIEWICZ, H. BLAND, R. P. EVERSHED, G. POSSNERT, AND S. PÄÄBO. 1998. Molecular coproscopy: dung and diet of the extinct ground sloth *Nothrotheriops shastensis*. Science, **281**: 402–406.
POINAR, H. N., M. HOSS, J. L. BADA, AND S. PÄÄBO. 1996. Amino acid racemization and the preservation of ancient DNA. Science, **272**: 864–866.
PORTER, C. A., S. L. PAGE, J. CZELUSNIAK, H. SCHNEIDER, M. P. C. SCHNEIDER, I. SAMPIO, AND M. GOODMAN. 1997. Phylogeny and evolution of selected primates as determined by sequences of the ε-globin locus and 5′ flanking regions. International Journal of Primatology, **18**: 261–295.
PORTER, C. A., I. SAMPAIO, H. SCHNEIDER, M. P. C. SCHNEIDER, J. CZELUSNIAK, AND M. GOODMAN. 1995. Evidence on primate phylogeny from ε-globin gene sequences and flanking regions. Journal of Molecular Evolution, **40**: 30–55.
RABINOWITZ, P. D., M. F. COFFIN, AND D. FALVEY. 1983. The separation of Madagascar and Africa. Science, **220**: 67–69.
RICHARD, A. F., AND R. E. DEWAR. 1991. Lemur ecology. Annual Review of Ecology and Systematics, **22**: 145–75.
RICHARDS, M. B., B. C. SYKES, AND R. E. M. HEDGES. 1995. Authenticating DNA extracted from ancient skeletal remains. Journal of Archaeological Science, **22**: 291–299.
SCHMIDT, T., S. HUMMEL, AND B. HERRMANN. 1995. Evidence of contamination in PCR laboratory disposables. Naturwissenschaften, **82**: 423–431.
SCHWARTZ, J. H., AND I. TATTERSALL. 1985. Evolutionary relationships of living lemurs and lorises (Mammalia, Primates) and their potential affinities with European Eocene Adapidae. Anthropological Papers of the American Museum of Natural History, **60**: 1–100.
SELIGSOHN, D., AND F. S. SZALAY. 1974. Dental occlusion and the masticatory apparatus in *Lemur* and *Varecia*: their bearing on the systematics of living and fossil primates, pp. 543–562. *In* Martin, R. D., G. A. Doyle, and A. C. Walker, eds., Prosimian Biology. Duckworth, London.
SIMONS, E. L. 1994. The giant aye-aye *Daubentonia robusta*. Folia Primatologica, **62**: 14–21.
———. Lemurs: old and new, pp. 142–166. *In* Goodman, S. M., and B. D. Patterson, eds., Natural Change and Human Impact in Madagascar. Smithsonian Institution Press, Washington, D.C.
———, D. A. BURNEY, P. S. CHATRATH, L. R. GODFREY, W. L. JUNGERS, AND B. RAKOTOSAMIMANANA. 1995a. AMS ^{14}C Dates for extinct lemurs from caves in the Ankarana Massif, northern Mdagascar. Quaternary Research, **43**: 249–254.
SIMONS, E. L., L. R. GODFREY, W. L. JUNGERS, P. S. CHATRATH, AND B. RAKOTOSAMIMANANA. 1992. A new giant subfossil lemur, *Babakotia*, and the evolution of the sloth lemurs. Folia Primatologica, **58**: 197–203.
SIMONS, E. L., L. R. GODFREY, W. L. JUNGERS, P. S. CHATRATH, AND J. RAVAOARISOA. 1995b. A new species of *Mesopropithecus* (Primates, Palaeopropithecidae) from northern Madagascar. International Journal of Primatology, **16**: 653–682.
STANDING, H. F. 1908. On recently discovered subfossil primates from Madagascar. Transactions of the Zoological Society of London, **18**: 69–162.
STANGER-HALL, K., AND C. W. CUNNINGHAM. 1998. Support for a monophyletic lemuriformes: overcoming incongruence between data partitions. Molecular Biology and Evolution, **15**: 1572–1577.
STONE, A. C., AND M. STONEKING. 1993. Ancient DNA from a pre-columbian Amerindian population. American Journal of Physical Anthropology, **92**: 463–471.
SWOFFORD, D. L. 1998. PAUP*. Phylogenetic Analysis Using Parsimony (*and Other Methods). Sunderland, Massachusetts, Sinauer Associates.

Sykes, B. 1993. The Anglo-Saxon invasion of England—bloody conquest or just dirty fingers. Ancient DNA: 2nd International Conference, Smithsonian Institution, Washington, D.C.

Tuross, N. 1994. The biochemistry of ancient DNA in bone. Experientia, **50**: 530–535.

———. 1995. Changes in DNA in a series of naturally weathered bones. Ancient DNA III, Oxford, U.K.

Vuillaume-Randriamanantena, M. 1988. The taxonomic attributions of giant sub-fossil bones from Ampasambazimba: *Archaeoindris* and *Lemuridotherium*. Journal of Human Evolution, **17**: 379–91.

Walker, A. C. 1974. Locomotor adaptations in past and present prosimian primates, pp. 349–381. *In* Jenkins, F. A., ed., Primate Locomotion. Academic Press, New York.

Woodward, S. R., N. J. Weyand, And M. Bunnel. 1994. DNA sequence from Cretaceous period bone fragments. Science, **266**: 1229–1232.

Yang, H., E. M. Golenberg, And J. Shoshani. 1996. Phylogenetic resolution within the Elephantidae using fossil DNA sequence from the American mastodon (*Mammut americanum*) as an outgroup. Proceedings of the National Academy of Sciences, U.S.A., **93**: 1190–1194.

Yoder, A. D. 1994. Relative position of the Cheirogaleidae in strepsirhine phylogeny: a comparison of morphological and molecular methods and results. American Journal of Physical Anthropology, **94**: 25–46.

———. Back to the future: a synthesis of strepsirrhine systematics. Evolutionary Anthropology, **6**: 11–22.

———, M. Cartmill, M. Ruvolo, K. Smith, And R. Vilgalys. 1996a. Ancient single origin of Malagasy primates. Proceedings of the National Academy of Sciences, **93**: 5122–5126.

Yoder, A. D., M. Ruvolo, And R. Vilgalys. 1996b. Molecular evolutionary dynamics of cytochrome *b* in strepsirrhine primates: the phylogenetic significance of third position transversions. Molecular Biology and Evolution, **13**: 1339–1350.

Young, D. L., Y. Huyen, And M. W. Allard. 1995. Testing the vailidity of the cytochrome *b* sequence from the Cretaceous period bone fragments as dinosaur DNA. Cladistics, **11**: 199–209.

Zierdt, H., S. Hummel, And B. Herrmann. 1996. Amplification of human short tandem repeats from medieval teeth and bone samples. Human Biology, **68**: 185–199.

Zischler, H., M. Hoss, O. Handt, A. von Haeseler, A. C. van der Kuyl, J. Goudsmit, And S. Pääbo. 1995. Detecting dinosaur DNA. Science, **268**: 1191–1193.

PAST AND PRESENT DISTRIBUTIONS OF LEMURS IN MADAGASCAR

Laurie R. Godfrey,[1] William L. Jungers,[2] Elwyn L. Simons,[3] Prithijit S. Chatrath,[3] and Berthe Rakotosamimanana[4]

[1] Department of Anthropology
Machmer Hall, Box 34805
University of Massachusetts
Amherst, Massachusetts 01003-4805, USA
lgodfrey@anthro.umass.edu (corresponding author)
[2] Department of Anatomical Sciences
Health Sciences Center
State University of New York at Stony Brook
Stony Brook, New York 11794-8081, USA
[3] Duke University Primate Center
3705 Erwin Road
Durham, North Carolina 27705-5000, USA
[4] Département de Paléontologie et d'Anthropologie Biologique
BP 906, Faculté des Sciences
Université d'Antananarivo
Antananarivo (101), Madagascar

ABSTRACT

Holocene cave, marsh, and stream deposits on the island of Madagascar have yielded thousands of "subfossil" specimens that document recent megafaunal extinctions. Excavations conducted during the past 15 years of archaeological and paleontological sites in northern, northwestern and southwestern Madagascar have unearthed, in addition to new specimens of extinct lemurs and other megafauna, an abundance of bones of still-extant lemur species. These specimens, as well as specimens of extant lemurs from subfossil sites excavated in the early and mid-1900's, prove that living lemur species once had much broader geographic ranges than they have today, and they help to explain the currently disjunct distributions of a number of species. This paper examines the pattern of distribution of extant primate species at subfossil sites, and compares recent to modern primate communities.

RÉSUMÉ

Les milliers d'échantillons subfossiles retrouvés dans les gisements des grottes, des marais et des rivières de l'Holocène apportent des éclaircissements sur les extinctions récentes de la mégafaune. Les fouilles faites pendant les 15 dernières années de sites archéologiques et paléontologiques du nord, du nord-ouest et du sud-ouest de Madagascar ont mis à jour, outre de nombreux échantillons de lémuriens disparus et d'autres mégafaunes, une multitude d'os d'espèces de lémuriens encore existants. Ces échantillons, tout comme ceux des lémuriens actuels provenant de sites subfossiles fouillés avant, prouvent que la répartition des lémuriens vivants était autrefois beaucoup plus vaste qu'actuellement. Et ils peuvent aider à la compréhension de la disjonction entre les distributions actuelles de nombreuses espèces. Cet article étudie le modèle de distribution d'espèces de primates actuels dans les sites subfossiles et compare les communautés de primates récentes aux communautés modernes.

INTRODUCTION

"Subfossil" (Late Pleistocene and Holocene) sites containing the bones of at least 17 species of extinct lemurs and other fauna have been found in all regions of Madagascar except the eastern rain forest and the Sambirano. Specimens of at least 21 (and perhaps as many as 23) species of extant lemurs have been found at these sites alongside the bones of extinct taxa. Little attention has been paid to extant lemur specimens in subfossil collections. These bones, however, comprise our only window into prehistoric changes in the geographic ranges of living lemurs, and they may help us better explain the current disjunct distributions of some. They also provide a framework within which prehistoric changes in primate community composition, structure, and ecology can be evaluated. Extant taxa have been found to occur in abundance at recently excavated subfossil sites (including the caves of the Mahajanga Plateau in the northwest, the Ankarana Massif in the extreme north, and the Manamby Plateau in the southwest). At some of the sites that were excavated during the early 20th century, screen washing was not employed, and the bones of extant lemurs are under represented. Collector bias was not complete, however, and the remains of extant lemur taxa do occur in the early collections (Standing, 1906; Lamberton, 1939a, 1939b).

There are 10 recognized genera and 33 recognized species of living lemurs (Mittermeier et al., 1994; Zimmermann et al., 1998). They belong to five families (Jenkins, 1987): the Indridae, Lemuridae, Lepilemuridae, Cheirogaleidae, and Daubentoniidae. The Cheirogaleidae are generally smaller in body size than other lemurs, although the largest cheirogaleid (*Cheirogaleus major*, ca. 400 g) and the smallest lepilemurid (*Lepilemur leucopus*, ca. 600 g) may overlap in body mass at their range extremes (see Smith & Jungers, 1997). For a variety of taphonomic reasons (for example, the vulnerability of bones of small-bodied species to pre-depositional destruction by carnivores and scavengers and to increased post-depositional acidic dissolution due to their high surface-to-volume ratios), there is a size bias against small-bodied species at fossil sites (see Shipman, 1981; Lyman, 1994, for a review). Nevertheless, specimens of some of the smallest-bodied lemur species have been found in subfossil deposits. Of the 10 recognized extant lemur genera, only *Allocebus*, *Mirza*, and *Phaner* have yet to be found alongside the bones of Madagascar's extinct lemurs.

This paper documents the Holocene distributions of extant lemurs as revealed by skeletal remains at subfossil sites. Many subfossils that formerly belonged to the Académie Malgache (and that were recently acquired by the Laboratoire de Paléontologie des Vertébrés, Université d'Antananarivo) are now catalogued for the first time. Virtually all associations for bones in this collection, if they ever existed, have been lost. Site information has not been lost because many specimens (including those belonging to extant species) have their sites and, sometimes, years of discovery, inscribed directly on them. Dates for specimens of extinct and extant lemurs, as well as other materials from subfossil sites (e.g., wood, charcoal, speleothems; MacPhee et al., 1985; MacPhee & Burney, 1991; Simons et al., 1995a; Simons, 1997; Burney et al., 1997; Burney, in press), confirm their recency as well as the synchronicity of extinct and extant taxa. *Hapalemur simus* is known to have lived at Ankarana 4560 ± 70 years ago (Simons et al., 1995a). The oldest date thus far obtained for a subfossil primate specimen is $26,150 \pm 400$ B.P. for a *Megaladapis* at Ankarana (Simons et al., 1995a); the most recent are 630 ± 50 B.P. and 510 ± 80 B.P. for a *M. madagascariensis* and a *Palaeopropithecus ingens*, respectively, from Ankilitelo (Simons, 1997). Many extinct lemur species are known to have survived the advent of humans in Madagascar (at ca. 2000 years ago) by at least 500 to 1500 years (see Burney, in press, for a review).

During the past several thousand years, while the larger-bodied lemurs and other megafauna disappeared entirely, the geographic ranges of many of the smaller-bodied lemurs contracted (Lamberton, 1939b; Vuillaume-Randriamanantena et al., 1985; Simons et al., 1990; Jungers et al., 1995; Godfrey et al., 1997a). Establishing a detailed chronology for these changes requires a degree of stratigraphic control that has been lost and cannot be reconstructed for most subfossil sites. Nevertheless, site-specific species rosters can be compiled, and these provide important insights into recent changes in lemur biogeography. We ask: 1) Are modern patterns of regional endemicity reflected in the primate communities of the recent past? How similar are the primate species compositions of modern and pre-colonization forests? 2) Was the Holocene primate community of the now nearly-barren central Madagascar "eastern" (wet forest) in character or did central Madagascar harbor a mixed-species assemblage? 3) What do the compositions of Holocene primate communities tell us about past routes of dispersal of lemur taxa? 4) Did the Holocene primate communities exhibit a "distance effect" (i.e., an inverse relationship between similarity in community composition and geographic distance) as has been demonstrated by Ganzhorn (1998a) for modern dry-forest communities of Madagascar? Finally, 5) do the extinct and extant taxa at subfossil sites yield similar biogeographic signals? (For a discussion of the geographic context of primate extinctions on Madagascar and the ecological implications of the loss of Madagascar's primate megafauna, see Godfrey et al., 1997a).

The data presented here can serve as a springboard for the analysis of changes in community structure ("rules of assembly", "niche ecospace", the differential vulnerability of taxa to extinction, "nestedness", etc.) in Madagascar (see Patterson & Atmar, 1986; Patterson, 1987; Atmar & Patterson, 1993; Ganzhorn, 1997, 1998; Ganzhorn et al., 1997; Godfrey et al., 1997a). Because the modern primate communities of Madagascar cannot be considered ecologically intact, these data should also inform studies of the "convergence" (or "non-convergence") of Malagasy and other primate communities (see Terborgh & van Schaik, 1987; Reed & Fleagle, 1995; Fleagle

& Reed, 1996; Kappeler & Heymann, 1996). Such analyses are beyond the scope of the current paper. We do, however, propose possible avenues for research along these lines.

METHODS

We surveyed the extensive subfossil collections of the Laboratoire de Paléontologie des Vertébrés, Université d'Antananarivo [UA], (Antananarivo, Madagascar), and the Duke University Primate Center [DUPC], (Durham, NC), identifying to the species level (whenever possible) all specimens of extant lemurs. Additional specimens of extant subfossil lemurs belonging to a number of different museums (see Acknowledgments) were also examined and identified. Identifications made on the basis of morphology were verified metrically; postcranial elements of species that are difficult to distinguish morphologically can often be identified by their size. The literature was consulted for additional references to subfossil specimens that may have been lost; these were included when adequately described. Comparative data on modern taxa were collected by LRG and WLJ. Published descriptive statistics for modern taxa were also examined (e.g., Albrecht et al., 1990).

Six subfossil sites were selected on the basis of their geographic location and the quality of our data on the extant lemurs found there (Table 1, Fig. 1). The marsh deposits of Ampasambazimba, as well as the caves of the Ankarana Massif, are among the richest of subfossil primate sites on Madagascar; they have also yielded numerous extant lemur specimens. Specimens of 10 extant lemur species have been found at Ampasambazimba (alongside those of seven or eight extinct primate taxa), and 11 extant lemur species have been found in the subfossil deposits of Ankarana (alongside those of six or seven extinct lemur species). Recent excavations at Anjohibe (northwest Madagascar) as well as at Ankilitelo (a pit 150 m in depth located on the karst landscape of the Mikoboka Plateau, north of Toliara, in southwest Madagascar) have also yielded huge numbers of specimens of extant species (belonging to seven species at Anjohibe, six at Ankilitelo). A fragmentary ulna from Anjohibe, recently allocated to *Hapalemur* sp. (cf. *griseus*) (Burney et al., 1997, Appendix), appears to have belonged, instead, to *Lepilemur edwardsi*. In the extreme north of Madagascar, two

Table 1. List of selected subfossil sites, region, references to geology, exploration, ecology

Site code and name	Geographic location	Key references
1. Ampasambazimba	Center	Jully and Standing (1904); Grandidier (1905); Standing (1906, 1908, 1909, 1910); Perrier de la Bâthie (1927); Lamberton (1939b); Battistini and Vérin (1967); Tattersall (1973); MacPhee et al. (1985).
2. Montagne des Français	North	Dewar and Rakotovololona (1986, 1992).
3. Ankarana	North	Duflos (1966, 1968); Decary and Kiener (1970); Radofilao (1977); Andre et al. (1986); Wilson et al. (1989); Simons et al. (1990, 1995a,b); Godfrey et al. (1990a); Jungers et al. (1991, 1995).
4. Anjohibe/Anjohikely	West (NW)	Decary (1934, 1939), de Saint-Ours (1953), Decary and Kiener (1970); Mahé (1976), Laumanns et al. (1991), Burney et al. (1997); for nearby sites, see also Mahé (1965).
5. Tsirave	West (SW)	Lamberton (1932, 1934, 1937, 1939a), Tattersall (1973).
6. Ankilitelo	West (SW)	Lapaire et al. (1975); Andre et al. (1986); Bonnardin (1988); unpublished data.

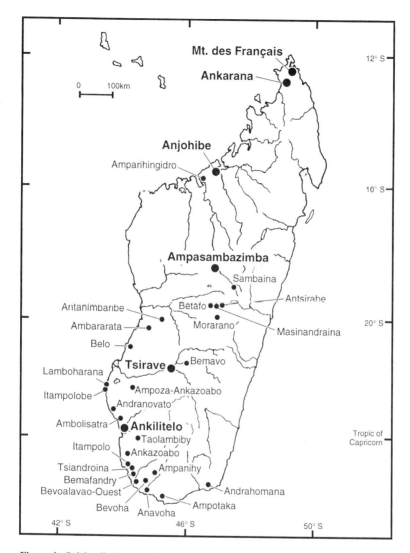

Figure 1. Subfossil site map highlighting sites selected for analysis (large circles).

dozen archaeological sites in small caverns or rockshelters within a narrow, steep-sided canyon (called the Gorge d'Andavakoera) at the Mt. des Français were found to contain charcoal, artifacts, shell, and bones of fish, tortoises, giant extinct lemurs, extant lemurs, cattle, and sheep or goats (Dewar & Rakotovololona, 1986, 1992). A total of seven species of extant lemurs have been found here. Finally, the stream beds of Tsirave on the Mangoky River in southwest Madagascar have yielded specimens of five species of extant lemurs.

Twenty-five modern forest sites were selected for comparative purposes (Table 2, Fig. 2); these span the range of ecological diversity of modern forest sites in Madagascar, from very dry to very wet, and from sea level to high mountains. One site (Andohahela, in southeast Madagascar) is actually comprised of two very distinct primate communities—one on the eastern (or wet forest) and the other on the southwestern (or dry, spiny forest) side of the wet/dry divide (Feistner & Schmid, 1999;

Table 2. List of selected modern forests and their ecological characteristics

Site code and name	Geographic region	Forest type
7. Analamazaotra	East	Wet
8. Zahamena	East	Wet
9. Ranomafana	East	Wet
10. Mantady	East	Wet
11. Ambatovaky	East	Wet
12. Anjanaharibe-Sud	East	Wet
13. Verezanantsoro	East	Wet
14. Betampona	East	Wet
15. Marojejy	East	Wet
16. Manongarivo	Sambirano	Wet
17. Manombo	East	Wet, degraded
18. Montagne d'Ambre	North	Wet
19. Ambohitantely	Center	Wet
20. Andringitra	East	Wet
21. Andohahela	East and South (on divide)	Mixed
22. Analamera	North	Dry
23. Ankarana today	North	Mixed
24. Zombitse	West	Dry
25. Andranomena	West	Dry
27. Namoroka	West (NW)	Dry
28. Berenty	South	Dry
29. Beza Mahafaly	South	Dry
30. Tsimanampetsotsa	South	Dry
31. Isalo	South	Dry

see also Goodman et al., 1997, for a discussion of the bird communities). However, we treated the two sectors as a single "mixed" community because there is a strong probability that subfossil specimens from neighboring communities along an ecotone would appear mixed in their depositional environments. This problem would be exacerbated by temporal shifts in the exact location of the ecological boundary.

Presence (1) or absence (0) of each extant lemur species was recorded for each subfossil and modern forest site, as gleaned from our own data for fossil sites and from published sources for modern forests (Ganzhorn, 1998a; Thalmann et al., 1998). This resulted in a raw data matrix (1–0) of 31 rows (six subfossil and 25 modern forest sites) by 21 columns (the total number of extant lemur taxa analyzed). A presence/absence matrix of 6 rows (subfossil sites) by 15 columns (the total number of extinct lemur species found at these sites) was also constructed to test regional patterns of commonality of extinct primate taxa. Finally, a "total evidence" matrix of 6 rows (subfossil sites) by 36 taxa (extinct plus extant primates) was constructed for subfossil sites only.

Omitted from the extant-taxon matrix were all species of *Microcebus* as well as all taxa whose bones have not been identified at any subfossil site (*Allocebus trichotis*, *Phaner furcifer*, *Mirza coquereli*, *Hapalemur aureus*, *Eulemur rubriventer*, and *E. macaco*). Specimens of *Microcebus* have been found at many subfossil sites, including Andrahomana (in southeastern Madagascar), Ankazoabo Cave (in the southwest), Ankilitelo (in the southwest), Ampasambazimba (in central Madagascar), Anjohibe (in the northwest), and Ankarana (in the extreme north); *Microcebus* is also the most ubiquitous of lemurs in modern (including degraded) forests. Nevertheless, *Microcebus* was excluded from our analysis for two reasons. First, given the taphonomic bias against

Figure 2. Map of Madagascar showing locations of the 25 selected modern forests (closed circles) in relation to the 6 subfossil sites selected for analysis (closed squares). One site (Ankarana) has both modern and subfossil extant-primate communities.

small species, we cannot take the absence of specimens of *Microcebus* at subfossil sites as diagnostic of their past absence from the surrounding habitats. Secondly, *Microcebus* is currently undergoing taxonomic revision (Rasoloarison, in prep.); thus, current species allocations must be treated as suspect. We did include in the matrix six species of *Lepilemur*. The two eastern forest forms, *L. mustelinus* and *L. microdon*, were treated as conspecific (we cannot distinguish their postcrania). Eastern and western forms of *Avahi* were also treated as conspecific. *E. fulvus* was considered absent at Berenty because it was introduced there by humans; a recent report of the possible occurrence of *Hapalemur simus* at Zahamena has been found to be dubious (see Ganzhorn, 1998a).

We used Jaccard's coefficients of community similarity to assess the commonality of species compositions of subfossil and modern forests (NTSYS-pc; Rohlf, 1992). Jaccard's coefficient differs from the "simple matching" coefficient in that the former scores as "matches" only shared presences of taxa. Jaccard's coefficient is calculated as the number of 1-1 matches divided by the sum of these matches plus all mismatches (0-1 and 1-0). Jaccard's coefficients were calculated for all pairwise comparisons of sites, using both extant and extinct species raw-data matrices. This generated a 31 × 31 symmetrical matrix of Jaccard's coefficients for extant species comparisons (for all sites), and 6 × 6 symmetrical matrices of Jaccard's coefficients for subfossil-site comparisons based on extinct taxa only, extant taxa only, and the total evidence (extinct plus extant taxa). A geographic distance matrix was also generated; this is the complete set of pairwise distances (in km) between sites (measured linearly, or "as the crow flies", from mapped site coordinates).

The Jaccard similarity matrices were analyzed for biogeographic content. We calculated mean similarity values for modern "wet" and "dry" forest communities and for communities in different geographic regions. "Mixed" habitat communities were excluded from both "wet" and "dry" forest comparisons. The relationship between Jaccard coefficients of similarity and geographic distance was evaluated via matrix correlations and the Mantel Z statistic (with 1000 random permutations) for all forests and for given subsets: i.e., all "wet" forests, eastern wet forests only, all "dry" forests, and the subset of modern forests from all regions (north, west, center, and south) with subfossil primate sites. Average similarities of subfossil (extant-primate) communities to modern forest sites were also assessed by geographic region and by forest type. A UPGMA phenogram was constructed to summarize the hierarchy of site similarities suggested by the full extant-taxon similarity matrix, and the cophenetic correlation between the similarities implied by the phenogram and the original Jaccard matrix was calculated to assess the adequacy of the clustering summary of similarities (Rohlf, 1992). Phenograms were also constructed for the subfossil sites (taken alone), based on the commonality of their extant taxa and on the commonality of their extinct taxa. A "total evidence" phenogram for subfossil site similarities (based on extant plus extinct taxa) was then constructed.

RESULTS

Distribution of Extant Lemurs at Subfossil Sites

Table 3 shows our raw-data matrix for extant lemur taxa; the subfossil sites are listed in the first six rows. Far from being rare, extant taxa are the dominant faunal elements at some subfossil sites. In each of the two major cave systems at Ankarana (Andrafiabe, Antsiroandoha), as well as in myriad isolated caves in and near the mountain range, specimens of the still-extant *Hapalemur simus* rival in ubiquity and abundance those of the extinct species of *Archaeolemur* found there (see Table 4). At Ankilitelo, along with the extinct *Palaeopropithecus ingens*, cheirogaleids and *Lemur catta* are the dominant primate taxa. Often, the extant taxa whose bones have been found at subfossil sites still live near the site. Thus, for example, there are some 10 or 11 lemur species that occur today in the forests of Ankarana (Hawkins et al., 1990; Wilson et al., 1995), and the bony remains of eight of these [*Microcebus* sp., *Lepilemur* sp. (probably *septentrionalis*), *Eulemur coronatus*, *E. fulvus* (probably *sanfordi*), *Hapalemur griseus*, *Daubentonia madagascariensis*, *Avahi laniger*, and *Propithecus diadema* (probably *perrieri*)] have been found in the subfossil deposits of the mountain caves. All of the extant primate taxa whose bones have been found at Anjohibe, with the exception of *H. simus*, still live in the region of Mahajanga. The extant subfossil fauna at Ankilitelo and at Tsirave in the southwest largely matches the fauna that survives in the southwest today. It is noteworthy that, at these two sites where subfossil *Lemur catta* and *E. fulvus* (probably *rufus*) co-exist, bones of the former are more common at the more southerly site (Ankilitelo) and the opposite is true at Tsirave. The current geographic ranges of *L. catta* and *E. fulvus* overlap at a few forest localities (see Sussman, 1977; Petter et al., 1977; Mittermeier et al., 1994; Ganzhorn, 1994), but that of *L. catta* extends predominantly to the south of the range of *E. fulvus*.

Locally extinct taxa are not uncommon at subfossil sites. Specimens of *Indri indri*, *Hapalemur simus*, and *Propithecus tattersalli* have been found at Ankarana, for example.

Table 3. Extant taxon raw data matrix. L. = *Lepilemur*

SITE CODE	*Cheirogaleus major*	*Cheirogaleus medius*	*Daubentonia madagascariensis*	*Avahi laniger*	*Propithecus diadema*	*Propithecus verreauxi*	*Propithecus tattersalli*	*Indri indri*	*Lemur catta*	*Eulemur mongoz*	*Eulemur coronatus*	*Eulemur fulvus*	*Varecia variegata*	*Hapalemur simus*	*Hapalemur griseus*	*Lepilemur septentrionalis*	*Lepilemur dorsalis*	*Lepilemur leucopus*	*Lepilemur ruficaudatus*	*Lepilemur edwardsi*	*Lepilemur microdon / mustelinus*
1	1	0	0	1	1	1	0	1	0	1	0	1	1	1	0	0	0	0	0	0	1
2	0	0	0	1	1	0	0	0	0	0	1	1	0	1	1	1	0	0	0	0	0
3	0	0	1	1	1	0	1	1	0	0	1	1	0	1	1	1	0	0	0	0	0
4	0	1	0	0	0	1	0	0	0	1	0	1	0	1	0	0	0	0	0	1	0
5	0	0	0	0	0	1	0	0	1	0	0	1	0	0	0	0	0	1	1	0	0
6	0	1	0	0	0	1	0	0	1	0	0	1	0	0	0	0	0	1	0	0	0
7	1	0	1	1	1	0	0	1	0	0	0	1	1	0	1	0	0	0	0	0	1
8	1	0	1	1	1	0	0	1	0	0	0	1	1	0	1	0	0	0	0	0	1
9	1	0	1	1	1	0	0	0	0	0	0	1	1	1	1	0	0	0	0	0	1
10	1	0	1	1	1	0	0	1	0	0	0	1	1	0	1	0	0	0	0	0	1
11	1	0	1	1	1	0	0	1	0	0	0	1	1	0	1	0	0	0	0	0	1
12	1	0	1	1	1	0	0	1	0	0	0	1	0	0	1	0	0	0	0	0	1
13	1	0	1	1	1	0	0	1	0	0	0	1	1	0	1	0	0	0	0	0	1
14	1	0	1	1	1	0	0	1	0	0	0	1	1	0	1	0	0	0	0	0	1
15	1	0	1	1	1	0	0	0	0	0	0	1	0	0	1	0	0	0	0	0	1
16	1	0	1	0	0	0	0	0	0	0	0	1	0	0	1	0	1	0	0	0	0
17	1	0	0	0	0	0	0	0	0	0	0	1	1	0	1	0	0	0	0	0	1
18	1	0	1	0	0	0	0	0	0	0	1	1	0	0	0	1	0	0	0	0	0
19	0	0	0	1	0	1	0	0	0	0	0	1	0	0	0	0	0	0	0	0	0
20	1	0	1	1	1	0	0	0	1	0	0	1	1	1	1	0	0	0	0	0	1
21	1	1	1	1	1	1	0	0	1	0	0	1	0	0	1	0	0	1	0	0	1
22	0	0	1	0	1	0	0	0	0	0	1	1	0	0	0	1	0	0	0	0	0
23	0	1	1	1	1	0	0	0	0	0	1	1	0	0	1	1	0	0	0	0	0
24	0	1	0	0	0	1	0	0	1	0	0	1	0	0	0	0	0	0	1	0	0
25	0	1	0	0	0	1	0	0	0	0	0	1	0	0	0	0	0	1	0	0	0
26	0	1	0	1	0	1	0	0	0	1	0	1	0	0	0	0	0	0	0	1	0
27	0	1	1	0	0	1	0	0	0	0	0	1	0	0	1	0	0	0	1	0	0
28	0	1	0	0	0	1	0	0	1	0	0	0	0	0	0	0	0	1	0	0	0
29	0	1	0	0	0	1	0	0	1	0	0	0	0	0	0	0	0	1	0	0	0
30	0	0	0	0	0	1	0	0	1	0	0	0	0	0	0	0	0	1	0	0	0
31	0	0	0	0	0	1	0	0	1	0	0	1	0	0	0	0	0	0	0	0	0

Table 4. Distribution of specimens of extant taxa in the subfossil deposits of the caves of the Ankarana Massif; overall abundance as subfossils: +++ common; ++ moderate, + scarce

	Cave of the Lone Barefoot Stranger	Antsiroandoha	Galerie des Gours Secs	Andrafiabe	Entrance Andrafiabe	Antenoaka	Anjohin' olona	Grotte de Matsaborimanga	Grotte Nord de la Cassure de Milaintety	Andetobe	Grotte de la Cassure des Arcades	Ankoatra	Grotte de la Forêt Isolée
Hapalemur simus +++	1	1	1	1	1	0	1	0	1	1	1	1	1
Eulemur fulvus +++	1	1	1	1	0	0	0	1	0	0	0	0	0
Propithecus diadema ++	1	1	1	1	0	1	0	0	0	0	0	0	0
Indri indri ++	1	1	1	1	0	0	0	0	0	0	0	0	0
Eulemur coronatus ++	1	0	1	1	0	0	0	0	0	0	0	0	0
Lepilemur sp. ++	1	1	0	1	0	0	0	1	0	0	0	0	0
Daubentonia mad/sis +	1	0	0	1	0	0	0	0	0	0	0	0	0
Hapalemur griseus +	1	1	0	0	0	0	0	0	0	0	0	0	0
Avahi sp. +	1	0	0	1	0	0	0	0	0	0	0	0	0
Microcebus sp. +	0	1	0	0	0	0	0	0	0	0	0	0	0
Propithecus tattersalli +	0	1?	0	0	0	0	0	0	0	0	0	0	0

None of these live in the area today. All but two of the extant taxa whose bones have been found in the marsh deposits at Ampasambazimba no longer live in the region.

A general though not universal phenomenon is the greater size and robusticity of specimens of extant lemurs found at subfossil sites in comparison to samples taken from modern forests. In particular, greater size and robusticity characterizes most of the larger-bodied of extant species whose bones have been found at subfossil sites—*Indri indri*, *Propithecus verreauxi*, *Varecia variegata*, *Hapalemur simus*, and *Eulemur fulvus*. Table 5 shows selected comparative data that illustrate this point. Note that in every comparison the value for the subfossil mean is greater than that of its modern counterpart. Although not every comparison is significant statistically, an overall sign test strongly supports the inference that subfossils were predictably more robust and/or larger. These data suggest a recent and rapid secular trend towards gracilization of the larger-bodied extant lemur taxa, which in turn suggests that these taxa may be food- or activity (range?)-limited in ways that previously did not hold. Small changes in the pattern of seasonality in Madagascar might account for these differences, as growth in lemurs is strongly influenced by the relative lengths of the wet and dry seasons (Albrecht et al., 1990; Godfrey et al., 1990b; Pereira, 1993; Ravosa et al., 1993; Leigh & Terranova, 1998). Burney and others have shown that a drying trend anticipated the arrival ca. 2000 years ago of humans in Madagascar (see, for example, Burney et al., 1986; Burney, 1987a, 1988, 1993).

The subfossil site distributions of several taxa warrant special comment. *Hapalemur simus* is found today in four forests in a small portion of southeastern Madagascar (Vondrozo, Kianjavato, Andringitra, and Ranomafana). Museum collections made only a century ago show that, even at that time, its geographic range extended throughout the eastern rain forest, from Vondrozo in the southeast to the region of the Bay of Antongil in the northeast (Godfrey & Vuillaume-Randriamanantena, 1986). Specimens of *H. simus* have now been found at four subfossil localities (Mt. des Français, Ankarana, Anjohibe, and Ampasambazimba), each of

Table 5. Descriptive statistics for specimens of still-extant lemurs found at subfossil sites (including Ampasambazimba, Tsirave, Anjohibe, Ankilitelo, Ankarana, Mt. des Français, and Manombo-Toliara) with modern comparisons. Values are means ± standard deviations (S.D.)

		Subfossil		Modern		Subfossil > Modern?
		N	Mean ± S.D.	N	Mean ± S.D.	
Hapalemur	Femoral robusticity	22	20.1 ± 1.1	5	18.3 ± 0.5	Yes**
simus	Maximum femur length	22	143.5 ± 5.4	5	140.3 ± 6.6	Yes
	Femoral midshaft transverse	46	8.7 ± 0.4	5	7.8 ± 0.5	Yes***
	Maximum humerus length	16	83.3 ± 2.7	5	81.2 ± 3.2	Yes
	Humeral midshaft transverse	16	7.1 ± 0.6	5	6.9 ± 0.6	Yes
Propithecus	Femoral robusticity	5	17.1 ± 0.7	13	16.2 ± 1.2	Yes
verreauxi	Maximum femur length	5	189.9 ± 9.2	21	183.0 ± 6.6	Yes
	Femoral midshaft transverse	19	9.8 ± 0.9	21	8.7 ± 0.6	Yes***
	Maximum humerus length	4	102.8 ± 5.3	21	96.5 ± 4.3	Yes*
	Humeral midshaft transverse	6	8.1 ± 0.7	21	6.6 ± 0.5	Yes*
Propithecus	Femoral midshaft transverse	10	11.1 ± 0.6	10	10.9 ± 0.6	Yes
diadema	Femoral midshaft circumference	15	37.6 ± 2.7	10	35.7 ± 1.5	Yes
Indri indri	Femoral robusticity	1	20.9	18	15.6 ± 0.6	Yes
	Femoral midshaft transverse	2	15.4 ± 1.8	18	11.8 ± 0.5	Yes
Varecia variegata	Humeral midshaft transverse	3	8.6 ± 0.7	7	7.9 ± 0.3	Yes*
Eulemur fulvus	Femoral robusticity	12	20.9 ± 1.1	18	19.4 ± 0.1	Yes
	Maximum femur length	12	131.7 ± 5.4	29	126.0 ± 3.7	Yes***
	Femoral midshaft transverse	18	8.2 ± 0.3	18	7.4 ± 0.3	Yes***
	Maximum humerus length	9	87.3 ± 1.5	27	84.6 ± 3.0	Yes**
	Humeral midshaft transverse	13	6.5 ± 0.5	14	6.0 ± 0.4	Yes**

Femoral robusticity = (Femoral midshaft circumference × 100) / Maximum femur length. All measurements in mm except femoral robusticity, which is dimensionless.
*p < 0.05, **p < 0.01, ***p < 0.001.

which is located well outside this species' historic geographic range (Vuillaume-Randriamanantena et al., 1985; Simons et al., 1990; Godfrey et al., 1997a). The occurrence of postcrania of *H. simus* at Ampasambazimba has not previously been reported; we have identified three specimens (a complete femur, UA 8764, and two partial femora, UA 8792 and UA 8791) as belonging to this taxon. The complete femur was mistaken by Lamberton (1939b) for a specimen of *Varecia variegata* (his Plate Ic).

Lamberton (1939b) was apparently correct, however, in allocating a femur (UA 8812) from Ampasambazimba to *Eulemur mongoz*, a species which occurs today (outside the Comores) only in the northwest (on either side of the Betsiboka River, in the region southwest of the Mahavavy River, and north through the Ankarafantsika Reserve to the forests near Antsohihy; see Mittermeier et al. [1994]). Two femoral specimens (UA 8812 and UA 8782) stand out as significantly smaller than any in the substantial series of *E. fulvus* from Ampasambazimba. At 109.9 mm in maximum length, the smaller of the two falls outside the size range for modern samples of *E. fulvus*, but well within the size range for modern samples of the *E. mongoz* (Fig. 3).

Cranial and postcranial specimens of *Indri indri* have been found at the Ankarana Massif; see Jungers et al. (1995). Only postcranial elements of this taxon are known from Ampasambazimba (Fig. 4a). These include a complete femur (UA 8724), a femoral shaft (UA 8725), a proximal tibia (UA 1185), a complete humerus (UA 8726), and an ulna (UA 8732). The hindlimb specimens from Ampasambazimba are quite robust (see Table 5).

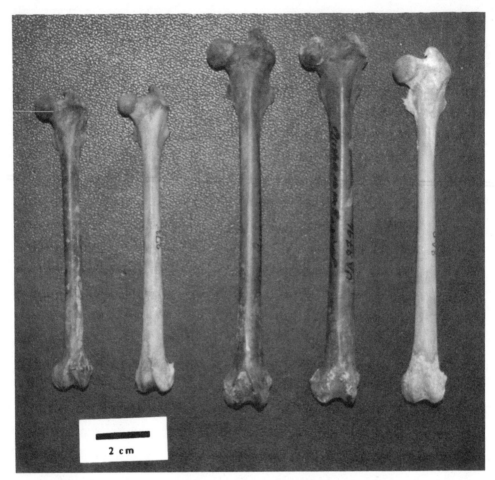

Figure 3. Femora of, from left to right, *Eulemur mongoz* from Ampasambazimba (UA 8812), modern *E. mongoz* (UA 373), *E. fulvus* from Ampasambazimba (UA 8793), *E. fulvus* from Ampasambazimba (UA 8794), modern *E. fulvus* (UA 840).

Several fragmentary humeri (UA 8765 and UA 8766) demonstrate that *Varecia variegata* was indeed present at Ampasambazimba. But rather than being gracile (as reported by Lamberton, 1939b, on the basis of a femur that actually belongs to *Hapalemur simus*), the subfossil *Varecia* from the Central Highlands was quite robust. Along with *Indri indri*, its presence at Ampasambazimba bears testimony to a former geographic range of "typically eastern" forms well to the west of their current ranges.

Eastern relict fauna and flora occur even today in the extreme west, in the region of Zombitse-Sakaraha, east of Morondava, and northeast of Toliara (Du Puy et al., 1994; Langrand & Goodman, 1997; Ganzhorn, 1998a). Indeed there exists in this region (west of Zombitse and north of Mahabobaka) a "mist oasis" called Analavelona that is a mixture of eastern and western in character (S. M. Goodman, pers. comm.). The primate species that live today in the forests near Morondava are western in character. However, some bones of "eastern" primate species occur at subfossil sites in this region. Two specimens that we have identified as *Varecia*—a femur of an immature individual (UA 8780) and a fragmentary humerus with intact distal end (UA 8781)—were collected in the early 1900's at "Manombo Tulear" (or Manombo Toliara). This site is

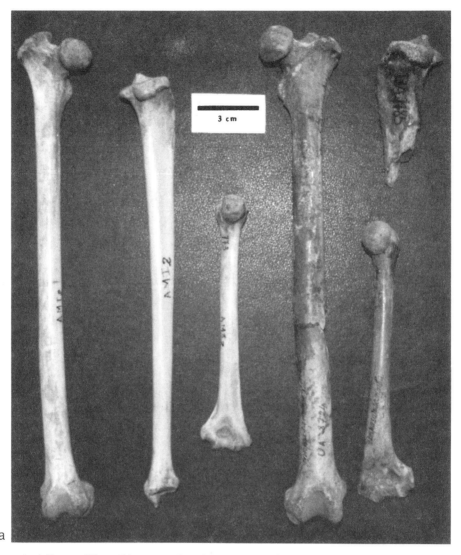

Figure 4. a) Femur, tibia, and humerus of modern *Indri indri* (UA-AMI2, right side) from the east, compared to subfossil *I. indri* (femur UA 8724, proximal tibia UA 1185, and humerus UA 8726, all left) from Ampasambazimba. b) From left to right, UA-AM 841 (left humerus of modern *Indri*), UA 8726 (left humerus of subfossil *Indri* from Ampasambazimba), and Field Museum of Natural History, field catalog E1-52 (left humerus of probable subfossil *Indri* from Ampoza).

located just north of Toliara, near Ambolisatra (Lamberton, 1934). Recently, a damaged humeral shaft that probably belonged to an *Indri indri* was also recovered at Ampoza (near Tsirave) by S. M. Goodman and A. D. Yoder (Field Museum of Natural History, field catalog # E1-52); see Fig. 4b. This specimen is missing both proximal and distal ends. It is indrid in morphology, significantly larger in size than *Propithecus* (modern and subfossil), and indeed larger than many modern *Indri*. We must await better specimens before making a definitive species allocation. However, it seems likely that the geographic ranges of both *Varecia* and *Indri* once extended far to the west and south of Ampasambazimba, into a region that today harbors relict eastern flora and fauna.

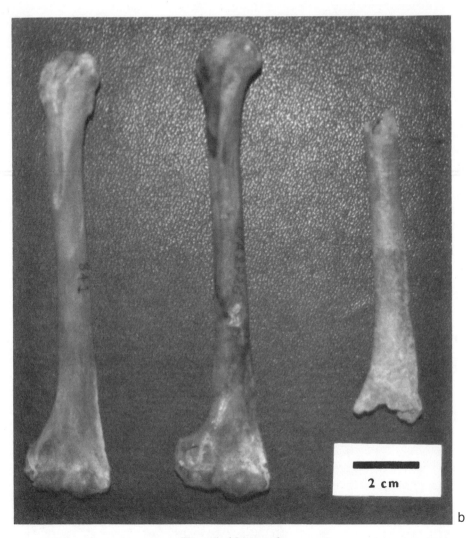

Figure 4. (*Continued*)

It furthermore appears that primate taxa that today live in western coastal forests once lived well to the east of their current ranges—i.e., in the Central Highlands. The example of *Eulemur mongoz* was discussed above. Another example is *Propithecus verreauxi*. Some specimens of *Propithecus* (e.g., femora UA 8737 and UA 8738, tibia UA 8739) from Ampasambazimba cannot be distinguished from subfossil *Propithecus* (presumably *P. verreauxi*) from the extreme west, while other *Propithecus* specimens from Ampasambazimba are considerably larger, and fall comfortably within the range of modern eastern *P. diadema*. Indeed, Jully and Standing (1904) noted the possible presence of two species of *Propithecus* in the deposits at Ampasambazimba (see also Chanudet, 1975, p. 59). In the historic past, *P. verreauxi* lived in the isolated rain forest of Ambohitantely, only 100 km north of Antananarivo, and not far from Ampasambazimba (Petter & Andriatsarafara, 1987). Today, the three species of *Propithecus* have allopatric distributions. *P. verreauxi* was reported to be sympatric with *P. diadema* in the eastern sector of Andohahela (Ganzhorn, 1998a), but a recent survey by Feistner and Schmid (in press) finds no overlap between the primate fauna on the western and eastern faces of the ecological divide between modern wet and dry forests at this mixed-habitat locality. *P. verreauxi* occupies the dry forests of southern and western Madagascar; *P. tattersalli* occupies the dry forests located between the Loky and Manambato Rivers in northeast Madagascar; and *P. diadema* occupies the rain forests of the east and the mixed forests of the extreme north.

Many subfossil sites in the southwest (including Ankazoabo Cave, Ambararata, Tsirave, Belo-sur-Mer, Ampoza, Anavoha, and Ankilitelo) have yielded subfossil *Propithecus*, and the postcrania of most of these are indeed at least slightly more robust than those of modern *P. verreauxi* from this region—sometimes markedly so (Table 5). Lamberton (1939a) had assigned a new species nomen, *P. verreauxoides*, to some of these specimens. Lamberton's new species was synonymized with *P. verreauxi* by Tattersall (1971). The craniodental specimens fall at the upper end of the range of variation of modern *P. verreauxi*. Most subfossil specimens of *Propithecus* from Ankarana fall comfortably within the size range of modern *P. diadema perrieri* (the Perrier's Sifaka at Ankarana is the smallest subspecies, and the subspecies at Ampasambazimba—probably *P. d. diadema*—is considerably larger). However, some specimens from the Ankarana (e.g., DUPC 6813) appear to be too small for *P. d. perrieri*, and were recently allocated to *P. tattersalli* (Jungers et al., 1995).

Subfossil specimens of *Lepilemur* from the region south of Morondava (from the Mangoky to the Onilahy Rivers) show an intriguing bimodality (Fig. 5). It appears that

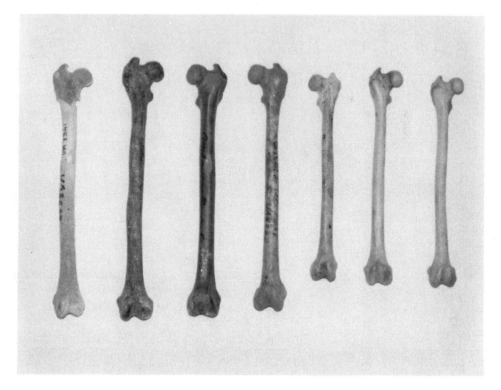

Figure 5. Subfossil femora of *Lepilemur*. Our species allocations are, from left to right, *L. edwardsi* (UA 8692 from Anjohibe), *L. mustelinus* (UA 8811 from Ampasambazimba), *L. mustelinus* (UA 8809 from Ampasambazimba), *L. ruficaudatus* (UA 8768 from Tsirave), *L. leucopus* (UA 8814 from Tsirave), *L. leucopus* (UA 8813 from Tsirave), *L. leucopus* (UA 9853 from Ankilitelo). For scale, UA 9853 is 83.7 mm in length.

there were two species of *Lepilemur* at Tsirave, the smaller of which matches the *Lepilemur* represented at Ankilitelo, and the larger of which more closely matches the *Lepilemur* represented at Anjohibe in the northwest. The *Lepilemur* from Ankilitelo and the smaller of the two apparent species at Tsirave were smaller even than individuals whose bones have been found at Ankazoabo Cave, which is situated well to the south of the Onilahy River.

Today, the smallest-bodied *Lepilemur* (*L. leucopus*) is reported to be limited to the north by the Onilahy River, where it is replaced by *L. ruficaudatus* (Mittermeier et al., 1994). The latter is in turn replaced further northward by *L. edwardsi* (see Thalmann et al., 1998, for comments of the current geographic limits of the range of *L. ruficaudatus*). The actual distribution of species of sportive lemurs in the southwest may be more complex than this; at least two may coexist, even today, in the region just north of the Onilahy River. Museum specimens collected there over the past hundred years exhibit striking coat color variation. Petter et al. (1977) considered the holotype of "*L. globiceps*" (BMNH 1892.11.6.1) from Ambolisatra, 25 km north of the Onilahy River, synonymous with *L. leucopus* (see Jenkins, 1987). Furthermore, individuals sighted in 1998 in the region of Ankilitelo had gray coats (with white ventra) as do *L. leucopus*; they lacked the red highlights and reddish tails that distinguish *L. ruficaudatus* (T. Rasmussen, pers. commun.). Various color morphs (including some with gray coats and no red in the tail) have also been observed recently in the forests of Kirindy (close to Morondava). They have been treated as color variants of *L. ruficaudatus* (J. U. Ganzhorn, pers. commun.; Tomiuk et al., 1997). However, all of the individuals collected almost 130 years ago at Morondava by van Dam (and now housed in the Nationaal Natuurhistorisch Museum, Leiden) show marked uniformity in coat characteristics, and appear to be typical *L. ruficaudatus*.

Figure 6 summarizes the similarities of the extant-species compositions for all sites (modern and subfossil). The phenogram displays two site clusters, the first of which (A) comprises eastern sites (as a subgroup, with its weakest link Manongarivo, a site from the Sambirano) and the forests of the extreme north (i.e., north and east of the Sambirano). The second major cluster (B) comprises the communities of the northwest (south and west of the Sambirano), southwest, and extreme south. The cophenetic correlation between the similarity implied by this phenogram and the original Jaccard matrix of similarity is 0.94. This means that the pairwise distances implied by the phenogram capture the lion's share of the variance contained in the original Jaccard matrix. The matrix correlation (or normalized Mantel statistic Z) between the Jaccard similarity matrix and the geographic distance matrix is -0.48 ($p = 0.001$). This implies strong island-wide geographic patterning.

The subfossil extant-primate communities show clear affinities with *nearby* modern primate communities (Table 6). Those at the Mt. des Français and Ankarana cluster with Ankarana's contemporary primate community and with two other modern communities in the extreme north—Mt. d'Ambre and Analamera. Their strongest affinities are with the wet and mixed forest communities of the north, followed closely by the rain forests of the east. Ampasambazimba clusters (as a distant member) with the communities of the east and with other modern wet forests. Anjohibe clusters with its nearest modern forest (Ankarafantsika); the same holds for Tsirave (with Isalo) and Ankilitelo (with Beza Mahafaly, as well as Berenty and Tsimanampetsotsa). Tsirave and Ankilitelo have strong "dry forest" signals.

The only apparent anomaly is that Ambohitantely, an isolated and strongly fragmented rain forest in the central part of Madagascar located slightly to the north and east of Ampasambazimba, clusters (weakly) with the communities of the north-

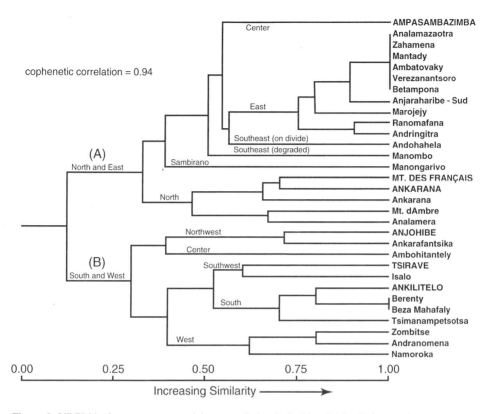

Figure 6. UPGMA phenogram summarizing overall site similarities. Subfossil sites are in upper case.

west (and thus with the west and south) rather than with the communities of the east. In part, this reflects our scoring of *Propithecus verreauxi* as present at Ambohitantely. (This species has only very recently disappeared from this forest; see above). More to the point, two of the three primate species that survive today at Ambohitantely (and both of the other species, *Avahi laniger* and *Eulemur fulvus*, that were considered in this analysis) occur in the northwest as well as the east. Had we considered species of *Microcebus* in our analysis, the linkage of Ambohitantely to the east would have been strengthened, as *Microcebus rufus*, which occurs at Ambohitantely, is characteristic of eastern communities. In fact, this odd mixture of taxa exemplifies the mixed nature of the taxa of the primate communities of central Madagascar.

Table 6. Mean Jaccard similarities of extant-taxon communities at subfossil sites to modern forest communities, by region and by forest type. Values are means ± standard deviations

Site	North	East	West	South	Dry	Wet
Montagne des Français	0.50 ± 0.17	0.35 ± 0.07	0.14 ± 0.05	0.03 ± 0.06	0.13 ± 0.16	0.33 ± 0.07
Ankarana	0.50 ± 0.14	0.43 ± 0.09	0.13 ± 0.07	0.02 ± 0.04	0.12 ± 0.16	0.40 ± 0.11
Ampasambazimba	0.17 ± 0.03	0.54 ± 0.08	0.20 ± 0.09	0.11 ± 0.05	0.15 ± 0.08	0.47 ± 0.16
Anjohibe	0.12 ± 0.04	0.09 ± 0.03	0.46 ± 0.17	0.20 ± 0.13	0.30 ± 0.20	0.11 ± 0.06
Tsirave	0.10 ± 0.02	0.09 ± 0.02	0.44 ± 0.19	0.55 ± 0.06	0.45 ± 0.18	0.11 ± 0.07
Ankilitelo	0.14 ± 0.04	0.09 ± 0.02	0.48 ± 0.14	0.70 ± 0.12	0.54 ± 0.22	0.11 ± 0.07

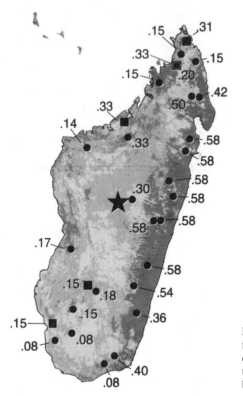

Figure 7. Map showing Jaccard similarity coefficients for all pairwise comparisons of Ampasambazimba's extant-primate community to extant-primate communities at all other sites. Ampasambazimba is represented by a solid star.

Ampasambazimba is another case in point. Figure 7 displays, in map form, this site's Jaccard similarities to all other communities (subfossil sites and modern forests) in our analysis. Whereas the extant-primate community at Ampasambazimba does indeed most closely resemble modern communities of the east, it is also moderately similar to modern forest communities of the west (especially the northwest) and the north. Note that Ampasambazimba shows stronger Jaccard similarities to the subfossil site communities in the north than to *equally distant* modern communities. This is due to the disappearance of *Hapalemur simus* and *Indri indri* from the north; both formerly occupied northern and central (as well as part of western) Madagascar, but survive today only in the east.

Table 7 shows the mean Jaccard similarities and correlations between Jaccard similarities and geographic distance for selected pairwise comparisons of primate communities, as well as the significance of the standardized Mantel statistic (=matrix correlation), generated using a random permutation test (Rohlf, 1992). For almost all comparisons, there is a significant "distance effect", or inverse correlation between geographic distance and community similarity, signalling that community similarity depends on geographic location. We found a weaker distance effect for wet forest comparisons than for dry forest comparisons, but, unlike Ganzhorn (1998a), who found a strong distance effect for dry forests but none at all for the wet forests of the east, we found statistically significant correlations between geographic distance and community similarity for both wet and dry forests. Like Ganzhorn, we found strong species commonality for wet (especially eastern) forest communities. Our mean Jaccard similarity was 0.60 for all wet forests and 0.80 for the subset of eastern wet forests. Our

Table 7. Correlations between Jaccard similarity coefficients (means ± standard deviation) and geographic distance for diverse pairwise comparisons. N = number of pairwise comparisons. Significance levels calculated using permutation tests (*$p < 0.05$, **$p < 0.01$, ***$p < 0.001$)

Sample	N	Matrix correlation (Mantel statistic)	Jaccard's coefficient ± S.D.
All pairwise comparisons of 6 subfossil sites to 9 modern dry forests	54	–0.78***	0.28 ± 0.23
All pairwise comparisons of 9 modern dry forests	36	–0.76***	0.35 ± 0.24
All pairwise comparisons of 13 modern forests in the south (including Andohahela), west, center, and north	78	–0.70***	0.30 ± 0.21
All pairwise comparisons of 6 subfossil sites to 14 modern wet forests	84	–0.30**	0.25 ± 0.18
All pairwise comparisons of 9 modern dry forests to 14 modern wet forests	126	–0.46***	0.11 ± 0.12
All pairwise comparisons of 6 subfossil sites to 11 eastern rain forests	66	–0.28*	0.26 ± 0.19
All pairwise comparisons of 13 modern forests in the south (including Andohahela), west, center, and north to 11 eastern rain forests	143	–0.12	0.11 ± 0.16
All pairwise comparisons of 14 modern wet forests	91	–0.34**	0.60 ± 0.29
All pairwise comparisons of 11 eastern rain forests	55	–0.62***	0.80 ± 0.17
All pairwise comparisons of 6 subfossil sites	15	–0.65**	0.26 ± 0.20
All pairwise comparisons of 31 sites	465	–0.48***	0.31 ± 0.26

mean Jaccard similarity for pairwise comparisons of modern dry forest communities was considerably lower (0.35). Dry forests (Fig. 8a, open squares) display a strong north-south distance effect combined with low mean Jaccard similarities, while wet forests show a moderate north-south distance effect combined with high mean Jaccard similarities. A weak east-west distance effect characterizes modern communities, as is evident when all wet forests (primarily in the east, but also in the north and center) are compared to all dry forests (primarily in the south and west, but also in the north). Only when all modern communities from the north, west, center and south were compared to all modern eastern rain forests did we find no distance effect (Fig. 8b, open squares).

The subfossil-site communities exhibit a strong north-south distance effect, paralleling that exhibited by modern communities from the same geographic regions (Fig. 8a, closed circles). When the extant primate communities at subfossil sites are compared to those occupying the eastern rain forests today (Fig. 8b, closed circles), a weak distance effect emerges. This distance effect is stronger than that derived for east-west comparisons of modern forests. It is, of course, impossible to compare subfossil communities across the east-west divide, since no primate subfossil sites are known from the east.

Distribution of Extinct Taxa at Subfossil Sites

Table 8 shows our extinct taxon raw-data matrix. The giant aye-aye, *Daubentonia robusta*, is recorded as present at Ampasambazimba. MacPhee and Raholimavo (1988) noted that Ekblom's (1953) "*D. robusta*" innominate from Masinandraina near Antsirabe (in Central Madagascar) actually belonged to *Plesiorycteropus madagas-*

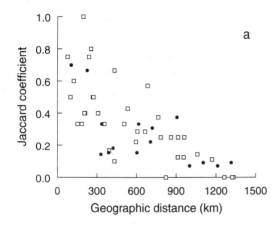

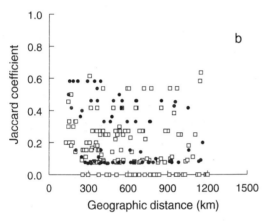

Figure 8. a) Scatter plot showing Jaccard similarity versus geographic distance for 36 pairwise comparisons of 9 modern dry forest communities (open squares, Mantel statistic = –0.76, p < 0.001), and for 15 pairwise comparisons of 6 subfossil extant-primate communities (closed circles, Mantel statistic = –0.65, p < 0.01). b) Scatterplot showing Jaccard similarity vs. geographic distance for 143 pairwise comparisons of 13 modern forest communities in the south, west, center, and north with 11 eastern rain forest communities (open squares, Mantel statistic = –0.12, NS), and for 66 pairwise comparisons of the extant-primate communites from the six subfossil sites with the same 11 forests of the east (closed circles, Mantel statistic = –0.28, p < 0.05).

Table 8. Extinct taxon raw data matrix. *M. = Mesopropithecus*

Site	*Pachylemur* sp.	*Megaladapis madagascariensis*	*Megaladapis grandidieri*	*Hadropithecus stenognathus*	*Archaeolemur majori*	*Archaeolemur edwardsi*	*Archaeolemur* sp.	*Palaeopropithecus ingens*	*Palaeopropithecus maximus*	*Palaeopropithecus* sp.	*Babakotia radofilai*	*M. pithecoides / globiceps*	*M. dolichobrachion*	*Archaeoindris fontoynontii*	*Daubentonia robusta*
Ampasambazimba	1	0	1	1	0	1	0	0	1	0	0	1	0	1	1
Montagne des Français	1	0	1	0	0	0	1	0	1	0	0	0	0	0	0
Ankarana	1	0	1	0	0	0	1	0	1	0	1	0	1	0	0
Anjohibe	0	1	0	0	0	0	1	0	0	1	1	0	0	0	0
Tsirave	1	1	0	1	1	0	0	1	0	0	0	1	0	0	1
Ankilitelo	1	1	0	0	1	0	0	1	0	0	0	0	0	0	1

cariensis. Jully and Standing (1904, p. 92) reported three lower jaw fragments of a large lemur from Ampasambazimba that they claimed resembled (in some unspecified manner) living *Daubentonia*. No anatomical descriptions were provided, and nobody has since been able to locate these specimens. Whereas some authors have speculated that these specimens may have belonged to *D. robusta* (see MacPhee et al., 1985; Vuillaume-Randriamanantena et al., 1985; MacPhee & Raholimavo, 1988), others have conservatively chosen to exclude *Daubentonia* entirely from rosters of primate taxa found at Ampasambazimba (Tattersall, 1973; Simons, 1994; Sterling, 1994; Godfrey et al., 1997a). However, we have recently discovered, in the collection of the Université d'Antananarivo (Laboratoire de Paléontologie des Vertébrés), a tibia from Ampasambazimba that can be definitively identified, despite some damage to its proximal and distal ends, as having belonged to *D. robusta* (UA 8754, Fig. 9). In addition, our 1997 expedition to Ankilitelo (in southwest Madagascar) recovered an incisor of this rare giant lemur (DUPC 17248). *D. robusta*, previously recorded at three subfossil sites in the southwest (Lamboharana, Tsirave, and Anavoha), is now known from five sites spanning central to southwestern Madagascar. It thus paralleled the paleodistribution of *Hadropithecus stenognathus* (Godfrey et al., 1997b).

Specimens of *Palaeopropithecus* from the northwest (Anjohibe, Amparihingidro) differ morphologically and metrically from those found at all other subfossil sites, and a new species based on these specimens is being described (Jungers et al., in prep.). Specimens of *Archaeolemur* from northwestern and northern Madagascar have been called "*Archaeolemur* sp. (cf. *edwardsi*)"; see Godfrey et al. (1997a). They are, in fact, quite similar to *A. edwardsi* from the Central Highlands. They nevertheless differ sufficiently from the latter to suggest possible independent species status (Rasoloharijaona, 1995). We followed Rasoloharijaona (1995) in treating samples of *Archaeolemur* from the extreme north and northwest as a distinct taxon.

The Late Pleistocene and Holocene distribution of species of *Archaeolemur* was certainly more complicated than is suggested by our six-site matrix. There is some evidence that the southern and western coastal form, *Archaeolemur majori*, periodically spread into central Madagascar (Ampasambazimba), and that the Central Highlands form, *A. edwardsi*, periodically spread into the west (Ampoza and Belo-sur-Mer); see Godfrey et al. (1997b). Indeed, multiple species of *Archaeolemur* may have also existed in the extreme north (at Ankarana) and in the southeast (at Andrahomana). The geographic variation and taxonomy of *Archaeolemur* is currently under study (Godfrey et al., in prep.). For our present purposes, we highlighted only the dominant regional forms.

The Central Highlands *Pachylemur* ("*P. jullyi*") is sometimes distinguished from coastal "*P. insignis*" on the basis of its greater size (see Lamberton, 1946; Albrecht et al., 1990; but see Szalay & Delson, 1979). However, there is considerable size variation in the postcrania of southern and coastal *Pachylemur*, and the geographic signal embedded in that variation needs evaluation. We see no clear morphological distinction between coastal and central forms of *Pachylemur* that warrants specific separation, and have thus included all *Pachylemur* in a single taxon. *Mesopropithecus globiceps* from the southwest is extremely similar to *Mesopropithecus pithecoides* from Ampasambazimba (Tattersall, 1971). We treat these, too, as conspecific, and as distinct from *M. dolichobrachion* (from the extreme north) (see Simons et al., 1995b). *Megaladapis* has recently been divided into two subgenera—*M.* (*Peloriadapis*) and *M.* (*Megaladapis*) (Vuillaume-Randriamanantena et al., 1992). The latter includes *M. madagascariensis* from the southwest and *M. grandidieri* from central Madagascar, as well as geographic

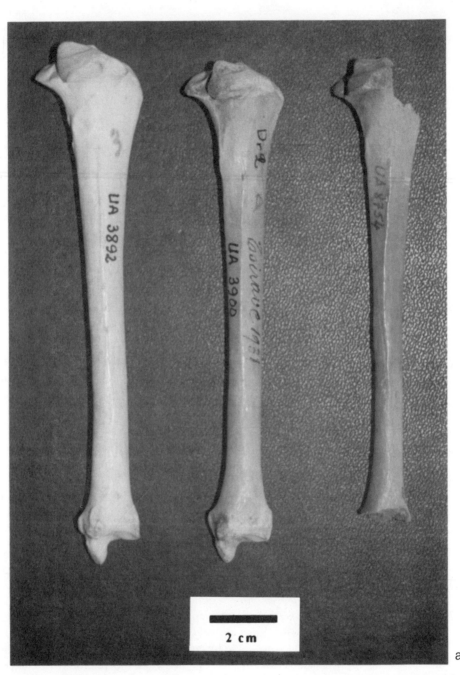

Figure 9. a) Lateral view of three right tibias of *Daubentonia robusta*, from left to right: UA 3892 from Beloha Anavoha, UA 3900 from Tsirave, and UA 8754 from Ampasambazimba. b) Medial view of the same three right tibias of *D. robusta*.

variants from the northwest and extreme north. Specimens of *Megaladapis* from the northwest (Anjohibe) are most similar to *M. madagascariensis*, whereas specimens from the extreme north (Ankarana, Mt. des Français) are closer to *M. grandidieri*. These allocations are uncertain, as the specimens from the northwest and north are actually

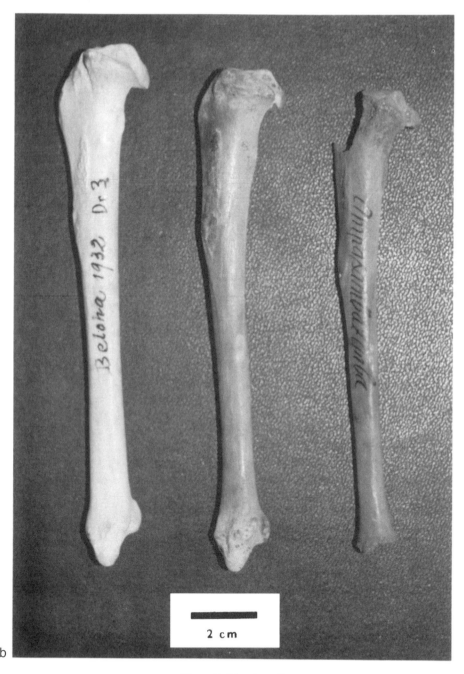

Figure 9. (*Continuted*)

intermediate in some characteristics between southwest coastal *M. madagascariensis* and *M. grandidieri* from the Central Highlands (Vuillaume-Randriamanantena et al., 1992.) A single mandible from Anjohibe with one damaged premolar appears to belong to *Babakotia radofilai* (Burney et al., 1997); this taxon is known elsewhere only from the caves of the Ankarana Massif (Godfrey et al., 1990a).

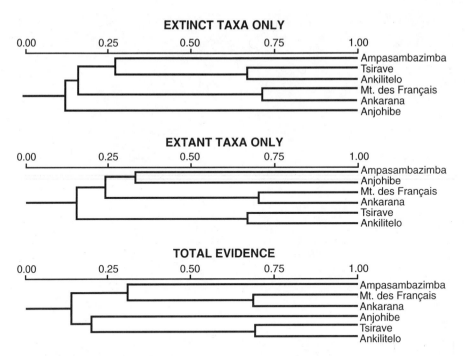

Figure 10. UPGMA phenograms comparing extant and extinct community signals for subfossil sites, plus total evidence.

Three phenograms were constructed to summarize the Jaccard similarities among subfossil sites; the first was based entirely on their extinct primate communites; the second on their extant primate communities, and the third on the total (combined extant and extinct taxon) evidence (Fig. 10). The three are consistent in displaying northern (Mt. des Français and Ankarana) and southern (Tsirave and Ankilitelo) clusters. They are also consistent in demonstrating relatively weak linkages between geographically intermediate sites (such as Anjohibe and Ampasambazimba) and both northern and southern clusters. The phenogram derived for extinct taxa places Ampasambazimba as a distant member of the Mt. des Français-Ankarana cluster, and Anjohibe as bearing a weak link to a cluster containing all other sites. The phenogram derived for extant taxa links both Ampasambazimba and Anjohibe (again, weakly) to the northern cluster. The total evidence phenogram links Ampasambazimba distantly with the northern site cluster, and Anjohibe distantly with the two sites in the southwest. The matrix correlation between Jaccard similarities for extinct and extant primate communities is 0.75 (for 15 pairwise comparisons; $p < 0.01$, using the random permutation test). Fig. 11 displays in map form, from the vantage of Tsirave, Jaccard similarity coefficients derived for comparisons of extant (Fig. 11a) and extinct (Fig. 11b) primate communities. Table 9 shows the Jaccard similarity coefficients based on the total evidence for all pairwise comparisons of primate communities at subfossil sites.

DISCUSSION

The primary goal of this paper was to reconstruct Madagascar's pre-human colonization (and early postcolonization) primate communities, and to contrast the dis-

Table 9. Jaccard's coefficients of similarity for comparisons of subfossil site primate communities, derived on the basis of their total evidence (extinct and extant taxa)

	Montagne des Français	Ankarana	Ampasambazimba	Anjohibe	Tsirave	Ankilitelo
Mt. des Français	1.00					
Ankarana	0.63	1.00				
Ampasambazimba	0.27	0.31	1.00			
Anjohibe	0.18	0.24	0.22	1.00		
Tsirave	0.05	0.08	0.25	0.22	1.00	
Ankilitelo	0.05	0.08	0.18	0.25	0.69	1.00

tributions of species of lemurs in the recent past with those of today. Constructing a detailed chronology of shifts and contractions in species' ranges requires careful and thorough dating of stratigraphically-controlled samples. Such a record is only beginning to be established. Yet, despite our inability to construct a detailed chronology of changes in lemur biogeography, our knowledge of the paleodistributions of Madagascar's lemurs has expanded dramatically over the past 15 years.

Ecologists have long studied biogeographic patterns and the processes (e.g., speciation, migration, extinction, vicariance) that create them. Some have attempted to use biogeographic patterning to reconstruct generative processes. Thus, for example,

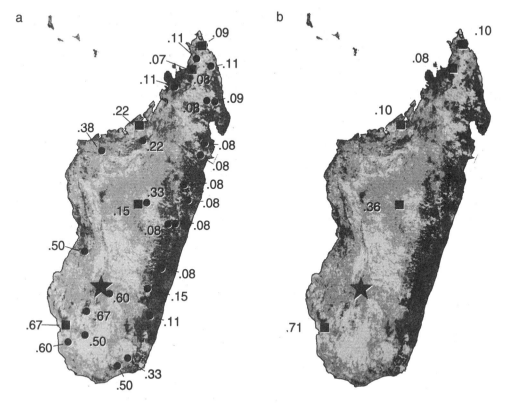

Figure 11. a) Map showing Jaccard similarity coefficients for all pairwise comparisons of the extant-primate community at Tsirave to those at all other sites. b) Map showing Jaccard similarities for all pairwise comparisons of the extinct-primate community at Tsirave to those at five other subfossil sites. Tsirave is represented by a solid star.

Patterson and Atmar (1986) suggested that observed differences in the strength of the relationship between similarity in community composition and the geographic distance between sites might be understood in terms of differences in processes of speciation, dispersal, and local extinction. Following Patterson and Atmar (1986), Ganzhorn (1998a) reasoned that, when forests with a common species pool lose taxa independently due to the differential sensitivity of individual taxa to forest degradation or to decreased fragment size, there should be no "distance effect" (or inverse relationship between community similarity and geographic distance). The selective disappearance of the more ecologically sensitive and larger-bodied taxa from increasingly degraded and fragmented Holocene forests should not, by itself, produce a distance effect. Little or no distance effect may also reflect sharp physical or ecological barriers to migration. Physical or ecological barriers to the dispersal of species between (at least some) forests in close proximity may be stronger than those separating (at least some) geographically distant forests.

In contrast, when: 1) species have different places of origin and different colonization potentials along the same ecogeographic gradient, and 2) at least intermittently, the ecological and physical barriers to dispersal are not prohibitive, then a significant distance effect may emerge. Whenever a distance effect exists, we can presume that the sampled forests were connected at times by corridors that permitted the dispersal (within the constraints of the differences in the colonization potential of the species and of temporal fluctuations in local climate and ecology) of species from one region to another.

Ganzhorn (1998a) examined the relationship between community similarity and geographic distance in Madagascar, as well as the pattern of distribution of species within regions, in an attempt to reconstruct the biogeographic history of the primates of eastern and western Madagascar. The distribution at subfossil sites of extant taxa, and indeed of extinct taxa, affords us an independent test of hypotheses derived from studies of the pattern of distribution of species in modern forests. The weak east-west distance effects that we observed for both modern and subfossil (extant community) comparisons bears testimony to the prior existence of faunal exchange routes that crossed the Central Highlands. The concordance of north-south distance effects for Holocene and modern primate communities (based on analysis of both extinct and extant taxa) bears testimony to the persistence of strong regional endemicity from the southern to the northern tips of the island. Our data confirm that the Central Highlands behaved as a crossroads for the past dispersal of both dry- and wet-loving species, providing both east-west and north-south corridors. They suggest that the ecological barriers between eastern and non-eastern communities were more porous in the past than they are today and that the strongly disjunct distributions exhibited by many lemur subspecies and species today may be a product of changes that occurred after human colonization.

Today's forests (particularly in the west) are highly fragmented; Smith (1997) estimates that only about 2.8% of the original (precolonization) western dry forest cover has survived into the 1990's. Thus, in some sense, every primate species that is found in several isolated forests can be said to exhibit a disjunct distribution. But only some of these species span the sharp ecological divide separating typical eastern wet from western dry flora (see, for example, Simons, 1993, 1994; Sterling, 1994; Mittermeier et al., 1994; Thalmann et al., 1998). Typical eastern-forest primates that occur in the west today include (in addition to *Eulemur fulvus fulvus* and *E. f. rufus*), *Daubentonia madagascariensis* (in the Tsingy of Namoroka as well as Bemaraha, across the Sambirano, and, in the historic past, in the vicinity of Mahajanga), *Hapalemur griseus* (at Bemaraha, Namoroka, the Sambirano, and in the extreme north), *Cheirogaleus major* (at Bemaraha and the Sambirano), and *Avahi laniger* (at Bemaraha and Ankarafantsika; the western *Avahi* is sometimes placed in its own species, *Avahi occidentalis;* see Mittermeier et al., 1994). Schmid and Kappeler (1994) suggest that the Pygmy Rufous Mouse Lemur (*Microcebus* sp. cf. *myoxinus*) that lives today in the forests of Kirindy (western Madagascar) alongside the Gray Mouse Lemur, *M. murinus*, is more closely related to the eastern rufous form, *M. rufus*, than to its sympatric congener. Species distributions at subfossil sites further demonstrate that other eastern species once lived not merely in the center but also in the extreme west. The latter include *Hapalemur simus* (at Anjohibe in the northwest), *Indri indri* (at Ampoza in the west), and *Varecia variegata* (at Manombo-Toliara near the west coast).

The converse is also true: Some "typical" western dry forest primate species occur today (or occurred during the recent past) in the east. *Phaner furcifer* is more widely distributed today in the west than in the east, but there is a population on the Masoala Peninsula just north and east of the Bay of Antongil on the east coast (Mittermeier et al., 1994). *Propithecus verreauxi* lived until recently in the rain forest at Ambohitantely. Some dry-loving cheirogaleids live on the wet side of the ecological divide separating Southern and Eastern Phytogeographic Domains, near Tolagnaro (Ganzhorn, 1998b). The evidence of the subfossil sites tells us that *Eulemur mongoz* once lived in Central Madagascar, and that both *P. verreauxi* and *P. diadema* occupied the Central Highlands—successively if not synchronously.

The Central Highlands has clearly functioned as a crossroads for species exchange. Palynological and speleothem research has demonstrated that northern and central Madagascar experienced intermittent periods of high and low aridity (MacPhee et al., 1985; Burney et al., 1986, 1997; Burney, 1987a,b, 1988; Matsumoto & Burney, 1994). The Central Highlands provided a mosaic of forest, woodland, and grassland habitats. Likely routes of dispersal for forest-limited species included gallery forests along rivers flowing westward from the region of the eastern escarpment (Ganzhorn, 1998a). Routes of dispersal from the north to the south (or vice versa) would have likely crossed the headwaters or the highland tributaries of rivers flowing east to west. Tributaries are far less likely to function as physical barriers to dispersal than are permanently-flowing waterways in the lowland regions nearer to the coasts (see, for a case in point, Thalmann & Rakotoarison, 1994).

The Central Highlands also provided likely routes for dispersal of taxa from the west into the north, and from the east into the Sambirano. Even during dry periods, the wet forests of the Sambirano, the montane forests of the Montagne d'Ambre, and the orographically moist eastern rain forest would have retained their evergreen forest character, providing refuge for species adapted to moist forest conditions. Passage through either the Sambirano (in the northwest) or into the eastern rain forest would have continued to be prohibitive for species adapted to dry conditions, but the drier Central Highlands may have provided suitable corridors for their dispersal. During periods of higher rainfall, dry-loving species in the north and west may have been more isolated, but densely forested corridors may have provided a means for the species of the eastern rain forest to spread into the Sambirano (through the northern highlands) and into the north. The limiting factors would have been the changing character of forested corridors through the highlands, and the ability of species to survive montane conditions.

Today the extreme north (excluding the Montagne d'Ambre) contains some of the driest habitats in all of Madagascar, and the vegetation is predominantly of western affinity (Koechlin et al., 1974). Nevertheless, the still-extant lemur species that once lived (*Hapalemur simus*, *Indri indri*) and that continue to live (*Propithecus diadema*, *Avahi laniger*, etc.) in the north are generally most similar to those that still live in the eastern rain forest as well as those that once lived in the center. The extant primate communities of the dry west have more in common with those of the south than with those of the north or east; this is particularly true of species inhabiting the western coastal forests to the south of Morondava. In the northwest, the western character of many species of lemurs, including *Propithecus verreauxi*, *Microcebus murinus*, and *Cheirogaleus medius*, is evident, but nevertheless the northwest exhibits some faunal affinities with the center, north, and east.

All currently recognized species of *Lepilemur* are presumed to be allopatric (but see Ausilio & Raveloanrinoro, 1998, and Sterling, 1998, for discussions of possible sympatric species or morphs in several forests). The occurrence of two species of *Lepilemur* at Tsirave does not prove their synchronicity, but it does bear testimony to possible changes in the biogeography of species belonging to this genus. If the small lepilemurid femora at Tsirave and Ankilitelo belong to *Lepilemur leucopus*, and if this species is indeed restricted to the region south of the Onilahy River, then this restriction is a recent phenomenon.

Our analysis of extinct species demonstrates that taxa typical for western or even southern Madagascar did at least intermittently penetrate central Madagascar; this is the case for *Hadropithecus stenognathus*, *Daubentonia robusta*, and possibly *Archaeolemur majori* (Godfrey et al., in prep.). Furthermore, if, as seems likely, many of the extinct species of central and northern Madagascar also occupied the east, then more taxa can be added to the list of "eastern" forms that lived at least periodically in the west. *Babakotia radofilai* was very common in the north; it also appears to have lived at Anjohibe in the northwest. *A. edwardsi* was common in the Central Highlands; it also occurred at Belo-sur-Mer (in the extreme west), Ampoza (in the southwest) and Andrahomana (in the extreme south); see Godfrey et al. (1997b, in prep.). *Mesopropithecus pithecoides* (from Ampasambazimba) strongly resembles *M. globiceps* from the south and west. Finally, even if the species of *Archaeolemur* that occupied the extreme north differed from that which occupied central Madagascar, it did not differ from the *Archaeolemur* of the northwest (Anjohibe).

The presence of primate taxa such as *Babakotia radofilai*, *Hapalemur simus*, and the large *Archaeolemur* at Anjohibe may suggest that the climate of the northwest was not always as dry as it is now; the presence of *Indri indri* at Ampoza and *Varecia variegata* at Manombo-Toliara (near Ambolisatra) suggest the same for the west. Pollen spectra document some changes in regional habitats from prehuman times to today, but those fluctuations are perhaps not as dramatic as one might expect from the paleodistributions of primate taxa. There is evidence that the dry southwest (in the region of Ambolisatra) was once wetter than it is today, and that climatic dessication beginning around 3000 BP followed by human disturbance resulted in a shift from forest woodland (with more mesic taxa) to the drier bushland/grassland that characterizes this region today (see especially Burney, 1993). Nevertheless, during the middle Holocene, the flora of the region of Ambolisatra would have been distinctly western in character, closely resembling the dry wooded habitats surrounding Morondava (in western Madagascar) today (Burney et al., 1998). Holocene pollen spectra from Ampasambazimba (in central Madagascar) reveal prehuman habitats somewhat like those in the vicinity of Lake Mahery (in the far north) today—ranging from closed forest to fully open grassland (Burney et al., 1998). The region of Lake Tritrivakely (near Antsirabe and Betafo, also in the Central Highlands) was similar to the Ankaratra

Massif today, with predominant montane heath bushland and riparian habitats, as well as grasslands, ecotones, and (rarely) closed forests (Burney et al., 1998). Finally, speleothem-derived pollen data from the northwest (Anjohibe) reveal a wooded grassland environment with satra palms (much like today's environment in the same region) over the past 40,000 years (Burney et al., 1997, 1998). Whereas the habitat of this region was sometimes more woody (and sometimes less so) than it is today, it did not resemble, during this time period, today's forests of the east or extreme north. Indeed, there is no evidence in pollen spectra that northwest, southwest, or indeed central Madagascar, was eastern in character during the recent past.

Species paleodistribution data can serve as a springboard for analysis of changes in lemur community structure. Ganzhorn (1997) demonstrated that most modern primate communities in Madagascar conform to Fox's "rule of assembly" in displaying balanced representation of species from different "functional" (or ecologically distinct) groups (e.g., folivores, frugivores, and omnivores). However, our data show that a disproportionate number of frugivorous and especially folivorous primate species have disappeared from the primate communities of Madagascar (Fig. 12). This suggests that these communities may not have always exhibited "favored" (or functionally balanced) states. Within a global context, even today, Malagasy primate communities are unusual in having a high percentage of folivorous and a low percentage of frugivorous species (Tattersall, 1982; Fleagle & Reed, 1996; Kappeler & Heymann, 1996; Godfrey et al., 1997a; Goodman & Ganzhorn, 1998). It is within the context of the entire Holocene

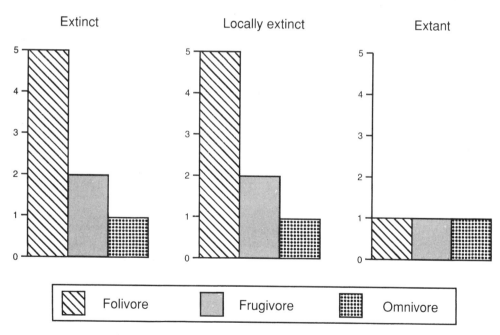

Figure 12. Number of primate folivores, frugivores, and omnivores that lived at Ampasambazimba during the past several thousand years but are now extinct (extreme left); that lived at Ampasambazimba during the past several thousand years but are now locally extinct (middle); that live in the region of Ampasambazimba (i.e., at Ambohitantely) today (extreme right). Some indrids, like colobines, prefer seeds over leaves; these are masticated rather than swallowed whole. The amount of fruit and seeds they consume varies seasonally. We classified all indrids as "folivorous" despite variation in the actual percentage of leaves vs. fruit (and seeds) consumed. In contrast to those lemurid species that swallow seeds whole or discard them, indrids are physiological folivores.

primate fauna that competing explanations for spatial and temporal variation in patterns of species assembly must be explored (see Fleagle, 1998).

CONCLUSIONS

Comparisons of extant primate community compositions at subfossil sites and modern forests demonstrate that there was, in the past, as today, a strong ecological distinction between eastern (moist forest) and western (dry forest) primate communities, and a strong geographic gradient (or distance effect) characterizing community composition from the southern to the northern tips of Madagascar. In other words, the regional endemism that characterizes today's primate faunas appears to have had some historic depth, as does the primary ecological division of Madagascar into eastern wet forest and western dry forest communities.

The precise boundaries of the geographic distributions of typically wet-loving and typically dry-loving species were quite fluid; fingers of "eastern" forest once spread across the island, perhaps all the way to the west coast. The Central Highlands functioned as a crossroads for dispersal of species in both east-west and north-south directions. The discovery of relict patches of eastern rain forest flora and fauna in the west (including the mountain ridge "mist oasis" of Analavelona that preserves eastern alongside western biota but is located to the west of Zombitse and northeast of Toliara), serves to underscore this fact. It also makes the existence of *Varecia variegata* at Manombo-Toliara and the apparent existence of *Indri* at Ampoza less enigmatic. This, coupled with recent fragmentation and degradation of forested habitats, explains the current disjunct distributions of many species and subspecies of lemurs.

Numerous extant species occur at subfossil sites in regions where they do not live today. These include *Cheirogaleus major* (at Ampasambazimba), *Hapalemur simus* (at Anjohibe, Ankarana, Montagne des Français, Ampasambazimba), *Indri indri* (at Ankarana, Ampasambazimba, Ampoza), *Eulemur mongoz* (at Ampasambazimba), *Propithecus diadema* (at Ampasambazimba), *P. tattersalli* (at Ankarana), *P. verreauxi* (at Ampasambazimba), *Varecia variegata* (at Ampasambazimba, Manombo-Toliara), and *Lepilemur* spp. (at Tsirave, Ampasambazimba). It appears certain that the geographic ranges of these taxa were considerably larger in the past than they are today.

ACKNOWLEDGMENTS

The research reported here was supported in part by funds from the National Science Foundation (Grant SBR-9630350 to ELS), the Margot Marsh Foundation (to WLJ), and the University of Massachusetts, Amherst (to LRG). Figures were prepared by Luci Betti-Nash (Figs. 1, 6, and 10) and Darren Godfrey (Figs. 2, 7, 8, 11, and 12). The subfossil specimens examined for this study are housed largely at the Laboratoire de Paléontologie des Vertébrés, Université d'Antananarivo, Madagascar, and at the Duke University Primate Center (Durham, North Carolina). Additional subfossil specimens housed at the Museum of Comparative Zoology (Cambridge, Massachusetts), The Natural History Museum (London), Field Museum of Natural History (Chicago, Illinois), and the Smithsonian Institution (Washington, D.C.), were also examined. Comparative data on extant lemurs were collected at the following museums: Museum of Comparative Zoology; American Museum of Natural History (New York, New York); Smithsonian Institution; Cleveland Museum of Natural History (Cleveland, Ohio); Yale

Peabody Museum (New Haven, Connecticut): Field Museum of Natural History; Duke University Primate Center; The Natural History Museum (London); Muséum National d'Historie Naturelle, Laboratoire d'Anatomie Comparée and Mammifères et Oiseaux; Naturhistoriska Riksmuseet; Nationaal Natuurhistorisch Museum (formerly the Rijksmuseum van Natuurlijke Historie, Leiden); Museum für Naturkunde (Berlin); Anthropologisches Institut und Museum (Zurich); University of Göttingen (H-J Kuhn Collection); and Institut Royal des Sciences Naturelles de Belgique (Brussels). We are deeply indebted to the curators of these collections for providing us access to these specimens. Comments on an earlier version of this manuscript by David Burney, Jörg Ganzhorn, Steven Goodman, Robert Sussman, and Ian Tattersall are greatly appreciated. This is Duke Primate Center Publication number 679.

REFERENCES

ALBRECHT, G. H., P. D. JENKINS, AND L. R. GODFREY. 1990. Ecogeographic size variation among the living and subfossil prosimians of Madagascar. American Journal of Primatology, **22**: 1–50.

ANDRE, P., D. BESSAGUET, T. CURTARELLI, M. MARGAILLAN, J. C. PEYRE, AND J. RADOFILAO. 1986. Exploration spéléologique dans l'ile de Madagascar (juillet à octobre 1985). La Société des Explorateurs Niçois, pp. 1–73, plus references.

ATMAR, W., AND B. D. PATTERSON. 1993. The measure of order and disorder in the distribution of species in fragmented habitats. Oecologia, **96**: 373–382.

AUSILIO, E., AND G. RAVELOANRINORO. 1998. Les lémuriens de la région de Bemaraha: Forêts de Tsimembo, de l'Antsingy et de la région de Tsiandro. Lemur News **3**: 4–7.

BATTISTINI, R., AND P. VÉRIN. 1967. Ecologic changes in protohistoric Madagascar, pp. 407–424. *In* Martin, P. S., and H. E. Wright, eds. Pleistocene Extinctions: The Search for a Cause. Yale University Press, New Haven.

BONNARDIN, P. 1988. Speleo tropique: une expédition spéléologique sous le tropique de Capricorn. Spelunca, **30**: 25–29.

BURNEY, D. A. 1987a. Late Holocene vegetational change in central Madagascar. Quaternary Research, **28**: 130–143.

———. 1987b. Pre-settlement vegetation changes at Lake Tritrivakely, Madagascar. Paleoecology of Africa, **18**: 357–381.

———. 1988. Modern pollen spectra from Madagascar. Palaeogeography, Palaeoclimatology, and Palaeoecology, **66**: 63–75.

———. 1993. Late Holocene environmental change in arid southwestern Madagascar. Quaternary Research, **40**: 98–106.

———. In press. Rates, patterns, and processes of landscape transformations and extinction in Madagascar. *In* MacPhee, R. D. E., ed., Extinctions in Near Time: Causes, Contexts, and Consequences. Plenum Press, New York.

BURNEY, D. A., H. F. JAMES, F. V. GRADY, J.-G. RAFAMANTANANATSOA, RAMILISONINA, H. T. WRIGHT, AND J. B. COWART. 1997. Environmental change, extinction, and human activity: Evidence from caves in NW Madagascar. Journal of Biogeography, **24**: 755–767.

BURNEY, D. A., R. D. E. MACPHEE, AND R. E. DEWAR. 1986. Cause, effect, and megafaunal extinction in Holocene Madagascar: pollen and charcoal influx at Lake Kavitaha. American Journal of Physical Anthropology, **69**: 182–183.

BURNEY, D. A., W. L. JUNGERS, H. F. JAMES, AND L. R. GODFREY. 1998. The paleoecology of *Archaeolemur*: An extinct prosimian for all seasons. Paper presented at the 17th Congress of the International Primatological Society, Antananarivo, August, 1998 (Abstract no. 219).

CHANUDET, C. 1975. Conditions géographiques et archéologiques des subfossiles malgaches. Mémoire de Maîtrise. Université de Bretagne Occidentale, Brest.

DECARY, R. 1934. Les grottes d'Anjohibe (Andranoboka). Revue de Madagascar, **8**: 81–85.

———. 1939 (for the year 1938). Les grottes d'Andranoboka (ou Anjohibe). Bulletin de l'Académie Malgache, **21**: 71–80.

DECARY, R., AND A. KIENER. 1970. Les cavités souterraines de Madagascar. Annales de Spéléologie, **25(2)**: 409–440.

DE SAINT-OURS, J. 1953. Etude des grottes d'Andranoboka. Travaux du Bureau Géologique, No. 43. Service Géologique, Antananarivo.
DEWAR, R. E., AND S. RAKOTOVOLOLONA. 1986. Hunting camps in northern Madagascar in the XIIth and XIIIth centuries. Paper presented at the International Congress "I primi uomini in ambiente insulare" (Early man in island environments). Oliena (Sardegna), Italy. Unpublished ms.
———. 1992. La chasse aux subfossiles: Les preuves du onzième siècle au treizième siècle. Taloha, **11**: 5–15.
DU PUY, B., J. P. ABRAHAM, AND A. J. COOKE. 1994. Les plantes, pp. 15–29. *In* Goodman, S. M., and O. Langrand, eds., Inventaire biologique Forêt de Zombitse. Recherches pour le développement, Série Sciences biologiques, No. Spécial, Centre d'Information et de Documentation Scientifique et Technique, Antananarivo.
DUFLOS, J. 1966. Bilan des explorations spéléologiques pour l'année 1965. Madagascar Revue de Géographie, **9**: 235–252.
———. 1968. Bilan des explorations spéléologiques de l'année 1966. Madagascar Revue de Géographie, **12**: 121–129.
EKBLOM, T. 1953. Studien über subfossile Lemuren von Madagaskar. Bulletin of the Geological Institute of Uppsala, **34**: 124–190.
FEISTNER, A. T. C., AND J. SCHMID. 1999. Lemurs of the Réserve Naturelle Intégrale d'Andohahela, Madagascar. Fieldiana: Zoology, new series, **94**: 269–283.
FLEAGLE, J. G. 1998. Primate Adaptation and Evolution (2nd edition). Academic Press, San Diego.
FLEAGLE, J. G., AND K. E. REED. 1996. Comparing primate communities: A multivariate approach. Journal of Human Evolution, **30**: 489–510.
GANZHORN, J. U. 1994. Les lémuriens, pp. 70–72. *In* Goodman, S. M. and O. Langrand, eds., Inventaire biologique Forêt de Zombitse. Recherches pour le développement, Série Sciences biologiques, No. Spécial, Centre d'Information et de Documentation Scientifique et Technique, Antananarivo.
———. 1997. Test of Fox's assembly rule for functional groups in lemur communities in Madagascar. Journal of Zoology, **241**: 533–542.
———. 1998a. Nested patterns of species composition and its implications for lemur biogeography in Madagascar. Folia Primatologica, **69**, Supplement **1**: 332–341.
———. 1998b. Progress report on the QMM faunal studies: lemurs in the littoral forest of southeast Madagascar. Lemur News, **3**: 22–23.
———, S. MALCOMBER, O. ANDRIANANTOAINA, AND S. M. GOODMAN. 1997. Habitat characteristics and lemur species richness in Madagascar. Biotropica, **29**: 331–343.
GODFREY, L. R., E. L. SIMONS, P. S. CHATRATH, AND B. RAKOTOSAMIMANANA. 1990a. A new fossil lemur (*Babakotia*, Primates) from northern Madagascar. Comptes Rendus de l'Académie des Sciences, Paris (série II), **310**: 81–87.
GODFREY, L. R., W. L. JUNGERS, K. E. REED, E. L. SIMONS, AND P. S. CHATRATH. 1997a. Subfossil lemurs: Inferences about past and present primate communities, pp. 218–256. *In* Goodman, S. M., and B. D. Patterson, eds., Natural Change and Human Impact in Madagascar. Smithsonian Institution Press, Washington, D. C.
GODFREY, L. R., W. L. JUNGERS, AND S. RASOLOHARIJAONA. In prep. Taxonomy and geographic distribution of *Archaeolemur* spp. in Madagascar.
GODFREY, L. R., W. L. JUNGERS, R. E. WUNDERLICH, AND B. G. RICHMOND. 1997b. Reappraisal of the postcranium of *Hadropithecus* (Primates, Indroidea). American Journal of Physical Anthropology, **103**: 529–556.
GODFREY, L. R., M. R. SUTHERLAND, A. J. PETTO, AND D. S. BOY. 1990b. Size, space, and adaptation in some subfossil lemurs from Madagascar. American Journal of Physical Anthropology, **81**: 45–66.
GODFREY, L. R., AND M. VUILLAUME-RANDRIAMANANTENA. 1986. *Hapalemur simus*: Endangered lemur once widespread. Primate Conservation, **7**: 92–96.
GOODMAN, S. M., AND J. U. GANZHORN. 1998. Rarity of figs (*Ficus*) on Madagascar and its relationship to a depauperate frugivore community. Revue d'Ecologie, **52**: 321–323.
GOODMAN, S. M., M. PIDGEON, A. F. A. HAWKINS, AND T. S. SCHULENBERG. 1997. The birds of southeastern Madagascar. Fieldiana: Zoology, new series, **87**: 1–132.
GRANDIDIER, G. 1905. Recherches sur les lémuriens disparus et en particulier sur ceux qui vivaient à Madagascar. Nouvelles Archives du Muséum, Paris (Série 4), **7**: 1–142, plus plates.
HAWKINS, A. F. A., P. CHAPMAN, J. U. GANZHORN, Q. M. C. BLOXAM, S. C. BARLOW, AND S. J. TONGE. 1990. Vertebrate conservation in Ankarana Special Reserve, northern Madagascar. Biological Conservation, **54**: 83–110.
JENKINS, P. D. 1987. Catalogue of the Primates in the British Museum (Natural History) and elsewhere in the

British Isles. Part IV. Suborder Strepsirrhini, including the subfossil Madagascan lemurs and Family Tarsiidae. British Museum (Natural History), London.

JULLY, A., AND H. F. STANDING. 1904. Les gisements fossilifères d'Ampasambazimba. Bulletin de l'Académie Malgache, **3/2**: 87–94.

JUNGERS, W. L., L. R. GODFREY, E. L. SIMONS, AND P. S. CHATRATH. 1995. Subfossil *Indri* from the Ankarana Massif of northern Madagascar. American Journal of Physical Anthropology, **97**: 357–366.

JUNGERS, W. L., L. R. GODFREY, E. L. SIMONS, P. S. CHATRATH, AND B. RAKOTOSAMIMANANA. 1991. Phylogenetic and functional affinities of *Babakotia* (Primates), a fossil lemur from northern Madagascar. Proceedings of the National Academy of Sciences, USA, **88**: 9082–9086.

JUNGERS, W. L., E. L. SIMONS, L. R. GODFREY, AND P. S. CHATRATH. In prep. New species of *Palaeopropithecus* (Indroidea, Palaeopropithecidae) from northwest Madagascar.

KAPPELER, P. M., AND E. W. HEYMANN. 1996. Nonconvergence in the evolution of primate life history and socio-ecology. Biological Journal of the Linnean Society, **59**: 297–326.

KOECHLIN, J., J-L. GUILLAUMET, AND P. MORAT. 1974. Flore et Végétation de Madagascar. J. Cramer, Vaduz.

LAMBERTON, C. 1932. Verbal report on excavations at Tsiravé, presented to the Académie Malgache on January 15, 1931. Bulletin de l'Académie Malgache, **14**: 21–22.

———. 1934. Contribution à la connaissance de la faune subfossile de Madagascar. Lémuriens et Ratites. *Chiromys robustus* sp. nov. Lamb. Mémoires de l'Académie Malgache, **17**: 40–46.

———. 1937 (for the year 1936). Fouilles paléontologiques faites en 1936. Bulletin de l'Académie Malgache (nouvelle série), **19**: 1–19.

———. 1939a. Contribution à la connaissance de la faune subfossile de Madagascar. Lémuriens et cryptoproctes. Note IV. Nouveaux lémuriens fossiles du groupe des Propithèques. Mémoires de l'Académie Malgache, **27**: 9–49 (plus 5 plates).

———. 1939b. Contribution à la connaissance de la faune subfossile de Madagascar. Lémuriens et cryptoproctes. Note V. Petits lémuriens subfossiles. Mémoires de l'Académie Malgache, **27**: 51–73 (plus 5 plates).

———. 1946. Contribution à la connaissance de la faune subfossile de Madagascar. Note XVII. Les Pachylemurs. Bulletin de l'Académie Malgache (nouvelle série), **27**: 7–22.

LAPAIRE, J.-P., G. ROSSI, AND OTHERS. 1975. Les plateaux de Mikoboka et Manamby D 56 et D 57, pp. 76–85, plus maps. *In* Prospection des Gisements de Guano des Bassins Sedimentaires de Majunga et de Morondava. Association des Géographes de Madagascar, Laboratoire de Géographie, Université de Madagascar, Antananarivo.

LEIGH, S. R., AND C. J. TERRANOVA. 1998. Comparative perspectives on bimaturism, ontogeny, and dimorphism in lemurid primates. International Journal of Primatology, **19**: 723–749.

LANGRAND, O., AND S. M. GOODMAN. 1997. Brève description biologique de la région des forêts de Vohibasia et d'Isoky-Vohimena, pp. 11–28. *In* Langrand, O., and S. M. Goodman, eds., Inventaire biologique Forêt de Vohibasia et d'Isoky-Vohimena. Recherches pour le développement, Série Sciences biologiques, No. 12, Centre d'Information et de Documentation Scientifique et Technique, Antananarivo.

LAUMANNS, M., J. BURGSMÜLLER, AND W. GEUCKE. 1991. Höhlenkundliche Activitäten der HuK Nordrhein nordöstlich Majunga und südlich Antsirabe. Mitteilungen des Verbandes der Deutschen Höhlen- und Karstforscher, **37**: 68–73.

LYMAN, R. L. 1994. Vertebrate Taphonomy. Cambridge University Press, Cambridge.

MACPHEE, R. D. E., AND D. A. BURNEY. 1991. Dating of modified femora of extinct dwarf *Hippopotamus* from southern Madagascar: implications for constraining human colonization and vertebrate extinction events. Journal of Archaeological Science, **18**: 695–706.

MACPHEE, R. D. E., D. A. BURNEY, AND N. A. WELLS. 1985. Early Holocene chronology and environment of Ampasambazimba, a Malagasy subfossil lemur site. International Journal of Primatology, **6**: 463–489.

MACPHEE, R. D. E., AND E. M. RAHOLIMAVO. 1988. Modified subfossil aye-aye incisors from southwestern Madagascar: Species allocation and paleoecological significance. Folia Primatologica, **51**: 126–142.

MAHÉ, J. 1965. Un gisement nouveau de subfossile à Madagascar. Comptes Rendus Sommaires des Séances de la Société Géologique de France, **2**: 66.

———. 1976. Craniométrie des lémuriens: Analyses multivariables-phylogénie. Mémoires du Muséum National d'Histoire Naturelle, Paris, (Série C) **32**: 1–342.

MATSUMOTO, K., AND D. A. BURNEY. 1994. Late Holocene environments at Lake Mitsinjo, northwestern Madagascar. The Holocene, **4**: 16–24.

MITTERMEIER, R. A., I. TATTERSALL, W. R. KONSTANT, D. M. MEYERS, AND R. B. MAST. 1994. Lemurs of Madagascar. Conservation International, Washington, D. C.

PATTERSON, B. D. 1987. The principle of nested subsets and its implication for biological conservation. Conservation Biology, **1**: 323–334.

——, AND W. ATMAR. 1986. Nested subsets and the structure of insular mammalian faunas. Biological Journal of the Linnean Society, **28**: 65–82.

PEREIRA, M. E. 1993. Seasonal adjustment of growth rate and adult body weight in ring-tailed lemurs, pp. 205–222. *In* Kappeler, P. M., and J. U. Ganzhorn, eds., Lemur Social Systems and their Ecological Basis. Plenum Press, New York.

PERRIER DE LA BÂTHIE, H. 1927. Fruits et graines du gisement de subfossiles d'Ampasambazimba. Bulletin de l'Académie Malgache **10**: 24–25.

PETTER, J.-J., AND S. ANDRIATSARAFARA. 1987. Les lémuriens de l'ouest de Madagascar, pp. 71–73. *In* Mittermeier, R. A., L. H. Rakotovao, V. Randrianasolo, E. J. Sterling, and D. Devitre, eds., Priorités en Matière de Conservation des Espèces à Madagascar. IUCN, Gland.

PETTER, J.-J., R. ALBIGNAC, AND Y. RUMPLER. 1977. Faune de Madagascar **44**: Mammifères Lémuriens (Primates Prosimiens). ORSTOM/CNRS, Paris.

RADOFILAO, J. 1977. Bilan des explorations spéléologiques dans l'Ankarana. Annales Université de Madagascar. Série Sciences de la Nature et Mathématiques **14**: 195–204.

RASOLOHARIJAONA, S. 1995. Contribution à l'étude du genre *Archaeolemur* sp. (Filhol 1895) de l'Ankarana: morphologie comparée et ostéométrie. Essai de reconstitution du paléoenvironnement de la région de l'Ankarana. Memoires Diplôme d'Etudes Approfondies, Université d'Antananarivo, Antananarivo.

RAVOSA, M. J., D. M. MEYERS, AND K. E. GLANDER. 1993. Relative growth of the limbs and trunk in sifakas: Heterochronic, ecological, and functional considerations. American Journal of Physical Anthropology, **92**: 499–520.

REED, K. E., AND J. G. FLEAGLE. 1995. Geographic and climatic control of primate diversity. Proceedings of the National Academy of Sciences, USA, **92**: 7874–7876.

ROHLF, F. J. 1992. NTSYS-pc. Numerical taxonomy and multivariate analysis system (version 1.70). Applied Biostatistics, Inc., New York.

SCHMID, J., AND P. KAPPELER. 1994. Sympatric mouse lemurs (*Microcebus* spp.) in western Madagascar. Folia Primatologica, **63**: 162–170.

SHIPMAN, P. 1981. Life History of a Fossil: An Introduction to Taphonomy and Paleoecology. Harvard University Press, Cambridge, Massachusetts.

SIMONS, E. L. 1993. Discovery of the western aye-aye. Lemur News, **1**: 6.

——. 1994. The giant aye-aye *Daubentonia robusta*. Folia Primatologica, **62**: 14–21.

——. 1997. Lemurs: Old and new, pp. 142–166. *In* Goodman, S. M., and B. Patterson, eds., Natural Change and Human Impact in Madagascar. Smithsonian Institution Press, Washington, D. C.

——, D. A. BURNEY, P. S. CHATRATH, L. R. GODFREY, W. L. JUNGERS, AND B. RAKOTOSAMIMANANA. 1995a. AMS[14] dates on extinct lemurs from caves in the Ankarana Massif of northern Madagascar. Quaternary Research, **43**: 249–254.

SIMONS, E. L., L. R. GODFREY, W. L. JUNGERS, P. S. CHATRATH, AND J. RAVAOARISOA. 1995b. A new species of *Mesopropithecus* (Primates, Palaeopropithecidae) from northern Madagascar. International Journal of Primatology, **16**: 653–682.

SIMONS, E. L., L. R. GODFREY, M. VUILLAUME-RANDRIAMANANTENA, P. S. CHATRATH, AND M. GAGNON. 1990. Discovery of new giant subfossil lemurs in the Ankarana Mountains of northern Madagascar. Journal of Human Evolution, **19**: 311–319.

SMITH, A. P. 1997. Deforestation, fragmentation, and reserve design in western Madagascar, pp. 415–441, plus separate plates and book-end references. *In* Laurance, W. F., and R. O. Bierregaard, Jr., eds., Tropical Forest Remnants: Ecology, Management, and Conservation of Fragmented Communities. The University of Chicago Press, Chicago.

SMITH, R. J., AND W. L. JUNGERS. 1997. Body mass in comparative primatology. Journal of Human Evolution, **32**: 523–559.

STANDING, H.-F. 1906 (for the year 1905). Rapport sur les ossements subfossiles provenant d'Ampasambazimba. Bullétin de l'Académie Malgache, **4**: 95–100.

——. 1908. On recently discovered subfossil primates from Madagascar. Transactions of the Zoological Society of London, **18**: 59–162.

——. 1909. Subfossiles provenant des fouilles d'Ampasambazimba. Bulletin de l'Académie Malgache, **6**: 9–11.

——. 1910. Note sur les ossements subfossiles provenant des fouilles d'Ampasambazimba. Bulletin de l'Académie Malgache, **7**: 61–64.

STERLING, E. 1994. Taxonomy and distribution of *Daubentonia*: A historical perspective. Folia Primatologica, **62**: 8–13.

———. 1998. Preliminary report on a survey for *Daubentonia madagascariensis* and other primate species in the west of Madagascar, June-August 1994. Lemur News, **3**: 7–8.
SUSSMAN, R. 1977. Distribution of the Malagasy lemurs. Part 2: *Lemur catta* and *Lemur fulvus* in southern and western Madagascar. Annals of the New York Academy of Sciences, **293**: 170–184.
SZALAY, F. S., AND E. DELSON. 1979. Evolutionary History of the Primates. Academic Press, New York.
TATTERSALL, I. 1971. Revision of the subfossil Indriinae. Folia Primatologica, **16**: 257–269.
———. 1973. Cranial anatomy of the Archaeolemurinae (Lemuroidea, Primates). Anthropological Papers of the American Museum of Natural History, **52**: 1–110.
———. 1982. The Primates of Madagascar. Columbia University Press, New York.
TERBORGH, J., AND C. P. VAN SCHAIK. 1987. Convergence vs. nonconvergence in primate communities, pp. 205–226. *In* Gee, J. H. R., and P. S. Giller, eds., Organization of Communities: Past and Present. Blackwell Scientific, Oxford.
THALMANN, U., P. KERLOCH, A. E. MÜLLER, AND A. ZARAMODY. 1998. A visit to the Reserve Tsingy de Namoroka (NW Madagascar). Poster presented at the 17th Congress of the International Primatological Society, Antananarivo, August, 1998 (Abstract no. 370).
THALMANN, U., AND N. RAKOTOARISON. 1994. Distribution of lemurs in central western Madagascar, with a regional distribution hypothesis. Folia Primatologica, **63**: 156–161.
TOMIUK, J., L. BACHMANN, M. LEIPOLDT, J. U. GANZHORN, R. RIES, M. WEIS, AND V. LOESCHCKE. 1997. Genetic diversity of *Lepilemur mustelinus ruficaudatus*, a nocturnal lemur of Madagascar. Conservation Biology, **11**: 491–497.
VUILLAUME-RANDRIAMANANTENA, M., L. R. GODFREY, AND M. R. SUTHERLAND. 1985. Revision of *Hapalemur* (*Prohapalemur*) *gallieni* (Standing 1905). Folia Primatologica, **45**: 89–116.
VUILLAUME-RANDRIAMANANTENA, M., L. R. GODFREY, W. L. JUNGERS, AND E. L. SIMONS. 1992. Morphology, taxonomy and distribution of *Megaladapis*—giant subfossil lemur from Madagascar. Comptes Rendus de l'Académie des Sciences, Paris, (Série II) **315**: 1835–1842.
WILSON, J. M., L. R. GODFREY, E. L. SIMONS, P. D. STEWART, AND M. VUILLAUME-RANDRIAMANANTENA. 1995. Past and present lemur fauna at Ankarana, north Madagascar. Primate Conservation, **16**: 47–52.
WILSON J. M., P. D. STEWART, G.-S. RAMANGASON, A. M. DENNING, AND M. S. HUTCHINGS. 1989. Ecology and conservation of the crowned lemur *Lemur coronatus* at Ankarana, Madagascar with notes on Sanford's lemur, other sympatrics and subfossil lemurs. Folia Primatologica, **52**: 1–26.
ZIMMERMANN, E., S. CEPOK, N. RAKOTOARISON, V. ZIETEMANN, AND U. RADESPIEL. 1998. Sympatric mouse lemurs in north-west Madagascar: A new rufous mouse lemur species (*Microcebus ravelobensis*). Folia Primatologica, **69**: 106–114.

SKELETAL MORPHOLOGY AND THE PHYLOGENY OF THE LEMURIDAE

A Cladistic Analysis

Gisèle Francine Noro Randria[1]

[1] Département de Paléontologie et d'Anthropologie Biologique
BP 906, Faculté des Sciences
Université d'Antananarivo, Antananarivo (101), Madagascar
brakoto@syfed.refer.mg

ABSTRACT

This paper reports the results of fourteen parsimony analyses of skeletal morphological characteristics of members of the Family Lemuridae. These treatments differed in the data set (quantitative or qualitative) chosen for analysis, as well as the selected outgroup. Outgroup species included members of the families Lepilemuridae, Cheirogaleidae, and Indridae. The data presented here, particularly when viewed in conjunction with those of other studies based on independent data sets, support the following hypothesis of lemurid relationships: that *Hapalemur* and *Lemur* are sister taxa, that *Eulemur* is the sister taxon to the *Hapalemur-Lemur* clade, and that *Varecia* is the sister taxon to the former three.

RÉSUMÉ

Dans le but de déterminer leurs rapports phylogénétiques, les caractéristiques anatomiques du squelette des Lemuridae ont été comparées avec celles de quelques genres et espèces de Lepilemuridae, de Cheirogaleidae, et d'Indridae. En tenant compte des résultats de 14 analyses cladistiques (basées sur les caractères qualitatifs et quantitatifs), et en comparaison aves les analyses réalisées par d'autres auteurs, nous retenons comme hypothèse finale: les genres *Hapalemur* et *Lemur* sont groupes-frères et, ensembles, ils forment le groupe-frère du genre *Eulemur*. Enfin, des deux derniers ensembles composent le groupe-frère de *Varecia*.

INTRODUCTION

Extensive homoplasy has plagued prior studies of the phylogeny of the Lemuridae. Researchers applying different methodologies or examining different kinds of data have obtained contradictory evolutionary trees, and the evolutionary history of this important group of lemurs remains poorly understood. Lemurid taxonomy continues to be debated; for example, brown lemurs, red-bellied lemurs, crowned lemurs, mongoose lemurs, and black lemurs were transferred from the genus *Lemur* to the genus *Eulemur* by Simons and Rumpler (1988), but retained, with *Varecia*, in the genus *Lemur* by Tattersall and Schwartz (1991). A new species of bamboo lemur from southeast Madagascar, *Hapalemur aureus*, was described in 1987 (Meier et al., 1987), and a poorly known subspecies of *Eulemur macaco* was rediscovered only a couple of years earlier (Koenders et al., 1985). Our knowledge of the geographic distributions of living lemur species is being revised even today as forests in remote regions of the island are increasingly explored (see, for example, Sterling, 1998, on western populations of *Daubentonia madagascariensis*). Given the vast size and ecological diversity of the island of Madagascar, and given the growing roster of recently discovered or rediscovered lemur species (e.g., *Propithecus tattersalli*, *Allocebus trichotis*, *Microcebus* cf. *myoxinus*, *M. ravelobensis*, and more), there is good reason to believe that new extant lemuriform taxa (including, perhaps, new lemurids) await discovery (Simons, 1988; Meier & Albignac, 1991; Schmid & Kappeler, 1994; Zimmermann et al., 1998).

We recognize five genera (with 12 species) in the Family Lemuridae (Table 1 generally from Mittermeier et al., 1994). One of these genera (*Pachylemur* Lamberton, 1946) is entirely extinct. *Pachylemur* is usually considered the sister taxon to *Varecia* because of apparent dental and craniofacial synapomorphies (Seligsohn & Szalay, 1974; Walker, 1974; Mahé, 1976; Szalay & Delson, 1979; Tattersall, 1982); but Randria (1990), in a cladistic analysis of variation in lemur vertebral morphology, found *Pachylemur* to be the sister taxon to all other lemurids. Two species of *Pachylemur* (*P. insignis* and

Table 1. Taxonomy of the Lemuridae

Order Primates Linnaeus, 1758
  Suborder Strepsirrhini (E. Geoffroy, 1812)
    Superfamily Lemuroidea (Mivart, 1864)
      Family Lemuridae Gray, 1821
        Genus *Varecia* Gray, 1863
          V. variegata (Kerr, 1792)
        Genus *Lemur* Linnaeus, 1758
          L. catta Linnaeus, 1758
        Genus *Eulemur* Simons and Rumpler, 1988
          E. fulvus (E. Geoffroy, 1796)
          E. rubriventer (I. Geoffroy, 1850)
          E. mongoz (Linnaeus, 1766)
          E. coronatus (Gray, 1842)
          E. macaco (Linnaeus, 1766)
        Genus *Hapalemur* I. Geoffroy, 1851
          H. griseus (Link, 1795)
          H. simus Gray, 1842
          H. aureus Meier et al., 1987
        Genus *Pachylemur* Lamberton, 1946
          P. insignis Lamberton, 1946
          P. jullyi Lamberton, 1946

P. jullyi) are recognized, but their distinctiveness is not certain. *Pachylemur* will not be considered further in this chapter (because associated skeletons do not exist); nevertheless, the current analysis establishes the background required for its eventual analysis.

This chapter presents a comprehensive cladistic analysis of the phylogeny of extant members of the Family Lemuridae, based on their skeletal morphology. My primary goal was to conduct the most complete skeletal analysis possible, utilizing characters scored for skull, mandible, teeth, vertebrae, scapulae, pelves, humeri, radii, ulnae, femora, tibiae, and fibulae. A secondary goal was to compare the optimal trees generated by detailed analyses of skeletal morphology to trees generated by other researchers—with particular reference to the results of studies of the genetics (molecular biology and chromosomes) of the Lemuridae. The optimal trees generated here show some of the same inconsistencies and contradictions that characterize trees generated by earlier studies. The most parsimonious cladograms differ depending on the outgroup selected and the subset of characters (qualitative, quantitative, or both) chosen for analysis. Thus, in some sense, the results of this study serve only to underscore the failure of the data to yield a clear solution. Nevertheless, the concordance of some of the most parsimonious cladograms found here with cladograms derived from recent studies of lemurid molecular biology is quite remarkable—particularly given the entirely independent nature of the data upon which these analyses are based. I would therefore argue that the present results succeed in bringing lemur biologists closer to discovering the true phylogeny of the Family Lemuridae.

MATERIALS AND METHODS

Table 2 lists sample sizes for individuals belonging to the species examined here; samples are listed by museum. Only adult individuals were included in the analysis. Whole skeletons were selected whenever available. Individuals from known localities

Table 2. Samples used in this study

Genus and species	Total sample	MNHN	FMNH	AMNH	MCZ
Varecia variegata	20	2		14	4
Lemur catta	6		1	5	
Eulemur fulvus	29	1	2	12	14
E. rubriventer	9		1	2	6
E. mongoz	5	2		2	1
E. coronatus	5	1		4	
E. macaco	2			2	
Hapalemur griseus	21			9	12
H. simus	1				1
Outgroup: *Propithecus verreauxi*	15	2	2	11	
Outgroup: *P. diadema*	4	1		3	
Outgroup: *Cheirogaleus medius*	1		1		
Outgroup: *C. major*	5		1	1	3
Outgroup: *Lepilemur leucopus*	10			10	
Outgroup: *L. mustelinus*	1			1	

MNHN = Laboratoire d'Anatomie Comparée, Muséum National d'Histoire Naturelle, Paris.
FMNH = Field Museum of Natural History, Chicago.
AMNH = American Museum of Natural History, New York City.
MCZ = Museum of Comparative Zoology, Harvard University, Cambridge, Massachusetts.

were preferred. Outgroup taxa included *Propithecus* (*verreauxi* and *diadema*), *Cheirogaleus* (*major* and *medius*), and *Lepilemur* (*leucopus* and *mustelinus*). In all, 134 individuals belonging to seven genera were included in the analysis; four of the seven belong to the Family Lemuridae. Of the 10 recognized extant lemurid species, nine were included in the analysis (n = 98 individuals). Due to the lack of skeletal samples *Hapalemur aureus* was excluded from this study.

Both qualitative and quantitative characters were hypothesized. Qualitative characters were either binary or multistate; some were ordered (i.e., treated as directional) while others were not. Qualitative characters were used to score genera. In all, twenty-six qualitative characters were constructed for genus-level comparisons.

Quantitative characters were simple indices (or ratios of metric traits). Character states were high, medium, or low values for ratios of lengths, widths, or heights of metric traits. In some cases, as many as five character states were defined for single characters; others had as few as three. The cut-points for character states depended on the range and distribution of values across taxa. All quantitative characters were coded as ordered and multistate. Only when ratios were found to yield statistically significant differences among the taxa considered here were they selected for use in the cladistic analyses of quantitative traits. Quantitative character states were coded separately for genera and for species, and separate genus-level and species-level transformation matrices were constructed. Thirty-eight quantitative characters were examined for both genus- and species-level analyses.

All data were first entered into MacClade; cladograms were then evaluated using PAUP Version 3.1 (see Wiley, 1981; Wiley et al., 1991; Swofford, 1993). Polarity of characters was determined by rooting the tree according to outgroups. ACCTRAN was employed for character optimization. [Character optimization is the process of distributing homoplasies to the nodes of any given tree.] ACCTRAN pushes evolutionary transformations down the tree as far as possible, thus favoring reversals over para lelisms when the choice is equally parsimonious. The branch-and-bound algorithm was used to find those trees that minimize evolutionary transformations among character states (i.e., the most parsimonious trees) and a variety of tree statistics for these optimal trees. These statistics included tree length (the total number of implied evolutionary steps, or independent changes in character state), character consistency indices (i.e., the degree to which homoplasy, for any given character, is required by a particular tree topology), and the overall Consistency Index (which reveals how much homoplasy is required for an entire data matrix given a particular tree topology; see Kluge & Farris, 1969).

Separate parsimony analyses were performed for (1) qualitative characters, (2) quantitative characters, and (3) combined character sets. Separate analyses were also performed taking *Cheirogaleus* or *Lepilemur* as outgroups, in isolation or together (and in association with *Propithecus*). In all, eleven parsimony analyses (differing in the number and type of characters treated and in the outgroup selected) were performed. These produced 14 optimal trees—three for species-level analyses and 11 for genus-level analyses. The 14 optimal trees were then compared, taking into account the fact that both overall Consistency Indices and tree lengths are affected by the number of characters and taxa analyzed.

RESULTS

Table 3 lists the indices used in the construction of quantitative characters, and provides descriptive statistics for these indices for the extant Lemuridae. Qualitative characters include some used by Tattersall and Schwartz (1991) (for example, the positions of cranial foramina, the development of a frontal sinus, the morphology of the orbital rim, etc.), but also postcranial characters (such as the form of the entepicondylar foramen on the humerus, the development and position of the third trochanter of the femur, and so on). A full list of the qualitative characters as well as the character states for each qualitative or quantitative character, as well as the (PAUP-generated) hypothetical character states at each critical node for each of the 14 optimal trees generated by the various analyses, can be found in Randria (1998).

No single sister-taxon relationship for lemurid genera was supported by all 14 optimal trees; indeed, there was no node supported by strict consensus for the eleven genus-level analyses (with only four ingroup taxa). As might be expected given the fact that the genus-level analyses treated only five to seven taxa (with four ingroup genera and one to three outgroup genera) while species-level analyses treated a minimum of 11 taxa, genus-level analyses yielded trees with considerably higher consistency indices (and lower tree lengths) than did species-level analyses. Combining qualitative and quantitative characters yielded trees of greater length, but only slightly depressed overall consistency; however, increasing the number of outgroup taxa from one to three did markedly affect the overall consistency indices (as well as tree length). Fig. 1 shows the topologies for the five trees (four genus-level trees and one species-level tree) with the highest overall Consistency Indices (controlling for number of taxa and characters analyzed), and Table 4 gives the summary statistics for this subset of optimal trees.

Table 5 illustrates the lack of consensus across all 14 parsimony analyses. There were, however, a few recurrent signals. For genus-level comparisons, three optimal trees were repeatedly produced (Fig. 1, Cladograms A–C). Cladograms A and B were supported by three independent parsimony analyses each, and Cladogram C was produced by two. In no solution was *Varecia* found to be the sister taxon to *Hapalemur*. *Lemur* and *Varecia* were most frequently found to be sister taxa, but sister-taxon relationships for *Varecia* and *Eulemur*, *Lemur* and *Hapalemur*, and (once) for *Lemur* and *Eulemur*, were also obtained.

Table 3. Indices used in the construction of quantitative characters, with mean values ± standard deviations for the extant lemurid species; for sample sizes, see Table 2[1]

Index[2]	Eulemur fulvus	Eulemur coronatus	Eulemur mongoz	Eulemur macaco	Eulemur rubriventer	Lemur catta	Hapalemur griseus	Hapalemur simus	Varecia variegata
1	75.1 ± 3.4	70.5 ± 1.7	75.0 ± 4.1	74.7 ± 1.0	70.0 ± 3.6	70.3 ± 2.8	61.0 ± 3.3	55.0	81.8 ± 3.0
2	26.7 ± 2.7	25.8 ± 1.7	24.6 ± 2.9	28.7 ± 2.7	26.9 ± 3.0	23.7 ± 1.1	27.5 ± 2.2	22.1	31.6 ± 1.9
3	54.8 ± 3.5	58.6 ± 3.7	56.6 ± 6.1	56.3 ± 2.8	57.6 ± 3.8	59.1 ± 3.7	63.0 ± 4.0	62.6	49.5 ± 3.1
4	72.8 ± 8.6	66.3 ± 3.6	66.2 ± 10.5	74.2 ± 10.0	68.4 ± 8.7	61.4 ± 4.7	63.9 ± 6.4	52.0	94.7 ± 8.0
5	72.4 ± 9.5	78.2 ± 6.7	75.2 ± 18.1	72.5 ± 0.8	72.4 ± 8.4	77.1 ± 3.2	50.8 ± 5.9	90.7	52.8 ± 4.2
6	47.9 ± 3.5	48.6 ± 3.5	48.5 ± 4.7	42.3 ± 7.1	49.0 ± 3.9	47.9 ± 3.2	49.9 ± 3.5	64.4	39.6 ± 3.0
7	93.9 ± 9.2	87.7 ± 10.4	91.1 ± 14.8	95.8 ± 9.6	83.9 ± 5.6	85.5 ± 4.2	82.1 ± 6.7	60.5	115.5 ± 6.7
8	103.8 ± 5.7	105.7 ± 7.4	103.2 ± 7.2	104.0 ± 0.8	107.0 ± 9.3	105.9 ± 8.3	75.5 ± 5.3	88.5	108.1 ± 5.2
9	28.4 ± 2.3	27.9 ± 1.9	27.2 ± 1.4	26.6 ± 0.4	30.2 ± 3.0	26.8 ± 2.6	35.5 ± 3.0	32.0	25.0 ± 2.3
10	60.8 ± 6.0	64.0 ± 2.9	61.2 ± 4.1	64.7 ± 6.1	43.1 ± 4.9	57.3 ± 6.3	43.7 ± 3.8	45.3	58.5 ± 4.3
11	84.3 ± 3.0	88.4 ± 0.9	87.4 ± 5.1	81.5 ± 0.1	86.5 ± 2.6	84.8 ± 2.3	98.2 ± 3.6	104.3	76.7 ± 1.8
12	43.7 ± 2.9	46.6 ± 2.5	44.0 ± 4.3	45.2 ± 1.9	46.8 ± 2.5	46.1 ± 2.5	54.4 ± 2.8	53.5	38.5 ± 2.7
13	75.8 ± 2.6	71.5 ± 2.0	71.1 ± 5.2	76.4 ± 0.4	71.1 ± 2.6	71.5 ± 2.9	69.0 ± 4.5	60.7	84.0 ± 3.2
14	49.6 ± 5.3	44.5 ± 2.8	47.3 ± 11.6	49.9 ± 4.2	44.3 ± 5.7	41.4 ± 2.3	46.8 ± 4.5	34.1	58.1 ± 3.2
15	62.7 ± 2.7	65.6 ± 2.2	62.7 ± 2.6	64.9 ± 0.7	67.9 ± 2.7	64.9 ± 3.1	72.6 ± 3.2	98.7	62.5 ± 2.3
16	58.0 ± 2.4	58.7 ± 0.9	56.7 ± 1.8	58.8 ± 1.1	62.1 ± 2.8	58.4 ± 2.0	65.6 ± 1.8	71.5	56.0 ± 2.0
17	37.6 ± 1.4	37.3 ± 0.7	34.6 ± 1.8	37.2 ± 1.4	38.7 ± 1.7	39.9 ± 1.2	43.3 ± 1.3	47.5	35.6 ± 1.3

(continued)

Table 3. (*Continued*)

Index[2]	*Eulemur fulvus*	*Eulemur coronatus*	*Eulemur mongoz*	*Eulemur macaco*	*Eulemur rubriventer*	*Lemur catta*	*Hapalemur griseus*	*Hapalemur simus*	*Varecia variegata*
18	56.3 ± 2.1	56.1 ± 1.6	52.7 ± 4.3	54.0 ± 1.6	56.6 ± 3.1	61.2 ± 3.0	62.5 ± 3.3	69.9	52.2 ± 1.9
19	86.6 ± 3.8	88.2 ± 2.2	85.6 ± 4.2	85.5 ± 0.9	90.8 ± 5.2	88.3 ± 2.6	94.9 ± 3.5	105.4	82.0 ± 3.5
20	61.0 ± 9.4	69.6 ± 11.3	67.4		58.0 ± 12.0	63.7 ± 0.9	58.5 ± 5.3	56.9	77.9 ± 7.3
21	44.4 ± 1.8	44.7 ± 1.1	40.3 ± 1.7	43.3 ± 1.6	46.0 ± 2.0	47.6 ± 2.3	54.5 ± 1.9	59.8	40.6 ± 1.4
22	96.7 ± 8.6	104.1 ± 19.7	105.2		87.1 ± 9.7	97.4 ± 6.6	85.3 ± 6.8	81.9	118.6 ± 9.5
23	68.8 ± 5.8	64.0 ± 0.6	63.9		77.5 ± 9.6	80.2 ± 9.3	84.3 ± 6.5	81.1	56.5 ± 5.4
24	78.1 ± 9.7	78.0 ± 4.3	79.3 ± 4.9	78.4 ± 1.6	77.8 ± 11.2	93.2 ± 3.3	86.9 ± 4.9	100.0	69.1 ± 7.9
25	80.6 ± 5.6	82.8 ± 4.0	75.7 ± 1.6	80.4 ± 1.5	79.8 ± 2.3	78.3 ± 6.8	76.4 ± 8.5	72.7	90.2 ± 5.4
26	103.5 ± 5.2	102.6 ± 1.0	91.3 ± 2.8	104.6 ± 9.2	107.6 ± 6.5	98.2 ± 5.1	93.9 ± 3.8	98.3	114.3 ± 6.3
27	57.4 ± 2.7	60.8 ± 4.0	56.8 ± 0.3		57.8	55.6 ± 1.3	56.9 ± 4.4	56.4	52.2 ± 1.5
28	104.1 ± 9.8	107.8 ± 0.1	108.5		97.7 ± 7.4	109.8 ± 4.0	113.0 ± 4.0	94.7	89.5 ± 2.4
29	83.1 ± 6.9	94.4			88.2	80.1 ± 2.6	84.1 ± 6.0	93.0	77.2 ± 6.4
30	28.5 ± 4.5	26.4			28.1 ± 0.2	24.8 ± 4.0	37.6 ± 6.6	31.8	25.1 ± 4.2
31	66.6 ± 2.6	63.2 ± 0.1			63.5 ± 0.0	62.7 ± 3.1	66.4 ± 2.5	66.8	72.8 ± 2.4
32	69.8 ± 6.8	60.5 ± 1.2			71.6 ± 9.9	77.2 ± 5.0	64.1 ± 10.5	75.8	83.3 ± 8.5
33	62.0 ± 14.3	65.7			57.2 ± 6.6	49.1 ± 9.0	98.5 ± 22.8	68.0	50.9 ± 10.1
34	86.5 ± 4.9	84.2 ± 4.6	88.9 ± 7.5		80.0 ± 16.6	91.9 ± 8.7	90.4 ± 4.8	93.0	75.7 ± 5.1
35	55.0 ± 4.3	57.5 ± 5.9	52.5 ± 4.3		57.2 ± 3.1	46.8 ± 1.5	46.8 ± 3.4	41.1	71.2 ± 5.8
36	54.7 ± 4.0	57.5 ± 5.9	52.2 ± 3.1		56.0 ± 2.2	46.5 ± 1.1	46.4 ± 3.4	41.5	70.9 ± 6.3
37	42.6 ± 5.2	45.3 ± 12.2	33.5 ± 3.0		43.4 ± 3.0	36.7 ± 0.7	38.5 ± 4.2	32.7	56.7 ± 7.6
38	87.0 ± 5.2	90.9 ± 6.9	85.7 ± 11.4		88.9 ± 5.1	74.4 ± 1.8	74.8 ± 4.8	66.9	105.3 ± 10.5

[1]Some individuals had damaged or missing parts, causing reduced sample sizes for some indices.
[2]All of the following are measured as percentages (i.e., multiplied by 100). For full descriptions, see Randria (1998):
Index 1: Palate length / Bizygomatic breadth
Index 2: Nasal length / Cranial length
Index 3: Cranial height / Mandibular length
Index 4: Nasal length / Cranial height
Index 5: Interorbital breadth / Nasal length
Index 6: Height of the postorbital bar / Length of the dental arcade
Index 7: Nasal length / Posterior nasal height
Index 8: Maximum maxillary breadth / Height of the mandibular coronoid process
Index 9: Mandibular corpus height / Maximum maxillary breadth
Index 10: Maximum breadth across choanae / Maximum inter-pterygoid breadth
Index 11: Nasion-opisthocranion / Basicranial length
Index 12: Cranial height / Basicranial length
Index 13: Palate length / Biorbital breadth
Index 14: Nasal length / Biorbital breadth
Index 15: Maximum maxillary breadth / Maximum
Index 16: Bizygomatic breadth / Cranial length
Index 17: Bi-tympanic bullar breadth / Cranial length
Index 18: Bi-tympanic bullar breadth / Mandibular length
Index 19: Bizygomatic breadth / Mandibular length
Index 20: Length of the lumbar laminae / Bi-occipital condyle breadth
Index 21: Bi-tympanic bullar breadth / Basicranial length
Index 22: Lumbar laminae length / Occipital foramen breadth
Index 23: Lumbar laminae breadth / Lumbar laminae length
Index 24: M^3 mesiodistal length / M^2 mesiodistal length
Index 25: M^2 mesiodistal length / M^2 buccolingual breadth
Index 26: M_1 mesiodistal length / M_2 mesiodistal length
Index 27: Scapular glenoid fossa width / Scapular glenoid fossa length
Index 28: Transverse diameter of the radial head / Anteroposterior diameter of the radial head
Index 29: Scapular glenoid fossa width / Anteroposterior diameter of the humeral shaft below the head
Index 30: Width of the brachialis flange / Breadth of the supraspinous fossa (scapula)
Index 31: Scapula length / Radius length
Index 32: Breadth of the sigmoid cavity (or trochlear notch of the ulna) / Width of the humeral trochlea
Index 33: Width of the brachialis flange / Breadth of the sigmoid cavity
Index 34: Scapular glenoid fossa width / Width of the humeral trochlea maxillary length
Index 35: Transverse diameter of femoral head / Iliac breadth
Index 36: Anteroposterior diameter of femoral head / Iliac breadth
Index 37: Transverse diameter of femoral neck / Iliac breadth
Index 38: Femoral bicondylar breadth / Iliac breadth

Table 4. Tree statistics for five most parsimonious trees. Analyses that produced unique, alternative cladograms are not shown

Outgroup	Character set	Level of analysis	Tree length (No. of steps)	Overall consistency index	Cladogram supported
Lepilemur	Quantitative	Genus	80	0.86	A
Cheirogaleus	Quantitative	Genus	77	0.90	C
Lepilemur	Qualitative	Genus	53	0.89	B
Cheirogaleus	Qualitative (tree I)	Genus	57	0.90	B
Cheirogaleus	Qualitative (tree II)	Genus	57	0.90	D
Cheirogaleus-Propithecus-Lepilemur	Qualitative	Genus	74	0.82	B
Lepilemur	Combined (tree I)	Genus	135	0.86	A
Cheirogaleus	Combined	Genus	136	0.88	C
Cheirogaleus-Propithecus-Lepilemur	Combined	Genus	194	0.75	A
Cheirogaleus	Quantitative	Species	121	0.80	E

Table 5. Consistency of phylogenetic signals across all parsimony analyses performed here[1]

		Monophyly of						
Outgroup	Character set	*Varecia* and *Hapalemur*	*Lemur* and *Hapalemur*	*Varecia* and *Eulemur*	*Varecia* and *Lemur*	*Eulemur* and *Lemur*	*Hapalemur*	*Eulemur*
Lepilemur	Quantitative	no	no	yes	no	no	NA	NA
Cheirogaleus	Quantitative	no	yes	no	no	no	NA	NA
Cheirogaleus-Propithecus-Lepilemur	Quantitative	no	no	no	no	no	NA	NA
Lepilemur	Qualitative	no	no	no	yes	no	NA	NA
Cheirogaleus	Qualitative I	no	no	no	yes	no	NA	NA
Cheirogaleus	Qualitative II	no	yes	no	no	no	NA	NA
Cheirogaleus-Propithecus-Lepilemur	Qualitative	no	no	no	yes	no	NA	NA
Lepilemur	Combined I	no	no	yes	no	no	NA	NA
Lepilemur	Combined II	no	no	no	yes	no	NA	NA
Cheirogaleus	Combined	no	yes	no	no	no	NA	NA
Cheirogaleus-Propithecus-Lepilemur	Combined	no	no	yes	no	no	NA	NA
Lepilemur	Quantitative I	no	no	no	yes	no	yes	no
Lepilemur	Quantitative II	no	no	no	yes	no	yes	no
Cheirogaleus	Quantitative	no	no	no	no	yes	yes	yes

[1] Note: Some analyses yielded two equally parsimonious optimal trees. NA = not applicable.

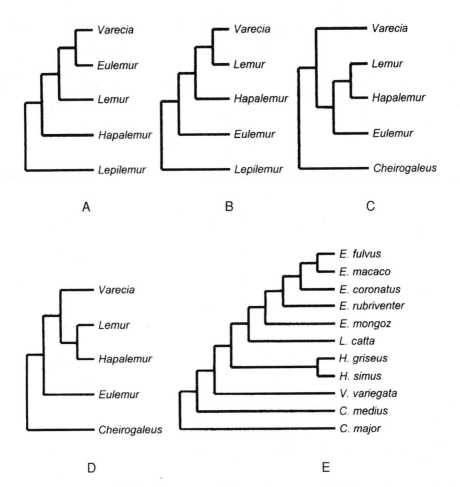

Figure 1. Principal types of cladograms generated (sometimes repeatedly) by parsimony analysis of different data sets, and with different selected outgroups. See text for explanation.

For species-level comparisons, a sister-taxon status for the two species of *Hapalemur* was very well-supported, and monophyly for species of the genus *Eulemur* was sometimes (but not always) supported. (When monophyly for *Eulemur* was violated, this was due to one slightly deviant species.) All three of the optimal cladograms derived for species-level comparisons supported a sister-taxon relationship for *Eulemur macaco* and *E. fulvus*.

Cladogram C (Fig. 1) is the genus-level solution that I favor. According to Cladogram C, *Lemur* and *Hapalemur* are sister taxa, *Eulemur* is the sister to the *Lemur-Hapalemur* clade, and *Varecia* is the sister to the *Lemur-Hapalemur-Eulemur* clade. Cladogram C was obtained through two parsimony analyses of genus-level variation, with *Cheirogaleus* selected as the outgroup. It emerged through an analysis of quantitative characters considered in isolation, and in combination with qualitative characters. The associated tree statistics are very strong. My main reason for preferring it over other trees, however, has nothing to do with the strong tree statistics (other solutions are equally impressive in this regard). Rather, I am impressed by its strong concordance with the results of cladistic analyses performed on entirely independent data sets, as described below.

DISCUSSION

The literature on the phylogeny of the Family Lemuridae continues to be contentious. Researchers employing different methods of analysis (some cladistic, some phenetic) and examining independent sets of data have reached little consensus even with regard to the relationships among the four extant genera, let alone relationships among species. Nevertheless, there is some concordance between the results obtained by other researchers and the results of the present study. Let us consider each of our five most optimal cladograms, in turn.

Cladogram A (Fig. 1) shows a sister taxon relationship for *Varecia* and *Eulemur*. *Lemur* is the sister to the *Varecia-Eulemur* clade, and *Hapalemur* the sister to the *Varecia-Eulemur-Lemur* clade. This cladogram resulted from a parsimony analysis of quantitative characters, taking *Lepilemur* as the outgroup. The same tree was generated by a parsimony analysis of quantitative and qualitative characters combined, again with *Lepilemur* as outgroup (the first of two equally parsimonious optimal trees), by a parsimony analysis of qualitative characters, and another of quantitative and qualitative characters combined, with *Cheirogaleus*, *Lepilemur*, and *Propithecus* as outgroups. Parallel results (with strong support, at least, for a *Varecia-Eulemur* clade) were obtained by Tattersall (1985, 1993) and Tattersall and Schwartz (1991) through a cladistic analysis of dental and cranial variation of the lemurids, and Yoder et al. (1996a), again through a treatment of morphological data. Another cladistic analysis of lemurid morphology (Yoder, 1994) supported a *Varecia-Lemur* clade, as in Cladogram B (Fig. 1) of our results. Cladogram B was generated by a parsimony analysis of qualitative characters, taking *Lepilemur* as the outgroup, a parsimony analysis of qualitative characters (the first of two equally parsimonious optimal trees), taking *Cheirogaleus* as the outgroup, and a parsimony analysis of qualitative characters, taking *Cheirogaleus*, *Lepilemur*, and *Propithecus* as outgroups.

Cladograms C and D were obtained by three parsimony analyses, each taking *Cheirogaleus* as the outgroup. Both support a *Lemur-Hapalemur* clade. Cladogram D was one of two equally parsimonious trees produced by parsimony analysis of qualitative characters (the other was Cladogram B, described above). Cladogram C was the sole optimal tree generated by parsimony analysis of quantitative characters, as well as by the combined set of quantitative and qualitative characters, again taking *Cheirogaleus* as the outgroup. According to Cladogram C: (1) *Lemur* and *Hapalemur* are probable sister taxa; (2) *Eulemur* is the probable sister taxon to the *Lemur-Hapalemur* clade; and (3) *Varecia* is the sister taxon to the *Eulemur-Lemur-Hapalemur* clade. The identical topology was obtained by Dutrillaux (1988) on the basis of cytogenetic data; Rumpler et al. (1991) on the basis of their analysis of the evolution of lemurid karyotypes (*Varecia* did not figure in this analysis); Crovella et al. (1995) on the basis of their analysis of lemurid molecular biology; Yoder (1996a), and Yoder and Irwin (in press) also on the basis of the molecular biology of the Lemuridae (see Yoder et al., this volume). Stanger-Hall (1997) defended a distinction of *Varecia* from all other living lemurids on morphological and behavioral data.

Nearly all genetic studies support the topology illustrated by Cladogram C: *Lemur* and *Hapalemur* are sister taxa (Rumpler et al., 1991; Jung et al., 1992; Adkins &

Honeycutt, 1994; Crovella et al., 1995; Yoder et al., 1996a,b; Yoder & Irwin, in press). This hypothesis was also defended by Groves and Eaglen (1988), Simons and Rumpler (1988), Tattersall (1988), and Macedonia and Stanger (1994) on the basis of their independent analyses of aspects of *Lemur* and *Hapalemur* soft anatomy and behavior. Indeed, the year 1988 saw a flurry of papers supporting a *Lemur-Hapalemur* sister taxon relationship, as well as considerable discussion as the proper generic allocation of all species of "*Lemur*" other than the type species, *L. catta*. If indeed Ring-tailed Lemurs and bamboo lemurs are sister taxa, then species that had previously been included with *L. catta* in the genus *Lemur* could no longer be included in the same genus with the type (see Groves & Trueman, 1995, for a summary of the publication sequence of alternative nomenclatural suggestions, and the ruling of the International Commission on Zoological Nomenclature favoring *Eulemur*). Both Ian Tattersall and Colin Groves have since rejected the argument that Ring-tailed Lemurs and *Lemur* "all-the-rest" must be separated at the generic level. This change of opinion resulted from new cladistic analyses of lemurid craniodental variation (Tattersall & Schwartz, 1991; Tattersall, 1993; Groves & Trueman, 1995).

In a "total evidence" treatment of both molecular and morphological characters, however, Yoder (1996a) again defended monophyly for a *Lemur-Hapalemur* clade. The data analyzed in that study also supported monophyly for a *Eulemur-Varecia* clade. None of the cladograms obtained here reveal monophyly for *both* of these pairs of taxa, although *Lemur-Hapalemur* monophyly and *Eulemur-Varecia* monophyly are both supported by three of the fourteen independently-generated optimal solutions (see Table 5). The *Lemur-Hapalemur* clade is also strongly supported by a wide variety of non-morphological characters examined by other researchers. It is furthermore the case that *Varecia* exhibits many morphological (as well as behavioral) characteristics that distinguish it from *Eulemur* as well as from *Lemur* and *Hapalemur*. Some of these morphological distinctions are shared only with *Pachylemur* (Seligsohn & Szalay, 1974; Walker, 1974; Szalay & Delson, 1979; Tattersall, 1982). They are generally interpreted as *Varecia-Pachylemur* synapomorphies. However, if *Pachylemur* is indeed the sister taxon to a *Varecia-Eulemur-Lemur-Hapalemur* clade rather than the sister taxon to *Varecia* alone (as argued by Randria, 1990, on the basis of the evidence of the vertebral column), then it would appear likely that the craniodental characters shared by *Varecia* and *Pachylemur* are primitive for the Lemuridae, and that characters shared by *Eulemur*, *Lemur*, and *Hapalemur* are derived. I favor Cladogram C over Cladogram D (Fig. 1), in part, for this reason.

The data analyzed for Cladogram E (Fig. 1) include multiple species for *Eulemur*, *Hapalemur* and the outgroup *Cheirogaleus*, in addition to the monotypic *Varecia* and *Lemur*. Cladogram E is one of three optimal trees generated for all ingroup species considered in this analysis (see Randria, 1998, for the other two). Produced by a parsimony analysis of quantitative characters, Cladogram E posits: (1) monophyly for *Eulemur*; (2) monophyly for *Hapalemur*; and (3) a sister-taxon relationship for *Varecia* to a *Hapalemur-Lemur-Eulemur* clade. It does not, however, support a *Hapalemur-Lemur* clade. Other morphology- and/or behavior-based studies support monophyly for *Eulemur* as well as a sister-taxon relationship of *Varecia* to all other extant lemurids (see Tattersall, 1992; Stanger-Hall, 1997; Tattersall & Sussman, 1997). Monophyly of *Hapalemur* is unequivocal. Monophyly of *Eulemur* is strongly supported by genetic studies as is a sister-taxon relationship for *Varecia* to a *Hapalemur-Lemur-Eulemur* clade (see reviews by Yoder & Irwin, in press; Yoder et al., this volume; see also Jung et al., 1992; Crovella et al., 1995).

Interestingly, Cladogram E is quite similar to the "phylogenetic" tree published by Mahé (1976, p. 131). Indeed, aside from the fact that Mahé regarded *Eulemur macaco* and *E. fulvus* as synonymous, and that neither *Hapalemur* nor *E. coronatus* were considered in his "phylogenetic" analysis of the Lemuridae, the two trees are identical. Mahé used cluster analysis of phenetic similarities to construct his tree; phenetic similarities and differences between taxa were summarized using Principal Component Analysis. The correspondence between Mahé's tree of lemurid relationships and Cladogram E is notable, given the differences in the methodological tools (phenetic versus cladistic) used to construct them. It suggests that this tree is well-supported by metric data, and is robust with respect to the particular analytical technique (and morphological data set) used to construct it.

It should be remembered that the cladistic analyses reported here are based not merely on characters of the skull, mandible, and dentitions, but also of the vertebrae, scapula, pelvis, and the long bones. Clearly, postcranial characters can hold the same value to phylogenetic analysis as can the craniodental characters that have been favored by most previous researchers. In fact, a recent study of Sanchez-Villagra and Williams (1998) demonstrated no difference in levels of homoplasy inherent in analyses of postcranial and craniodental morphological characters.

CONCLUSIONS

Cladistic analysis of skeletal variation in the Lemuridae supports a number of sister-taxon relationships. *Hapalemur griseus* and *H. simus* are probable sister taxa, as are *Eulemur macaco* and *E. fulvus* (these are the only nodes upheld by the strict consensus of competing optimal trees, regardless of whether *Cheirogaleus* or *Lepilemur* is selected as the outgroup).

Genus-level analyses yielded no nodes supported by strict consensus. Majority rule would favor a sister-taxon relationship for *Varecia* and *Lemur*; this is the sister-taxon relationship most frequently observed when the 14 optimal trees produced by 11 independent analyses are simultaneously considered. Reasons to prefer an alternative solution include the strong concordance of one of the other optimal solutions found here with optimal trees derived from analyses of entirely independent sets of data—in particular, genetic data. The latter optimal cladogram is supported by both quantitative characters (taken separately) and by quantitative and qualitative characters taken in combination, with *Cheirogaleus* taken as the outgroup.

The lack of consensus across diverse analyses of skeletal morphological characters, and the strong dependence of the outcome on the data set analyzed and (especially) the outgroup selected, suggest that morphological data offer only a weak phylogenetic signal (despite the high Consistency Indices of competing trees). Under such circumstances, perhaps the strongest argument one can make for any given tree topology is that it is supported by independent data sets. Concordance of morphology-based optimal trees with trees derived from molecular studies might be used as a heuristic tool to find the "best" outgroup for morphological parsimony analysis (and with that selection, the most likely sequence of morphological character-state evolutionary change). In this case, the "best" outgroup for analyzing the morphological evolution of the family Lemuridae appears to be *Cheirogaleus*.

Several questions remain open—in particular, despite its strong support from other data sets, the monophyly of *Eulemur* was not strongly supported by the morphological data presented here, and the relationship of *Varecia* to the other lemurids remains uncertain. A detailed study of the skeletal morphology of *Pachylemur*, the only subfossil member of the Lemuridae, may shed more light on the relationship of *Varecia* to other lemurids. It is now possible to include subfossil lemurs in cladistic analyses of molecular (as well as morphological) data. Thus, "total evidence" cladograms need not exclude the subfossil forms. This is because, using PCR (Polymerase Chain Reaction) techniques, researchers can now extract ancient DNA from the bones and teeth of subfossil lemurs (see review by Yoder et al., this volume). Preliminary analysis of the molecular biology of *Pachylemur* (using DNA hybridization techniques) supports a *Pachylemur-Varecia* clade (Crovella et al., 1994). Preliminary phenetic and cladistic analyses of the dental and skeletal morphology of *Pachylemur* and other lemurids have yielded diverse results (cf. Seligsohn & Szalay, 1974; Mahé, 1976; Randria, 1990). Whereas substantially complete, associated skeletons of *Pachylemur* have not yet been found, *Pachylemur* is well-represented by most of the elements of its skeletal anatomy (skulls, mandibles, vertebrae, scapulae, pelves, and all of the long bones, as well as some of the elements of its hand and foot). It is now known from subfossil sites from the northern to the southern tips of the island of Madagascar (see Godfrey et al., this volume), and it is extremely well represented at Tsirave and at other sites in southwestern Madagascar, as well as at Ampasambazimba in central Madagascar. Data on its postcranial and craniodental morphology must be considered (in conjunction with the evidence of the molecular biology) in deriving a more comprehensive hypothesis regarding the phylogenetic history of the Lemuridae.

In conclusion, whereas parsimony analysis of morphological data can be invoked to support various competing interpretations of lemurid phylogeny, the combination of morphological and genetic data argues in favor of the hypothesis that: (1) *Lemur* and *Hapalemur* are sister taxa; (2) *Eulemur* is the sister taxon to the *Lemur-Hapalemur* clade; and (3) *Varecia* is the sister taxon to the *Eulemur-Lemur-Hapalemur* clade.

ACKNOWLEDGMENTS

I thank the editors of this volume for inviting me to submit this chapter. I thank Berthe Rakotosamimanana, Anne Yoder, William Jungers, Laurie Godfrey, and Michael Sutherland, for their advice and guidance through various aspects of this research, and for comments on my dissertation research and/or earlier versions of this manuscript. For facilitating my dissertation research, I thank Steven Goodman, Ian Tattersall, Russ Mittermeier, and Eleanor Sterling. The support of the Ranaivoson family, the Ramanantsoa family, the Godfrey family, and Jean Claude Razafimahaimodison is also greatly appreciated. B. Scott Leon provided technical help with MacClade and PAUP. The following institutions provided access to skeletal materials and/or facilities (computers, computer software, office space): Muséum National d'Histoire Naturelle, Paris; Field Museum of Natural History, Chicago; Museum of Comparative Zoology, Harvard University, Cambridge, Massachusetts; Department of Anthropology, University of Massachusetts, Amherst, Massachusetts; and the American Museum of Natural History, New York. I thank the staff and curators of specimens in these institutions for providing me access to specimens under their care. Special thanks go to François Jouffroy, Steven Goodman, Ross MacPhee, Ian Tattersall, and Maria Rutzmoser. Financial support from the MacArthur Foundation, Conservation International, the Université d'Antananarivo, and the American Museum of Natural

History is greatly appreciated. Laurie Godfrey translated and edited the original French version of this manuscript.

REFERENCES

ADKINS, R. M., AND R. L. HONEYCUTT. 1994. Evolution of the primate cytochrome *c* oxidase subunit II gene. Journal of Molecular Evolution, **38**: 215–231.

CROVELLA, S., D. MONTAGNON, B. RAKOTOSAMIMANANA, AND Y. RUMPLER. 1994. Molecular biology and systematics of an extinct lemur: *Pachylemur insignis*. Primates, **35**: 519–522.

CROVELLA, S., D. MONTAGNON, AND Y. RUMPLER. 1995. Highly repeated DNA sequences and systematics of Malagasy Primates. Human Evolution, **10**: 35–44.

DUTRILLAUX, B. 1988. Chromosome evolution in Primates. Folia Primatologica, **50**: 134–135.

GROVES, C. P., AND R. H. EAGLEN. 1988. Systematics of the Lemuridae (Primates, Strepsirhini). Journal of Human Evolution, **17**: 513–538.

GROVES, C. P., AND J. W. H. TRUEMAN. 1995. Lemurid systematics revisited. Journal of Human Evolution, **28**: 427–437.

JUNG, K. Y., S. CROVELLA, AND Y. RUMPLER. 1992. Phylogenetic relationships among lemuriform species determined from restriction genomic DNA banding patterns. Folia Primatologica, **58**: 224–229.

KLUGE, A. G., AND J. S. FARRIS, 1969. Quantitative phyletics and the evolution of anurans. Systematic Zoology, **18**: 1–32.

KOENDERS, L., Y. RUMPLER, J. RATSIRARSON, AND A. PEYRIÉRAS. 1985. *Lemur macaco flavifrons* (Gray, 1867): a rediscovered subspecies of primate. Folia Primatologica, **44**: 210–215.

LAMBERTON, C. 1946. Contribution à la connaissance de la faune subfossile de Madagascar. Note XVII Les Pachylemurs. Bulletin de l'Académie Malgache, Nouvelle série **27**: 7–22.

MACEDONIA, J. M., AND K. F. STRANGER. 1994. Phylogeny of the Lemuridae revisited: Evidence from communication signals. Folia Primatologica, **63**: 1–43.

MAHÉ, J. 1976. Crâniométrie des Lémuriens: Analyses Multivariables. Phylogénie. Mémoires du Muséum National d'Histoire Naturelle, Paris, Série C, **32**: 1–342.

MEIER, B., AND R. ALBIGNAC. 1991. Rediscovery of *Allocebus trichotis* Günther 1875 (Primates) in northeast Madagascar. Folia Primatologica, **56**: 57–63.

———, A. PEYRIÉRAS, Y. RUMPLER, AND P. C. WRIGHT. 1987. A new species of *Hapalemur* (Primates) from south east Madagascar. Folia Primatologica, **48**: 211–215.

MITTERMEIER, R. A., I. TATTERSALL, W. R. KONSTANT, D. M. MEYERS, AND R. B. MAST. 1994. Lemurs of Madagascar. Conservation International, Washington, D.C.

RANDRIA, G. F. N. 1990. Contribution à l'Etude de la Colonne Vertébrale du genre *Pachylemur* (Lamberton, 1946): Anatomie et Analyse Cladistique. Thèse de Doctorat de 3ème Cycle, Université d'Antananarivo, Madagascar.

———. 1998. Morphologie, Analyses Phenetique et Cladistique des Lemuridae (Gray, 1821). Thèse de Doctorat d'Etat des Sciences, Université d'Antananarivo, Madagascar.

RUMPLER, Y., S. WARTER, M. HAUWY, V. RANDRIANASOLO, AND B. DUTRILLAUX. 1991. Cytogenetic study of *Hapalemur aureus*. American Journal of Physical Anthropology, **86**: 81–84.

SANCHEZ-VILLAGRA, M. R., AND B. A. WILLIAMS. 1998. Levels of homoplasy in the evolution of the mammalian skeleton. Journal of Mammalian Evolution, **5**: 113–126.

SCHMID, J., AND P. M. KAPPELER. 1994. Sympatric mouse lemurs (*Microcebus* spp.) in western Madagascar. Folia Primatologica, **63**: 162–170.

SELIGSOHN, D., AND F. S. SZALAY. 1974. Dental occlusion and the masticatory apparatus in *Lemur* and *Varecia*: their bearing on the systematics of living and fossil Primates, pp. 543–561. *In* Martin, R. D., G. A. Doyle, and A. C. Walker, eds., Prosimian Biology. Duckworth, London.

SIMONS, E. L. 1988. A new species of *Propithecus* (Primates) from northeast Madagascar. Folia Primatologica, **50**: 143–151.

———, AND Y. RUMPLER. 1988. *Eulemur*: new generic name for species of *Lemur* other than *Lemur catta*. Comptes Rendus de l'Académie des Sciences, Paris, **307**: 547–551.

STANGER-HALL, K. F. 1997. Phylogenetic affinities among the extant Malagasy lemurs (Lemuriformes) based on morphology and behavior. Journal of Mammalian Evolution, **4**: 163–194.

STERLING, E. 1998. Preliminary report on a survey for *Daubentonia madagascariensis* and other primate species in the west of Madagascar, June-August 1994. Lemur News, **3**: 7–8.

SZALAY, F. S., AND E. DELSON. 1979. Evolutionary History of the Primates. Academic Press, New York.
SWOFFORD, D. L. 1993. PAUP: Phylogenetic Analysis Using Parsimony, Version 3.1. Computer program distributed by the Illinois Natural History Survey, Champaign.
TATTERSALL, I. 1982. The Primates of Madagascar. Columbia University Press, New York.
———. 1985. Systematics of the Malagasy strepsirhine primates, pp. 43–72 in Swindler, D., and J. Erwin, eds., Comparative Primate Biology Vol. 1. Alan R. Liss, New York.
———. 1988. A note on the nomenclature in Lemuridae. Physical Anthropology News, **7**: 14.
———. 1992. Systematic versus ecological diversity: The example of the Malagasy primates, pp. 25–39. *In* Eldredge, N., ed., Systematics, Ecology, and the Biodiversity Crisis. Columbia University Press, New York.
———. 1993. Speciation and morphological differentiation in the genus *Lemur*, pp. 163–176. *In* Kimbel, W. H., and L. B. Martin, eds., Species, Species Concepts, and Primate Evolution. Plenum Press, New York.
———, AND J. H. SCHWARTZ. 1991. Phylogeny and nomenclature in the *Lemur*-group of Malagasy strepsirhine Primates. Anthropological Papers of the American Museum of Natural History, **69**: 3–18.
TATTERSALL, I., AND R. W. SUSSMAN. 1998. 'Little brown lemurs' of northern Madagascar: Phylogeny and ecological role in resource partitioning. Folia Primatologica, **69**, Supplement 1: 379–388.
WALKER, A. C. 1974. Locomotor adaptations in past and present prosimian primates, pp. 349–382. *In* Jenkins, F. A., ed., Primate Locomotion. Academic Press, New York.
WILEY, E. O. 1981. Phylogenetics: The Theory and Practice of Phylogenetic Systematics. John Wiley and Sons, New York.
———, D. SIEGEL-CAUSEY, D. R. BROOKS, AND V. A. FUNK. 1991. The Complete Cladist: A Primer of Phylogenetic Procedures. Museum of Natural History, University of Kansas, Lawrence, Kansas.
YODER, A. D. 1994. Relative position of the Cheirogaleidae in Strepsirhine phylogeny: a comparison of morphological and molecular methods and results. American Journal of Physical Anthropology, **94**: 25–46.
———, M. CARTMILL, M. RUVOLO, K. SMITH, AND R. VILGALYS. 1996a. Ancient single origin for Malagasy primates. Proceedings of the National Academy of Sciences, USA, **93**: 5122–5126.
YODER, A. D., AND J. A. IRWIN. In press. Phylogeny of the Lemuridae: Effects of character and taxon sampling on resolution of species relationships within *Eulemur*. Cladistics.
YODER, A. D., M. RUVOLO, AND R. VILGALYS. 1996b. Molecular evolutionary dynamics of cytochrome **b** in strepsirrhine primates: the phylogenetic significance of third position transversions. Molecular Biology and Evolution, **13**: 1339–1350.
ZIMMERMANN, E., S. CEPOK, N. RAKOTOARISON, V. ZIETEMANN, AND U. RADESPIEL. 1998. Sympatric mouse lemurs in north-west Madagascar: A new rufous mouse lemur species (*Microcebus ravelobensis*). Folia Primatologica, **69**: 106–114.

SUPPORT UTILIZATION BY TWO SYMPATRIC LEMUR SPECIES

Propithecus verreauxi verreauxi **and** *Eulemur fulvus rufus*

Léonard Razafimanantsoa[1]

[1] Département de Paléontologie et d'Anthropologie Biologique
BP 906, Faculté des Sciences
Université d'Antananarivo, Antananarivo (101), Madagascar

ABSTRACT

The supports utilized by *Propithecus verreauxi verreauxi* and *Eulemur fulvus rufus* for displacement and feeding were studied in the Kirindy Forest/CFPF of southwestern Madagascar. Support dimensions were classified in four categories (small, medium, large, and enormous). Support orientation was classified in four categories (horizontal, oblique, angled, and vertical). Comparisons between the structures available in the forest and the supports used by these lemurs showed that *P. verreauxi* used mainly intermediate and large sized supports and vertical structures for locomotion; while *E. fulvus* used the different sized supports according to their availability in the forest, but moved disproportionally more often quadrupedally on horizontal and oblique supports. Both species used small and horizontal supports to feed from. Comparing the two lemur species, *P. verreauxi* used larger trunks and branches that were more often vertical or angular than the supports used by *E. fulvus* for locomotion and feeding. In contrast, *E. fulvus* used horizontal branches more frequently for locomotion and especially when feeding.

RÉSUMÉ

Les supports utilisés par *Propithecus verreauxi verreauxi* et *Eulemur fulvus rufus* pour assurer leurs déplacements et leurs activités d'alimentation ont été étudiés dans la Forêt Kirindy/CFPF dans le sud-ouest de Madagascar. Les dimensions des supports ont été classés en quatre catégories (petit, moyen, grand et énorme). L'orientation du support a été classée en quatre catégories (horizontal, oblique, anguleux et vertical). La comparaison entre la végétation disponible et celle utilisée par les lémuriens montre que *P. verreauxi* préfère se mouvoir sur des supports de dimensions intermédiaire et large, d'une orientation verticale alors que *E. fulvus* ne montre une préférence que dans l'orientation, favorisant les supports horizontaux et obliques. Une préférence pour les supports de petites tailles orientés horizontalement est montrée par les deux espèces lors des activités d'alimentation.

Les supports privilégiés pour les deux activités étaient caractéristiques du point de vue de la taille comme de l'orientation pour les deux espèces. En même temps, le support exploité pour les déplacements s'avérait être utilisé simultanément pour les activités alimentaires, *P. verreauxi* fréquentant les arbres de tailles intermédiaire et grande avec des orientations anguleuse et verticale alors que *E. fulvus* utilisait les supports de petite taille orientés obliquement ou horizontalement. En plus d'une variation dans l'orientation du support exploité pour les déplacements, une différence dans le type de locomotion a été relevée.

INTRODUCTION

Locomotion, its underlying morphology and the influence of substrate use, are all important aspects of primate biology. Forests are complex three dimensional environments in which to move. Primates have evolved a variety of locomotion modes which enable them to exploit this habitat (Napier & Napier, 1967). Different locomotor modes have different energetic requirements and these have been linked to body mass and the utilization of different food types (e.g., Fleagle, 1984; Demes & Günther, 1989; Pounds, 1991; Crompton et al., 1993; Demes et al., 1995; Preuschoft et al., 1995, 1998). In addition, different forms of locomotion in the forest can reduce interspecific competition by allowing access to different food sources or to similar food resources in different places, such as flowers accessible from branch tips or from main trunks (e.g., Grand, 1984; McGraw, 1996). Understanding the links between behavior, locomotion, and morphology might allow us to reconstruct the lifestyle of extinct primates in relation to their environment and give us more information on primate evolution (e.g., Jungers, 1978; Jouffroy et al., 1982; Schwartz & Tattersall, 1985; Oxnard et al., 1990; Godfrey et al., 1997). However, while many studies have described interspecific differences in the types of locomotion, very few attempts have been made to determine actual preferences or avoidances of free ranging arboreal primates for specific substrates (e.g., Cannon & Leighton, 1994; Warren, 1997). This, however, is a prerequisite to assess the adaptive value of morphological adaptations and to understand the constraints acting upon different primate species.

A number of studies described the locomotion of Malagasy prosimians and classified the different species into different locomotor categories based on morphological and behavioral characteristics (e.g., Petter et al., 1977; Gebo, 1987; Oxnard et al., 1990; Dagosto, 1994, 1995; Warren, 1997 and references quoted therein). However, Warren's study was the first that quantified the supports available in the natural environment of lemurs and thus provided a basis for comparisons between frequencies of locomotor behavior and the general frequency of supports available in the forest. Her data showed that two lemur species, *Avahi laniger* and *Lepilemur edwardsi*, used horizontal branches more often than would be predicted by the availability of these substrates in the forest. In the literature and also based on casual observations in the forest, these two species are classified as vertical clingers and leapers. Assuming that morphology and behavior evolved in parallel to optimize energy budgets, the discrepancy between the animals' behavior observed in Warren's (1997) study and what would have been expected based on the morphological characteristics, raises the question whether or not the traits we observe today are adaptations to the present environmental situation or whether they reflect constraints that operated in the past (Richard & Dewar, 1991; van Schaik & Kappeler, 1996; Warren & Crompton, 1996).

The goal of this study was to extend the data base initiated by Warren (1997) to allow further comparisons of the locomotion of various lemur species in forests with different structure or under different competitive regimes. In order to extend the data base, I first quantify the structures available in a dry deciduous forest. Then the frequency of these structures was compared with the supports used by two sympatric lemur species, *Propithecus verreauxi verreauxi* and *Eulemur fulvus rufus*.

METHODS

Study Site

The study was carried out in plot CS7 in near primary forest of the forestry concession of the Centre de Formation Professionnelle Forestière de Morondava (CFPF) during the wet season between October 1996 and April 1997 (Fig. 1). The plot is part of a large (>12,000 ha) tract of deciduous dry forest in western Madagascar (Kirindy Forest/CFPF), 60 km north of Morondava (44°39′E, 20°03′S). The study area is dissected by a grid of trails spaced at about 25 m intervals. More details are given by Abraham et al. (1996), and Ganzhorn and Sorg (1996). The climate is highly seasonal with about eight rainless months per year. During the dry season, most trees shed their leaves. Fruit is available year-round (Sorg & Rohner, 1996).

The structure of the forest can be subdivided into three groups (Rakotonirina, 1996):

1) a dense undergrowth between 1 and 5 m in height and 6000–8000 stems smaller than 10 cm diameter at breast height (DBH) per ha;
2) an intermediate layer between 6 and 12 m in height with up to 1500 stems of 10–25 cm DBH per ha;
3) a discontinuous canopy at a height of about 10–20 m with some emergents reaching up to 25 m.

Estimating the Supports Available in the Forest

Quantification of the supports available in the forest follows the technique applied by Warren (1997). For this a transect of $200 \times 2\,m^2$ was subdivided in cubes of $2 \times 2 \times 2\,m^3$. Bamboo poles and flagging tape was used to mark the boundaries of the cubes. The different trunks and branches within each cube were counted and classified into different categories based on their size and orientation (Table 1). Classification was based on estimations, checked by estimations of structures of known diameter and orientation. The cubes were only used to subdivide the forest and to facilitate regis-

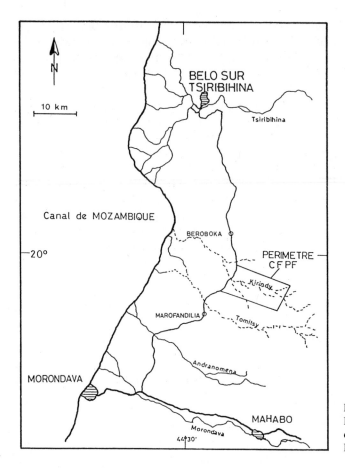

Figure 1. Location of the Kirindy Forest/concession of the Centre de Formation Professionnelle Forestière de Morondava.

tration of supports. For the analyses the total number of supports within the $200 \times 2\,m^2$ strip were used.

Animals and Sampling Procedure

Study animals were adults from two groups of *Propithecus verreauxi* and two groups of *Eulemur fulvus*. For *P. verreauxi* the first group contained one female and two males and the second group two females and two males. For *E. fulvus* one group

Table 1. Classification of supports available in the Kirindy Forest/CFPF

Support variable	Support class	Class details
Orientation	vertical	81–90°
	angled	46–80°
	oblique	11–45°
	horizontal	0–10°
Diameter	small	≤ 5 cm
	medium	5.1–10 cm
	large	10.1–15 cm
	enormous	> 15 cm

consisted of two adult females and four adult males. The other group was composed of three females and five males. All groups contained subadults that were not included in this study. *P. verreauxi* is classified as vertical clinger and leaper with adult body mass of about 3 kg while *E. fulvus* is considered quadrupedal with a body mass of 2–2.5 kg (Kappeler & Heymann, 1996). All groups were habituated to the presence of observers.

Each group was followed for four days per month between October 1996 and April 1997 from dawn to dusk. For each group focal animals were followed when they were moving, according to the technique originally applied by Fleagle (1978). The terminology and classification of locomotion follow Fleagle (1978), Gebo (1987), Dagosto (1994), and Gebo and Chapman (1995). For this study a locomotor bout was used as the unit of analyses. For example, if an animal climbed down a tree, leaped to another tree, and then climbed up, three different locomotor bouts would be recorded: two for climbing and one for leaping. If an animal stopped moving for more than one minute, another individual became the focal animal. The diameter and orientation of the supports used by the animals were estimated.

The following categories of locomotion and posture were distinguished (after Fleagle & Mittermeier, 1980; Dagosto 1994, p. 193; Gebo & Chapman, 1995, p. 53):

Leaping—a movement in which the hindlimbs are used to propel an animal across a gap.
Climbing—a movement up or down an oblique or vertical support.
Clinging—the animal clings to a substrate without supporting any of its weight on other supports.
Quadrupedalism—a movement in which all four limbs move in a regular pattern above a horizontal or oblique support.
Bipedalism—movement in which only the hind feet are used to take a short walk.
Types of locomotion and supports were recorded for moving or feeding animals separately.

Chi-square tests were applied for the comparisons between supports available and supports used. For this, all records were treated as independent samples. This may not be correct and overestimate chi-square values. But I found it impossible to define "independence" of data in this type of study.

The difference between the observed frequency and the frequency expected based on chi-square tests was expressed as percentage deviation from the expected value. This percentage was then plotted to illustrate deviations of the utilization of supports of different size and orientation from the availability of these support types in the forest. The associated p - values are given in the text. Since they may be based on inflated chi-square values, they have to be considered as indications for significance rather that taken as correct probabilities.

RESULTS

Forest Structure

The supports available within the transect of 200 m × 2 m are listed in Table 2. The number of supports remained more or less constant up to a height of 6 m. This represents the understory with many vertical structures of regenerating trees that grow towards the light. Between 6 and 12 m the number of supports was also constant. Above

Table 2. Number of structural supports available in the Kirindy Forest/CFPF at different heights within a transect of 200 m × 2 m and their relative representation (in percentages) in different classes of dimensions and orientations. Except for sample size (n) values are percentages; S small, I intermediate, L large, E extra large; V vertical, A angular, O oblique, H horizontal

Height (m)	n	Dimension (% of n per row)				Orientation (% of n per row)			
		S	I	L	E	V	A	O	H
0–2	445	58.4	30.4	5.8	5.4	65.2	24.9	7.0	2.9
2–4	460	60.2	29.4	5.2	5.2	48.5	39.6	17.5	3.9
4–6	417	60.1	19.2	6.7	6.0	34.8	43.4	17.5	4.3
6–8	255	48.6	36.1	9.8	5.5	29.8	46.6	16.9	6.7
8–10	196	46.4	32.1	12.8	8.7	21.9	46.4	17.9	13.8
10–12	216	49.1	30.1	16.2	4.6	24.5	49.1	14.4	12.0
12–14	104	45.2	48.1	4.2	1.9	21.2	50.0	19.2	9.6
14–16	82	73.2	24.4	2.4	0.0	19.5	32.9	30.4	18.2
TOTAL	2175	57.4	29.4	7.8	5.4	39.9	39.9	13.6	6.6

12 m, supports are represented by the branches and stems of overstory trees. Since the distribution of these larger trees is patchy, the overall number of supports above 12 m drops. Across all heights, vertical, and oblique supports were the most abundant, each accounting for 39.9% of all available supports. Oblique and horizontal supports followed with 13.6% and 6.6%, respectively. Horizontal supports are most abundant in the layer between eight and 12 m, representing the main canopy of the forest and between 14–16 m, representing the crowns of the largest trees in the forest.

Supports Used by *Propithecus verreauxi* and *Eulemur fulvus* for Locomotion

The supports used by the two species are listed in Tables 3 and 4. For *Propithecus verreauxi* the frequency of different sized supports and their orientations differed significantly from the supports available in the forest (chi-square > 31.5; df = 3, p < 0.001; see summary in Table 5). This species used intermediate and large sized supports more often than would be expected based on their availability. Small supports were used less often than expected (Fig. 2). Vertical and horizontal supports were used more often than expected by chance, while angular and oblique branches were under represented as supports during locomotion. The differences between supports available and

Table 3. Supports used by *Propithecus verreauxi* during locomotion. Except for sample size (n) values are percentages; for abbreviations see Table 2

	n	%	Dimension (% of n per row)				Orientation (% of n per row)			
			S	I	L	E	V	A	O	H
Leaping	2265	57.0	48.1	35.8	10.7	5.3	56.0	25.0	11.0	8.0
Climbing	489	12.2	45.4	34.6	12.2	7.8	60.1	35.2	23.7	1.0
Clinging	918	23.1	49.9	34.0	11.7	4.4	62.9	34.0	2.6	0.5
Quadrupedal	111	2.7	60.4	23.4	3.6	12.6	0	0	45.0	55.0
Bipedal	27	0.6	74.0	18.6	0	7.4	0	0	48.1	51.9
Others	163	4.1	88.3	9.2	1.8	0.7	8.6	15.3	26.4	49.7

Table 4. Supports used by *Eulemur fulvus rufus* for locomotion. Except for sample size (n) values are percentages; for abbreviations see Table 2

	n	%	Dimension (% of n per row)				Orientation (% of n per row)			
			S	I	L	E	V	A	O	H
Leaping	1589	48.4	55.9	32.7	7.7	3.7	49.7	24.6	12.7	13.3
Climbing	577	17.6	51.1	35.2	9.7	4.0	61.7	35.9	1.9	0.5
Clinging	325	9.9	68.9	24.9	3.1	3.1	57.2	39.7	2.2	0.9
Quadrupedal	669	20.4	60.0	21.7	10.8	7.5	0.2	12.1	36.0	51.7
Bipedal	2	0.1	100	0.0	0.0	0.0	0.0	0.0	0.0	100
Others	115	3.5	88.7	7.9	2.6	0.7	8.7	21.7	26.1	43.5

supports used seem small for the different categories, but due to the large sample size these differences are significant.

Eulemur fulvus used mainly vertical and angular supports for jumps, but horizontal and oblique branches for quadrupedal movements (Table 4). Most supports used were small or intermediate in size. The frequency of different supports used by *E. fulvus*, irrespective of height, did not differ from the frequency of supports available in the forest with respect to support size (chi-square = 3.11, df = 3; $p > 0.05$), but the animals used horizontal and vertical supports more often than expected based on their availability in the forest (chi-square = 227.03; df = 3; $p < 0.001$; Fig. 2).

Supports Used by *Propithecus verreauxi* and *Eulemur fulvus* while Feeding

In both species, the supports used while feeding differed significantly from those occurring in the forest (*Propithecus verreauxi*: dimension and orientation: chi-square = 41.82 and 252.93, respectively; *Eulemur fulvus*: dimension and orientation: chi-square = 93.79 and 406.57, respectively; df = 3; $p < 0.001$ in both comparisons). These species used small and especially horizontal branches much more frequently than would be expected (Table 5; Fig. 3).

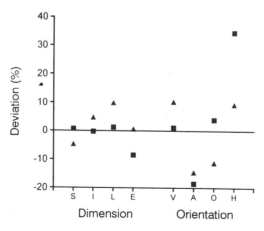

Figure 2. Deviations of the observed frequency of support utilization during locomotion by *Propithecus verreauxi* (triangles) and *Eulemur fulvus* (squares) from the frequency of supports available in the forest. Deviations are in % of the expected frequencies based on the calculations of chi-square values.

Table 5. Summary of supports available in the forest and used by *Propithecus verreauxi* and *Eulemur fulvus* for locomotion and while feeding. Except for sample size (n) values are percentages

	n	Dimension (% of n per row)				Orientation (% of n per row)			
		S	I	L	E	V	A	O	H
Available in the forest	2175	57.4	29.4	7.8	5.4	39.9	39.9	13.6	6.6
P. verreauxi									
Locomotion	3973	50.4	33.7	10.5	5.5	54.2	27.0	10.0	8.7
Feeding	617	71.8	19.3	5.5	3.4	30.8	22.9	18.1	28.2
E. fulvus									
Locomotion	3277	58.4	29.2	8.1	4.4	41.0	25.4	15.0	18.6
Feeding	360	83.1	13.4	3.2	0.3	19.2	18.4	22.4	40.0

Comparison of Supports Used by *Propithecus verreauxi* and *Eulemur fulvus* for Locomotion and while Feeding

The size of supports used for locomotion differed significantly from the size of supports used while feeding (chi-square = 99.63; df = 3; p < 0.001; Table 5). *Propithecus verreauxi* held on to small branches when they were feeding, while during locomotion they used larger supports. In addition the orientation of supports also differed between these two types of activity (chi-square = 271.34; df = 3, p < 0.001). Feeding took place on horizontal and oblique branches, while the animals used more vertical and angular supports for locomotion. *Eulemur fulvus* showed the same preferences for small horizontal branches for feeding and larger vertical structures for locomotion (chi-square = 90.40 and 124.82 for dimension and orientation, respectively; df = 3; p < 0.001; Figs. 2 and 3).

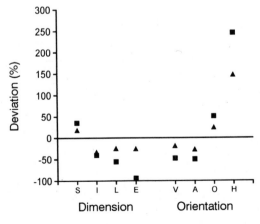

Figure 3. Deviations of the observed frequency of support utilization when feeding by *Propithecus verreauxi* (triangles) and *Eulemur fulvus* (squares) from the frequency of supports available in the forest. Deviations are in % of the expected frequencies based on the calculations of chi-square values.

Interspecific Comparison of the Locomotion Types for *Propithecus verreauxi* and *Eulemur fulvus*

According to the data presented in Tables 3 and 4, the frequency of the different types of locomotion differed significantly between the two species (chi-square = 777.88; df = 5; p < 0.001). Relative to *Eulemur fulvus*, *Propithecus verreauxi* jumped more using mainly vertical and angular supports, while *E. fulvus* were more quadrupedal on horizontal and oblique trunks and branches, though jumping using vertical supports was still relatively frequent.

Interspecific Comparison of the Supports Used by *Propithecus verreauxi* and *Eulemur fulvus* for Locomotion and while Feeding

For locomotion, *Propithecus verreauxi* used larger trunks and branches that were more often vertical or angular than the supports used by *Eulemur fulvus* (Fig. 2). Both characteristics (dimensions and orientations) differed significantly between the two species (chi-square = 48.29 and 236.82 for dimension and orientation, respectively; df = 3; p < 0.001).

These differences were less evident, though still significant when the animals were feeding. Again, *Propithecus verreauxi* used bigger trunks and branches that were more often vertical or angular than the supports used by *Eulemur fulvus*. In contrast, *E. fulvus* used horizontal branches more frequently for locomotion and especially when feeding. Both characteristics differed significantly between the two species (chi-square = 21.16 and 26.27 for dimension and orientation, respectively; df = 3; p < 0.001; Figs. 2 and 3).

DISCUSSION

Different postures and modes of locomotions have been suggested repeatedly as ways for sympatric primate species to exploit different resources and thus to reduce interspecific competition (e.g., Prosimians: Sussman, 1974; Crompton, 1984; Harcourt & Nash, 1986; Gebo 1987; Ganzhorn, 1989; Dagosto 1994, 1995; Overdorff, 1996; Terranova, 1996; Old World Monkeys: Fleagle, 1978; Gautier-Hion et al., 1981; McGraw, 1996; New World Monkeys: Fleagle & Mittermeier, 1980; Mittermeier & van Roosmalen, 1981; Fleagle, 1984).

In addition, the energetic costs associated with different modes of locomotion are considered as critically important for the understanding of primate adaptations (e.g., Charles-Dominique & Hladik, 1971; Demes & Günther, 1989; Crompton et al., 1993; Warren, 1994; Demes et al., 1995; Warren & Crompton, 1996, 1998; Drack et al., this volume). Yet, despite the important role attributed to primate locomotion, there are very few studies that attempted to quantify the supports actually available in a given forest and thus provide a null hypothesis that would allow to actually test which supports are selected or avoided for locomotion (e.g., Cannon & Leighton, 1994; Warren, 1997).

The present study describes utilization of supports by two sympatric lemur species, *Propithecus verreauxi* and *Eulemur fulvus* in a dry deciduous forest of western Madagascar. The trunks and branches used during locomotion and feeding were compared to the supports available in a random transect representative of the forest structure within the study area. The majority of supports (>80%) used by either species for locomotion or during feeding consisted of small (<5 cm diameter) or intermediate sized (5–10cm) trunks and branches. This matches the size of supports used by four rain forest species of similar size *(Eulemur fulvus rufus, E. rubriventer, Varecia variegata,* and *Propithecus diadema)* who also used mainly branches between 2.5 and 10cm (Dagosto, 1995). However, the larger species *(P. verreauxi)* also used larger supports than the smaller *E. fulvus.* The two species differed most with respect to the orientation of supports used for locomotion. Though both species used vertical supports for travel and movements between trees, *P. verreauxi* used more vertical while *E. fulvus* used more horizontal structures. When feeding, the two species looked for food mainly on small horizontal branches, i.e. where food is of higher quality (Ganzhorn, 1995) and possibly more abundant. Yet, again, *P. verreauxi* used more vertical supports while *E. fulvus* concentrated on horizontal branches. This also matches the situation found for their counterparts *(P. diadema edwardsi* and *E. f. rufus)* in the rain forest of the Parc National de Ranomafana (Dagosto, 1995).

Though *Propithecus verreauxi* used mainly small and intermediate sized structures for locomotion (>80%), the comparison with the size of structures available showed a much higher utilization of intermediate and large trunks and branches than would be expected based on the occurrence of these structures in the forest. *Eulemur fulvus* did not show much deviation with respect to the size of trunks and branches but seem to use whatever sized structure was available. However, even though size did not seem to play a major role in the utilization of supports in *E. fulvus,* both species showed substantial "preferences" for specific orientations of trunks and branches during travel. *P. verreauxi* used more vertical structures, while *E. fulvus* mainly traveled on horizontal supports.

The disproportionate utilization of small horizontal branches during feeding by both species is likely to reflect the distribution of food items. This is similar to the situation found in other primate communities (e.g., Fleagle & Mittermeier, 1980). However, despite the high concentration of feeding activities on small branches, the two species considered here differed considerably in the extent they actually used these supports. The heavier *P. verreauxi* still relied more on larger and more vertical structures than *E. fulvus.* In concert with the more folivorous diet of *P. verreauxi* and the emphasis on fruits of *E. fulvus* in the Kirindy Forest (Ganzhorn & Kappeler, 1996), these differences also help to reduce interspecific competition between the two species.

In addition to facilitate coexistence of the two species through resource partitioning, the different modes of locomotion and substrate utilization must also be seen in the light of different predation pressure on these species. In the Kirindy Forest the Fossa *(Cryptoprocta ferox)* is the main predator on large lemur species (Rasolonandrasana, 1994; Rasoloarison et al., 1995). This carnivore seems to be much more successful hunting *P. verreauxi* than *E. fulvus.* This seems logical because, with a body mass of about 10kg, *C. ferox* needs good sized supports to sustain its weight when hunting in trees. Being able to use small branches may represent an advantage to escape this carnivore. But, certainly, in order to be able to use small branches, body mass has to be low. Using the outer parts of trees makes the animals more vulnerable to aerial predators such as *Polyboroides radiatus* and *Accipiter henstii* (Goodman et al., 1993, 1998). Thus, the different types of support utilization in different lemur species has to be seen as a compromise between present day substrate and food availability, food acquisition, and predator pressure, with substantial phylogenetic heritage (Warren &

Crompton, 1996; Godfrey et al., 1997, this volume) and under different predation pressures (Goodman, 1994; van Schaik & Kappeler, 1996).

ACKNOWLEDGMENTS

I would like to thank Mme. B. Rakotosamimanana for her support throughout my studies. J. Ganzhorn initiated this project and translated this chapter. J. Schmid, P. Kappeler, R. Rasoloarison, and F. Ranaivo were there to help when needed. B. Demes and R. Warren provided very helpful comments on the manuscript. Special thanks go to my brother Rakotonirina and his family for their expert help and unflagging support throughout my studies. The study was conducted as part of the cooperation between the Deutsches Primatenzentrum, Göttingen, and the Laboratoire de Primatologie de l'Université d'Antananarivo.

REFERENCES

ABRAHAM, J.-P., B. RAKOTONIRINA, M. RANDRIANASOLO, J. U. GANZHORN, V. JEANNODA, AND E. G. LEIGH. 1996. Tree diversity on small plots in Madagascar: a preliminary review. Revue d'Ecologie, **51**: 93–116.

CANNON, C. H., AND M. LEIGHTON. 1994. Comparative locomotor ecology of gibbons and macaques: selection of canopy elements for crossing gaps. American Journal of Physical Anthropology, **93**: 505–524.

CHARLES-DOMINIQUE, P., AND C. M. HLADIK. 1971. Le *Lepilemur* du sud de Madagascar: écologie, alimentation et vie sociale. La Terre et la Vie, **25**: 3–66.

CROMPTON, R. H. 1984. Foraging, habitat structure, and locomotion in two species of *Galago*, pp. 73–111. *In* Rodman, P. S., and J. G. H. Cant, eds., Adaptations for Foraging in Nonhuman Primates. Contributions to an Organismal Biology of Prosimians, Monkeys, and Apes. Columbia University Press, New York.

CROMPTON, R. H., W. I. SELLERS, AND M. M. GÜNTHER. 1993. Energetic efficiency and ecology as selective factors in the saltatory adaptation of prosimian primates. Proceedings of the Royal Society of London B, **254**: 41–45.

DAGOSTO, M. 1994. Testing positional behavior of Malagasy lemurs: a randomization approach. American Journal of Physical Anthropology, **94**: 189–202.

———. 1995. Seasonal variation in positional behavior of Malagasy lemurs. International Journal of Primatology, **16**: 807–833.

DEMES, B., AND M. GÜNTHER. 1989. Biomechanics and allometric scaling in primate locomotion and morphology. Folia Primatologica, **53**: 125–141.

DEMES, B., W. L. JUNGERS, T. S. GROSS, AND J. G. FLEAGLE. 1995. Kinetics of leaping primates: Influence of substrate orientation and compliance. American Journal of Physical Anthropology, **95**: 85–99.

FLEAGLE, J. G. 1978. Locomotion, posture, and habitat utilization in two sympatric Malaysian Leaf Monkeys (*Presbytis obscura* and *Presbytis melalophus*), pp. 243–251. *In* Montgomery, G. G., ed., Ecology of Arboreal Folivores. Smithsonian Institution Press, Washington, D.C.

———. 1984. Primate locomotion and diet, pp. 105–117. *In* Chivers, D. J., B. A. Wood, and A. Bilsborough, eds., Food Acquisition and Processing in Primates. Plenum Press, New York.

———, AND R. A. MITTERMEIER. 1980. Locomotor behavior, body size, and comparative ecology of seven Surinam monkeys. American Journal of Physical Anthropology, **52**: 301–314.

GANZHORN, J. U. 1989. Niche separation of the seven lemur species in the eastern rainforest of Madagascar. Oecologia, **79**: 279–286.

———. 1995. Low level forest disturbance effects on primary production, leaf chemistry, and lemur populations. Ecology, **76**: 2084–2096.

———, AND P. M. KAPPELER. 1996. Lemurs of the Kirindy Forest, pp. 257–274. *In* Ganzhorn, J. U., and J.-P. Sorg, eds., Ecology and Economy of a Tropical Dry Forest in Madagascar. Primate Report, **46–1**: 1–382.

GANZHORN, J. U., AND J.-P. SORG [eds.]. 1996. Ecology and Economy of a Tropical Dry Forest in Madagascar. Primate Report, **46–1**: 1–382.

GAUTIER-HION, A., J. P. GAUTIER, AND R. QURIS. 1981. Forest structure and fruit availability as complementary factors influencing habitat use by a troop of monkeys (*Cercopithecus cephus*). La Terre et la Vie, **35**: 511–536.

GEBO, D. L. 1987. Locomotor diversity in prosimian primates. American Journal of Primatology, **13**: 271–281.

——, AND C. A. CHAPMAN. 1995. Positional behavior in five sympatric Old World Monkeys. American Journal of Physical Anthropology, **97**: 49–76.

GODFREY, L. R., W. L. JUNGERS, K. E. REED, E. L. SIMONS, AND P. S. CHATRATH. 1997. Inferences about past and present primate communities in Madagascar, pp. 218–256. *In* Goodman, S. M., and B. D. Patterson, eds., Natural Change and Human Impact in Madagascar. Smithsonian Institution Press, Washington, D.C.

GOODMAN, S. M. 1994. Description of a new species of subfossil eagle from Madagascar: *Stephanoaetus* (Aves: Falconiformes) from the deposits of Ampasambazimba. Proceedings of the Biological Society of Washington, **107**: 421–428.

——, S. O'CONNOR, AND O. LANGRAND. 1993. A review of predation on lemurs: implications for the evolution of social behavior in small nocturnal primates, pp. 51–66. *In* Kappeler, P. M., and J. U. Ganzhorn, eds., Lemur Social Systems and their Ecological Basis. Plenum Press, New York.

GOODMAN, S. M., L. A. RENE DE ROLAND, AND R. THORSTROM. 1998. Predation on the eastern Wholly Lemur (*Avahi laniger*) and other vertebrates by Henst's Goshawk (*Accipiter henstii*). Lemur News, **3**: 14–15.

GRAND, T. I. 1984. Motion economy within the canopy:four strategies for mobility, pp. 55–72. *In* Rodman, P. S., and J. H. Cant, eds., Adaptation for Foraging in Non-human Primates. Columbia University Press, New York.

HARCOURT, C. S., AND L. T. NASH. 1986. Species difference in substrate use, and diet between sympatric galagos in two Kenyan coastal forest. Primates, **27**: 41–52.

JOUFFROY, F. K., C. E. OXNARD, AND R. Z. GERMAN. 1982. Interpretation phylogenetique des proportions des membres des tarsiers, par comparison avec les autres prosimiens sauteurs. Analyse multivarée. Compte Rendus Académie Science, Paris, Série D, **295**: 315–320.

JUNGERS, W. L. 1978. The functional significance of skeletal allometry in *Megaladapis* in comparison to living prosimians. American Journal of Physical Anthropology, **49**: 303–314.

KAPPELER, P. M., AND E. W. HEYMANN. 1996. Nonconvergence in the evolution of primate life history and socio-ecology. Biological Journal of the Linnean Society, **59**: 297–326.

McGRAW, W. S. 1996. Cercopithecid locomotion, support use and support availability in the Tai Forest, Ivory Coast. American Journal of Physical Anthropology, **100**: 507–522.

MITTERMEIER, R. A., AND M. G. M. VAN ROOSMALEN. 1981. Preliminary observations on habitat utilization and diet in eight Surinam monkeys. Folia Primatologica, **36**: 1–39.

NAPIER, J. R., AND P. H. NAPIER. 1967. A Handbook of Living Primates. Academic Press, London.

OVERDORFF, D. J. 1996. Ecological correlates to social structure in two lemur species in Madagascar. American Journal of Physical Anthropology, **100**: 487–506.

OXNARD, C. E., R. H. CROMPTON, AND S. S. LIEBERMAN. 1990. Animal Lifestyles and Anatomies. University of Washington Press, Seattle.

PETTER, J. J., R. ALBIGNAC, AND Y. RUMPLER. 1977. Faune de Madagascar: Mammifères Lémuriens, Vol. 44. ORSTOM/CNRS, Paris.

POUNDS, J. A. 1991. Habitat structure and morphological patterns in arboreal vertebrates, pp. 109–119. *In* Bell, S., E. McCoy, AND H. Mushinsky, eds., Habitat Structure. The Physical Arrangements of Objects in Space. Chapman & Hall, New York.

PREUSCHOFT, H., H. WITTE, AND M. FISCHER. 1995. Locomotion in nocturnal prosimians, pp. 453–472. *In* Alterman, L., G. A. Doyle, and M. K. Izard, eds., Creatures of the Dark: The Nocturnal Prosimians. Plenum Press, New York.

PREUSCHOFT, H., M. M. GÜNTHER, AND A. CHRISTIAN. 1998. Size dependence in Prosimian locomotion and its implications for the distribution of body mass. Folia Primatologica, **69**, Supplement **1**: 60–81.

RAKOTONIRINA. 1996. Composition and structure of a dry forest on sandy soils near Morondava, pp. 81–87. *In* Ganzhorn, J. U., and J.-P. Sorg, eds., Ecology and Economy of a Tropical Dry Forest in Madagascar. Primate Report, **46–1**: 1–382.

RASOLOARISON, R. M., B. P. N. RASOLONANDRASANA, J. U. GANZHORN, AND S. M. GOODMAN. 1995. Predation on vertebrates in the Kirindy Forest, western Madagascar. Ecotropica, **1**: 59–65.

RASOLONANDRASANA, B. P. N. 1994. Contribution à l'étude de l'alimentation de *Cryptoprocta ferox* Bennet (1833) dans son milieu naturel. D.E.A. Thesis, Université d'Antananarivo, Antananarivo.

RICHARD, A. F., AND R. E. DEWAR. 1991. Lemur ecology. Annual Review of Ecology and Systematics, **22**: 145–175.

SCHWARTZ, J. H., AND I. TATTERSALL. 1985. Evolutionary relationships of the living lemurs and the lorises (Mammalia, Primates) and their potential affinities with European Eocene Adapidae. Anthropological Paper of the American Museum of Natural History, **1**: 1–100.

SORG, J.-P., AND U. ROHNER. 1996. Climate and phenology of the dry deciduous forest at Kirindy, pp. 57–80. *In* Ganzhorn, J. U., and J.-P. Sorg, eds., Ecology and Economy of a Tropical Dry Forest in Madagascar. Primate Report **46-1**: 1–382.

SUSSMAN, R. W. 1974. Ecological distinctions in sympatric species of *Lemur*, pp. 75–108. *In* Martin, R. D., G. A. Doyle, and A. C. Walker, eds., Prosimian Biology. Duckworth, London.

TERRANOVA, C. J. 1996. Variations in leaping of lemurs. American Journal of Primatology, **40**: 145–165.

VAN SCHAIK, C. P., AND P. M. KAPPELER. 1996. The social system of gregarious lemurs: lack of convergence with anthropoids due to evolutionary disequilibrium? Ethology, **102**: 915–941.

WARREN, R. D. 1994. Lazy leapers: a study of the locomotor ecology of two species of saltatory nocturnal lemur in sympatry at Ampijoroa, Madagascar. Ph.D. thesis, University of Liverpool, Liverpool.

——. 1997. Habitat use and support preference of two free-ranging saltatory lemurs (*Lepilemur edwardsi* and *Avahi occidentalis*). Journal of Zoology, London, **241**: 325–341.

——, AND R. H. CROMPTON. 1996. Lazy leapers: energetics, phylogenetic inertia, and the locomotor differentiation of the Malagasy primates, pp. 259–266. *In* Lourenço, W., ed., Biogeography of Madagascar. ORSTOM, Paris.

——. 1998. Diet, body size, and the energy costs of locomotion in saltatory primates. Folia Primatologica, **69**, Supplement **1**: 86–100.

FIELD METABOLIC RATE AND THE COST OF RANGING OF THE RED-TAILED SPORTIVE LEMUR (*LEPILEMUR RUFICAUDATUS*)

Sonja Drack,[1] Sylvia Ortmann,[1,2] Nathalie Bührmann,[3] Jutta Schmid,[4,5] Ruth D. Warren,[4,6] Gerhard Heldmaier,[1] and Jörg U. Ganzhorn[4,7]

[1] Tierphysiologie, Universität Marburg
PF 1929, 35032 Marburg, Germany
[2] present address: Deutsches Institut für Ernährungsforschung
Arthur Scheunert Allee 114–116, 14558 Bergholz-Rehbrücke
Germany
[3] Zoologie, Universität Giessen
Stephanstrasse 24, 35390 Giessen, Germany
[4] Deutsches Primatenzentrum
Kellnerweg 4, 37077 Göttingen, Germany
[5] present address: Department of Zoology
University of Aberdeen
Aberdeen, AB9 2TN, UK
[6] present address: Capel Chwarel Goch
Tregarth, Bangor Gwynedd LL57 4RD, UK
[7] present address and address for correspondence
Zoologisches Institut und Museum
Martin-Luther-King-Platz 3, 20146 Hamburg, Germany
ganzhorn@zoologie.uni-hamburg.de (corresponding author)

ABSTRACT

The goal of this study was to describe energy expenses of free ranging *Lepilemur ruficaudatus* during the dry season in the deciduous forest of western Madagascar. Since all lemur species measured so far have had very low resting metabolic rates (RMR) and some lemur species can go into daily or prolonged torpor, the question was, whether or not resting metabolic rates can be used to predict field metabolic rates (FMR) in lemurs. Doubly-labeled water measurements of FMR in 11 free-ranging *L. ruficaudatus* showed that these animals had FMR of about 65.6% of the values expected for FMR of eutherian herbivores of similar body mass (mean mass = 723 g). FMR was on average 2.6–3.1 times higher than RMR that had been measured previously as 63.6% of the expected value for RMR of eutherian mammals of similar body mass. The ratio of FMR/RMR matches the ratio found in other eutherian mammals and is consistent with the idea that FMR is a constant multiple of RMR for eutherian mammals over a wide range of body mass. FMR augmented with increasing range size. The present data do not provide evidence that the costs of locomotion were energetically limiting for *L. ruficaudatus*.

RÉSUMÉ

Le but de cette étude était de décrire les dépenses énergétiques de *Lepilemur ruficaudatus* dans un enclos libre au cours de la saison sèche dans la forêt décidue de l'ouest de Madagascar. Dans la mesure où toutes les espèces de lémuriens pour lesquels des mesures ont été effectuées montrent des taux de métabolisme au repos (TMR) très bas et que certaines espèces peuvent rentrer dans des états de torpeurs journaliers ou prolongés, la question était de savoir si oui ou non les taux de métabolisme au repos pouvaient être utilisés pour prédire les taux du métabolisme dans la nature (TMN) des lémuriens. Des TMR mesurés sur des molécules d'eau marquées sur 11 *Lepilemur ruficaudatus* dans un enclos ont montré que le TMN étaient de l'ordre de 65,6% des valeurs des TMR escomptées pour des herbivores placentaires de masse corporelle similaire (masse moyenne = 723 g). Le TMN étaient en moyenne 2,6 à 3,1 plus grand que le TMR qui avaient été mesuré antérieurement comme 63,6 % des valeurs de TMR de mammifères placentaires de masse corporelle similaire. Le rapport TMR/TMN correspond à celui qu'on trouve chez les autres mammifères placentaires et est compatible avec l'idée que le TMN est un multiple constant du TMR pour ces mammifères sur un échantillonnage important de masse corporelle. Le TMN augmentait en même temps que l'aire vitale. Les données actuelles ne permettent pas de montrer que les coûts de locomotion sont limitant pour *Lepilemur ruficaudatus* quant à l'énergie.

INTRODUCTION

Knowing the energetic cost of living represents a cornerstone for our understanding of animal biology, ranging from behavior to life history traits (Kleiber, 1961; McNab, 1986; Nagy, 1987; reviews by Koteja, 1991; Degen & Kam, 1995) to population biology, community assembly, and the distribution of body mass in animal taxa (McNab, 1980; Brown et al., 1993; Chown & Gaston, 1997). Due to methodological problems, the daily energy requirements of animals have been assumed to be described by basal or resting metabolic rates (BMR or RMR), measured by indirect calometry using O_2 consumption of inactive animals in metabolic chambers. It has been suggested that field metabolic rates (FMR) might actually be a constant multiple of BMR (Peterson et al., 1990; Koteja, 1991). Techniques using doubly labeled water have made it possible to measure the actual energy requirements of free ranging animals (i.e. their FMR). It became clear that the FMR can not simply be derived from BMR by multiplication by some constant factor, but rather that these factors vary in relation to phylogeny and body mass (Nagy, 1987; Degan & Kam, 1995; Nagy et al., 1995). For lemurs (as for all other taxa), this lack of predictability has important consequences for interpretations concerning the evolutionary basis of various life history traits.

In general, lemurs and other prosimians have low BMR when measured in resting animals by indirect calometry (reviews by Müller, 1985; Genoud et al., 1997). In addition, members of the Cheirogaleidae (*Microcebus* spp. and *Cheirogaleus* spp.) can go into daily or prolonged torpor (e.g., Petter, 1978; Schmid, 1996; Ortmann et al., 1997). Other lemur species seem to reduce their energy expenditure during times of food shortage (Pereira, 1993; Pereira et al., this volume). The lower than expected BMR in lemurs and seasonal changes in energy expenditures has invoked explanations as to why some prosimian primates, such as *Nycticebus* can tolerate toxic food items (Hladik, 1979), why lemurs can reach very high population densities despite extreme seasonality in parts of their ranges (Charles-Dominique & Hladik, 1971; Ganzhorn, submitted) and why females should be dominant to males (i.e. females should have priority access to food to raise their infants successfully; Jolly, 1966; Richard, 1987; for an extensive discussion see Kappeler, 1996; Pereira et al., this volume).

Recent measurements of RMR of *Lepilemur ruficaudatus* confirmed previous results obtained for other lemur species (Schmid & Ganzhorn, 1996). In this study, RMR were only 49.6% of the metabolic rate expected for eutherian mammals of similar body mass based on Kleiber's (1961) allometric relation when measured during the animals' resting phase. However, the RMR increased significantly to about 77.6% during the animals' active phase. Given this variability and the uncertainty whether or not BMR is a reliable indicator of an animal's FMR (pro: Peterson et al., 1990; Koteja, 1991; contra: Nagy et al., 1995), we measured the FMR of *L. ruficaudatus* in the dry deciduous forest of Kirindy/CFPF.

The specific questions addressed in this study were:

1. Is the lower than expected BMR paralleled by a lower than expected FMR in *L. ruficaudatus*?
2. How are nightly travel distances and home range sizes related to the FMR?

METHODS

Study Site

The study has been carried out in the Kirindy Forest/CFPF some 60 km north of Morondava at 44° 49'E, 20° 03'S. The forest is dry deciduous. Overstory and many understory trees and shrubs lose their leaves during the dry season that lasts from about April to October. More information on this forest can be found in Ganzhorn and Sorg (1996).

Animals

Lepilemur ruficaudatus is a nocturnal lemur species that feeds on leaves, flowers, and fruit. Body mass of adult animals averages about 800 g. They spend the day in tree holes. In the present study all animals were caught from their day shelters by hand early in the morning. Eleven different animals were used for the measurements. Their gender (F = female, M = male) and age class (a = adult, y = young, probably less than one year old) are listed in Table 1. Animals were injected with 1.6 ml $H_2^{18}O$ and 0.16 ml D_2O. Four

Table 1. Summary of field metabolic rate in *Lepilemur ruficaudatus* during the dry season (July and August 1996)

Animal[1]	Days[2]	Mean mass (g)[3]	Δ mass (% d^{-1})	FMR	Relative to predicted FMR[4]	FMR/ BMR[5]	Nightly travel distance (m)	Range (ha)[6]
aF2	7.9	877.5	−1.1	466.6	0.57	2.7		0.48
aF4	18.7 (1.6)	939.0	−0.0	762.7	0.88	4.2		0.39
aF4	7.1	941.5	0.6	428.8	0.50	2.3		
aF4 (x̄)		940.3	0.3	595.8	0.69	3.2		
aF7	2.1	659.5	5.1	218.9	0.33	1.6	607.5	0.10
aF9	6.6	831.5	0.1	852.7	1.08	5.1		
aF11	6.0 (1.0)	701.5	0.8	360.9	0.52	2.5		0.20
aF11	7.4	662.5	2.2	528.5	0.79	3.7	518.0	
aF11(x̄)		682.0	1.5	444.7	0.65	3.1		
aM12	3.6 (0.8)	760.5	−1.2	423.9	0.57	2.7		
aM14	7.7 (1.6)	733.5	1.6	270.9	0.38	1.8		
aM15	4.9 (0.9)	738.5	−1.4	822.0	1.14	5.4		
yF16	8.9	513.0	−0.5	401.3	0.72	3.4		
aM17	5.3 (0.9)	744.5	−1.7	459.7	0.63	3.0	717.4	0.10
yF18	7.2	470.0	0.1	241.8	0.46	2.2		
x̄ ± SD[7] n = 11		722.8 ± 141.9	0.3 ± 1.9	472.6 ± 211.8	0.66 ± 0.26	3.1 ± 1.2		

[1] a = adult, y = young; F = female, M = male; (x̄) marks the mean of repeated measures from the same individual.
[2] Number of days between blood samples with number of days kept in the enclosure in parentheses.
[3] Average of body masses measured across the time period of the experiment.
[4] Field metabolic rate was predicted using the regression for eutherian herbivores as: FMR [kJ d^{-1}] = 5.95 BM$^{0.727}$, where BM is body mass in g (Nagy, 1987).
[5] Basal metabolic rate was calculated as the mean of BMR measured by Schmid and Ganzhorn (1996) during the active (77.6% of the value predicted by Kleiber, 1961) and the resting phase (49.6% of the value predicted by Kleiber); i.e. BMR$_{predicted}$ (VO$_2$ in ml O$_2$/h) = 0.636 * 3.42 BM$^{0.75}$; BM in g). The resulting value was multiplied by 24 hours and 20.7 kJ/l O$_2$.
[6] Based on nine fixes per individual taken over three consecutive nights in at least one-hour intervals.
[7] Means and standard deviations based on mean values per individual (n = 11).

hours later the animals were anaesthetized for 10–20 min with 0.15–0.3 ml of 100 mg/ml Ketanest and Rompun (ratio 5:1) and 0.4–1 ml blood was taken. Four hours should be sufficient to allow uniform distribution of the labeled water in animals of about 1 kg (Nagy, 1987). The natural isotopic content, i.e., the concentrations of ^{18}O and deuterium occurring in the blood naturally was calculated from blood samples taken prior to injection. These background concentrations were assumed to remain constant during the experiment. Animals were recaptured for the second blood sample three to nine days after the first sample had been taken. Blood samples were stored in heparinized 4 ml vials (Greiner VACUETTE; LH Lithium Heparin). Preparation of blood samples and analyses were carried out at Marburg University on a gas isotope ratio mass spectrometer gIRMS; SIRA II; Fisons Instruments (Drack, 1997).

In captivity, *Lepilemur ruficaudatus* can maintain a stable body mass on a diet consisting of 80.0% carbohydrates, 12.8% proteins, and 7.2% fat (Petter-Rousseaux & Hladik, 1980). Since in our field study average changes in body mass between blood samples did not deviate significantly from 0 (mean ± standard deviation: Δ body mass/day = 0.39 ± 1.92%, Table 1), we assumed that the animals were in a steady state and that the dietary composition of free ranging *L. ruficaudatus* was comparable to the diet in captivity.

FMR, i.e., CO_2 production was calculated according to Lifson and McClintock

(1996), assuming that body water content averaged 70% of body mass. FMR was expressed in kilojoule per day (kJ/d). Conversion factors for 1 l of CO_2 are 20.8, 23.1 and 27.7 kJ, respectively, when feeding exclusively on carbohydrate, protein or fat (Nagy et al., 1995). Under the dietary conditions assumed above 1 l of CO_2 produced by *Lepilemur ruficaudatus* would be energetically equivalent to 21.6 kJ. The FMR measured was compared with the theoretical value for eutherian herbivores of this body mass from the equation: $kJ\,d^{-1} = 5.95\,BM^{0.727}$, where BM is body mass in g (Nagy, 1987).

RMR of *Lepilemur ruficaudatus* had been measured previously (Schmid & Ganzhorn, 1996). The oxygen consumption of resting animals scaled to the BMR predicted by Kleiber (1961) as O_2 (ml/h) = 0.496 * 3.42 $BM^{0.75}$ during the day (their normal time of inactivity) and as O_2 (ml/h) = 0.776 * 3.42 $BM^{0.75}$ during the night when they are active normally. BM was body mass in g. For the present study, the two values were averaged to predict the BMR as O_2 (ml/h) = 0.636 * 3.42 $BM^{0.75}$.

The amount of CO_2 produced in relation to O_2 consumed depends on dietary composition. A respiratory quotient (RQ = volume CO_2/O_2) of 1.0 is typical for carbohydrates, 0.81 for proteins, and 0.71 for fat (Schmidt-Nielsen, 1990). Under the dietary regime listed above, this corresponds to a mean RQ of 0.96. This conversion factor was then used to convert oxygen consumption to the associated heat equivalents as BMR $[kJ\,d^{-1}] = 0.636 * 3.42\,BM^{0.75} * 24\,h * 20.7\,kJ/1000$.

Blood samples of animals aF7, aM17 and yF18 could be run in duplicate. Values for these animals are means. Animals aF4 and aF11 were used twice for different experiments. Since their FMR were used for different aspects of the study (first: FMR of free ranging *Lepilemur ruficaudatus*; second: relation between FMR and nightly travel distance and/or home range), both values are listed in Table 1. Statistical tests were run with the help of Winstat or calculated by hand according to Siegel and Castellan (1988).

Radio Tracking

Animals were radio-collared and released the same day at their capture site. For a subset of animals the size of their home range was determined during three consecutive nights by radio-telemetry. For this, the location of each animal was recorded about once per hour from dusk to dawn. TW-4 button tags (Biotrack Ltd., Wareham, Dorset, UK) and a TRX-1000 receiver (Wildlife Materials Inc., Illinois, USA) were used for radio-tracking. Nightly travel distances and home ranges ("outer convex polygons") were calculated as with the help of TRACKER (Camponotus AB, Solna, Sweden).

RESULTS

Field Metabolic Rates

The FMR of 11 free ranging *Lepilemur ruficaudatus* are listed in Table 1. Animals aF4 and aF11 were used twice. Data are given for both measurements. To calculate the overall means for the 11 individuals, the two measurements were averaged for these two individuals. On average the animals did not change body mass between blood samples (mean ± standard deviation: Δ body mass/day = 0.3 ± 1.9%). Thus, it was assumed that animals were in a steady state. FMR of the 11 *L. ruficaudatus* was 65.6 ± 25.6% of the expected values for eutherian herbivorous mammals of the same body mass. Calculations of the expected theoretical FMR were based on the allometric

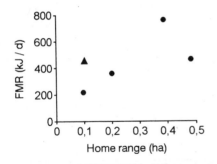

Figure 1. Relationship between home range and field metabolic rates. Dots = females; triangle = male.

regression for herbivorous eutherian species given by Nagy (1987) as FMR $[kJ\,d^{-1}]$ = $5.95\,BM^{0.727}$, where BM is body mass in g. The high FMR of animal aM15 was likely due to an error in the analysis (see Drack, 1997, for an extensive discussion). The high FMR of aF9 might have been due to unusual and very large distances this female travelled after she had been released. This value might be at the upper end of the range of field metabolic rates achieved by *L. ruficaudatus*. On average, the actual field metabolic rates were significantly below the FMR expected for eutherian herbivores of similar body mass (Sign test: p = 0.019). The FMR of adult females did not differ from the FMR of adult males (t-test: t = 0.14, p > 0.05).

The ratio of FMR to BMR was 3.1 ± 1.1 for all 11 animals and 2.6 ± 0.7 without animals aM15 and aF9 (Table 1). Individual field metabolic rates varied between 1.6 and 5.4 times their estimated BMR.

Relationship between FMR, Travel Distance, and Home Range Size

This analysis is restricted to the five animals studied simultaneously with doubly-labeled water and by radio-tracking. Continuous tracking data over two consecutive nights were available only for three animals. There was no obvious relation between FMR and the mean distance travelled per night (Table 1).

The calculation of the home range area had to be restricted to nine data points per individual taken over three consecutive nights in at least one-hour intervals. This reduction of the tracking data was necessary to have comparable sample sizes for all individuals that could be linked in a uniform way to field metabolic rates. The area covered does certainly not reflect the true home range area of the animals. They have to be considered as relative values. Despite the considerable reduction of data and the small sample size, home range size correlated well with FMR (Fig. 1; r_s = 0.80, p = 0.05, n = 5; Fig. 1).

DISCUSSION

The field metabolic rates of *Lepilemur ruficaudatus* averaged 65.6% of those expected for other eutherian herbivores based on Nagy's equation (Nagy, 1987). In a previous study RMR of *L. ruficaudatus* had been measured to average 63.6% of the value expected for the BMR of eutherian mammals of that size (Schmid & Ganzhorn, 1996). The proportion of 65.6% of expected FMR is very close to these 63.6% of the expected BMR. This result is consistent with the idea that field metabolic rates are constant multiples of BMR (Peterson et al., 1990; Koteja, 1991) and thus might actually be used to predict the energy expenditures of free ranging lemurs. Despite the low field and RMR of *L. ruficaudatus*, we observed an average ratio of FMR to RMR of 3.1 ± 1.1 for all animals and 2.6 ± 0.7 excluding animals aM15 and aF9. These ratios are well within biological limits and match the values calculated by Karasov (1992) with FMR: BMR = 2.65 (based on 17 mammalian species). However, at present, this is the only study of field metabolic rates in lemurs. Recent case studies and reviews showed that there are exceptions that do not match the idea of constant ratios of field metabolic rates to BMR (Nagy et al., 1995), and that this ratio is larger for small animals than for large animals, possibly due to different costs for thermoregulation (Degen & Kam, 1995).

In lemurs, energy expenditures is likely to vary between seasons. This is reflected most obviously by seasonal torpor in *Microcebus* spp. and *Cheirogaleus* spp. (e.g., Petter, 1978; Schmid, 1996; Ortmann et al., 1997). Surprisingly, other lemur species also exhibit pronounced seasonal changes in various traits that are likely to reflect energy requirements (Pereira, 1993; Pereira et al., this volume). It is unknown to what extent these seasonal changes are characteristic of the metabolism of all lemur species, how they are translated in actual energy requirements and which season has to be considered as the bottleneck that is likely to be effective in evolutionary terms. Recent studies on *L. ruficaudatus* provided evidence that these animals adapt their ranging pattern to optimize food availability during the wet season, when food is seemingly superabundant, while during the lean dry season the animals just "sit and wait" (Ganzhorn, submitted). Here, contrary to expectations, conditions during the rich wet season seemed to be more important than during the lean dry season. It is unknown how this is reflected in the energetic requirements of the animals.

Previously it has been argued that *Lepilemur* spp. were under energetic stress during the cool season (Nash, 1998) and that the costs of locomotion were an important factor limiting the suitability of different forests for *L. ruficaudatus* (Ganzhorn, 1993). In line with this argument, Warren and Crompton (1998) suggested that *Lepilemur* spp. maintain low costs of travel by very short nightly travel distances. The data at hand do not allow these ideas to be further developed. First, too few data were available to link FMR to nightly travel distances. Second, home range data are inconclusive. Recently it was shown that males and females *L. ruficaudatus* have similar monthly home ranges, but males travel longer distances per night than females, at least during the wet season (Pietsch, 1998). Nevertheless the present study illustrates the potential of the metabolic studies that are desperately needed to come to a better understanding of the energetic bases for lemur biology, ranging from conclusions about habitat suitability (e.g., Ganzhorn & Schmid, 1998) to the evolution of social systems (Kappeler, 1996; Pereira et al., this volume).

ACKNOWLEDGMENTS

We thank the Commission Tripartite, the Laboratoire de Primatologie et de Biologie Evolutive, the Direction des Eaux et Forêts and the Centre de Formation Professionnelle Forestière à Morondava for their permissions to perform this study. The study has been carried out under the Accord de Collaboration between the Laboratoire de Primatologie et de Biologie Evolutive, Université d'Antananarivo, and the Deutsches Primatenzentrum. As always, Mme. Berthe Rakotosamimanana and

Mme. Celestine Ravaoarinoromanga helped greatly on the administrative side of the project. The study received financial and logistical support through the Deutsche Forschungsgemeinschaft (DFG: Ga 342/3–1,2; SFB 305; the British Council/DFG Post-Doctoral Program), HSPII, and the World Wide Fund for Nature.

REFERENCES

Brown, J. H., P. A. Marquet, And M. L. Taper. 1993. Evolution of body size: consequences of an energetic definition of fitness. American Naturalist, **142**: 573–584.

Charles-Dominique, P., And C. M. Hladik. 1971. Le Lepilemur du sud de Madagascar: écologie, alimentation et vie sociale. La Terre et la Vie, **25**: 3–66.

Chown, S.-L., And K.-J. Gaston. 1997. The species-body size distribution: energy, fitness, and optimality. Functional Ecology, **11**: 365–375.

Degen, A. A., And M. Kam. 1995. Scaling of field metabolic rate to basal metabolic rate in homeotherms. Ecoscience, **2**: 48–54.

Drack, S. 1997. Messung des Energieumsatzes mittels der $D_2^{18}O$-Methode am freilebenden "KleinenWieselmaki" (*Lepilemur ruficaudatus*) an der Westküste Madagaskars. Diplom Thesis, Marburg University, Marburg.

Ganzhorn, J. U. 1993. Flexibility and constraints of *Lepilemur* ecology, pp. 153–165. *In* Kappeler, P. M., and J. U. Ganzhorn, eds., Lemur Social Systems and their Ecological Basis. Plenum Press, New York.

———. submitted. Implications of seasonal variation in food selection by folivorous lemurs for habitat carrying capacities.

———, And J. Schmid. 1998. Different population dynamics of *Microcebus murinus* in primary and secondary deciduous dry forests of Madagascar. International Journal of Primatology, **19**.

Ganzhorn, J. U., And J.-P. Sorg (eds.). 1996. Ecology and Economy of a Tropical Dry Forest. Primate Report, **46-1**: 1–382.

Genoud, M., R. D. Martin, And D. Glaser. 1997. Rate of metabolism in the smallest Simian primate, the Pygmy Marmoset (*Cebuella pygmaea*). American Journal of Primatology, **41**: 229–245.

Hladik, C. M. 1979. Diet and ecology of prosimians, pp. 307–357. *In* Doyle, G. A., and R. D. Martin, eds., The Study of Prosimian Behavior. Academic Press, London.

Jolly, A. 1966. Lemur Behaviour. The University of Chicago Press, Chicago.

Kappeler, P. M. 1996. Causes and consequences of life-history variation among Strepsirhine primates. American Naturalist, **148**: 868–891.

Karasov, W. H. 1992. Daily energy expenditure and cost of activity in mammals. American Zoologist, **32**: 238–248.

Kleiber, M. 1961. The Fire of Life. John Wiley, New York.

Koteja, P. 1991. On the relation between basal and field metabolic rates in birds and mammals. Functional Ecology, **5**: 56–64.

Lifson, N., And McClintock, R. 1966. Theory of use of the turnover rates of body water for measuring energy and material balance. Journal of Theoretical Biology, **12**: 46–74.

McNab, B. K. 1980. Food habits, energetics, and the population biology of mammals. American Naturalist, **116**: 106–124.

———. 1986. The influence of food habits on the energetics of eutherian mammals. Ecological Monographs, **56**: 1–19.

Müller, E. F. 1985. Basal metabolic rates in primates—the possible role of phylogenetic and ecological factors. Comparative Biochemical Physiology, **81A**: 707–711.

Nagy, K. A. 1987. Field metabolic rate and food requirement scaling in mammals and birds. Ecological Monographs, **57**: 111–128.

Nagy, K. A., C. Meienberger, S. D. Bradshaw, And R. D. Wooller. 1995. Field metabolic rate of a small marsupial mammal, the honey possum (*Tarsipes rostratus*). Journal of Mammalogy, **76**: 862–866.

Nash, L. T. 1998. Vertical clingers and sleepers: seasonal influence on the activities and substrate use of *Lepilemur leucopus* at Beza Mahafaly Special Reserve, Madagascar. Folia Primatologica, **69**, Supplement 1: 204–217.

Ortmann, S., G. Heldmaier, J. Schmid, And J. U. Ganzhorn. 1997. Spontaneous daily torpor in Malagasy mouse lemurs. Naturwissenschaften, **84**: 28–32.

PEREIRA, M. E. 1993. Seasonal adjustment of growth rate and adult body weight in ringtailed lemurs, pp. 205–221. *In* Kappeler, P. M., and J. U. Ganzhorn, eds., Lemur Social Systems and their Ecological Basis. Plenum Press, New York.

PETERSON, C. C., K. A. NAGY, AND J. DIAMOND. 1990. Sustained metabolic scope. Proceedings of the National Academy of Science USA, **87**: 2324–2328.

PETTER, J.-J. 1978. Ecological and physiological adaptations of five sympatric nocturnal lemurs to seasonal variations in food production, pp. 211–223. *In* Chivers, D. J., and J. Herbert, eds., Recent Advances in Primatology. Academic Press, New York.

PETTER-ROUSSEAUX, A., AND C. M. HLADIK. 1980. A comparative study of food intake in five nocturnal prosimians in simulated climatic conditions, pp. 169–179. *In* Charles-Dominique, P., H. M. Cooper, A. Hladik, C. M. Hladik, G. F. Pariente, A. Petter-Rousseaux, and A. Schilling, eds., Nocturnal Malagasy Primates. Academic Press, New York.

PIETSCH, T. 1998. Geschlechtsspezifische Unterschiede in der räumlichen Verteilung und Nahrungswahl von *Lepilemur ruficaudatus* im Trockenwald von Madagaskar. Diplom Thesis, Universität Hamburg, Hamburg.

RICHARD, A. F. 1987. Malagasy prosimians: female dominance, pp. 25–33. *In* Smuts, B. B., D. L. Cheney, R. M. Seyfarth, R. W. Wrangham, and T. T. Struhsaker, eds., Primate Societies. The University of Chicago Press, Chicago.

SCHMID, J. 1996. Oxygen consumption and torpor in mouse lemurs (*Microcebus murinus* and *M. myoxinus*): Preliminary results of a study in western Madagascar, pp. 47–54. *In* Geiser, F., A. J. Hulbert, and S. C. Nicol, eds., Adaptations to the Cold: the Tenth International Hibernation Symposium. University of New England Press, Armidale.

———, AND J. U. GANZHORN. 1996. Resting metabolic rates of *Lepilemur ruficaudatus*. American Journal of Primatology, **38**: 169–174.

SCHMIDT-NIELSEN, K. Animal Physiology. 4th ed. Cambridge University Press, Cambridge.

SIEGEL, S., AND N. J. CASTELLAN. 1988. Nonparametric statistics for the behavioral sciences. McGraw-Hill, New York.

WARREN, R. D., AND R. H. CROMPTON. 1998. Diet, body size, and the energy costs of locomotion in saltatory primates. Folia Primatologica, **69**, Supplement 1: 86–100.

6

METABOLIC STRATEGY AND SOCIAL BEHAVIOR IN LEMURIDAE

Michael E. Pereira,[1,2] Russ A. Strohecker,[1] Sonia A. Cavigelli,[3] Claude L. Hughes,[4] and David D. Pearson[1]

[1] Department of Biology
Bucknell University
Lewisburg, Pennsylvania 17837, USA
[2] Program in Animal Behavior
Bucknell University
Lewisburg, Pennsylvania 17837, USA
mpereira@bucknell.edu (corresponding author)
[3] Department of Psychology
Duke University
Durham, North Carolina 27710, USA
[4] Center for Women's Health
Cedars-Sinai Medical Center
Los Angeles, California 90048, USA

ABSTRACT

Research on lemurs contributes importantly to evaluation of hypotheses on primate development and evolution. A central question about social behavior has asked why adult females in many lemur species socially dominate males while this trait is rare among Anthropoids and other mammals. At present, the favored hypothesis—that female lemurs undertake unusually costly reproduction and dominance over males is necessary for females to access sufficient nutrition throughout Madagascar's dry season—has both supporters and detractors. We propose that confusion concerning female dominance in lemurs derives from failing to appreciate that adaptation should have minimized prospects for seasonal stress and that success during months of food abundance, not months of food scarcity, is most likely to be the primary foraging factor influencing relative fitness among today's lemurs. Analyses of hair growth, somatic growth, food intake, fatness, and two metabolically-active hormones (thyroxine and insulin-like growth factor 1) revealed that *Lemur catta* and *Eulemur fulvus rufus* prepare during the photoperiod of Madagascar's annual wet season for upcoming dry seasons and also make adjustments during dry seasons that make life less expensive.

Phenological data from three Malagasy ecosystems suggest that seasonal changes in fruit availability may be a less important selection pressure in the evolution of lemur life histories than previously thought, whereas changes in availabilities of leaves and flowers might be crucial. Detailed new fieldwork in nutritional ecology will be required to evaluate this hypothesis and also to help determine why *Eulemur fulvus* does not show female dominance. Ideally, such work would be accompanied by information on metabolic physiology, reproduction, and social behavior from a variety of lemurs. More generally, increased research on metabolic strategies in primates and other mammals would contribute as much to illuminating adaptive patterns of behavior as has research on sexual selection and mating strategies to date.

RÉSUMÉ

La recherche sur les lémuriens apporte une contribution importante à l'évaluation des hypothèses du développement et de l'évolution des primates. Une question cruciale abordée dans le comportement social est de savoir pourquoi les femelles adultes de nombreuses espèces de lémuriens dominent socialement les mâles alors que cet aspect est rare chez les primates et même chez les autres mammifères. L'hypothèse actuellement retenue qui voudrait que les femelles présentant une reproduction anormalement coûteuse domineraient les mâles pour assurer l'accès suffisant à la nourriture tout au long de la saison sèche à Madagascar, a autant de défenseurs que d'adversaires. Nous proposons que la confusion qui règne sur la dominance des femelles chez les lémuriens dérive du fait que l'adaptation montre de faibles perspectives, difficile à apprécier, lorsque la saison est contraignante et que le succès dans la recherche de nourriture au cours des mois d'abondance de nourriture, et non les mois de disette, est probablement le principal facteur influençant la santé relative au sein des lémuriens contemporains. Des analyses portant sur la pousse des poils, la croissance somatique, la consommation d'aliments, la corpulence et sur deux hormones actives dans le métabolisme (la thyroxine et une " insulin-like growth factor 1 ") ont montré que le Maki et le Lémur à front roux se préparent au cours de la photopériode de la saison humide annuelle de Madagascar pour les mois secs suivants et qu'ils opèrent aussi à des ajustements au cours de la saison sèche pour réduire les besoins vitaux.

Des données sur la phénologie provenant de trois écosystèmes malgaches suggèrent que les variations saisonnières dans la disponibilité de fruits opéreraient une pression sélective moins importante que celle avancée préalablement dans l'évolution de l'histoire de la vie des lémuriens alors que les changements dans les disponibilités de feuilles et de fleurs seraient décisifs. De nouveaux travaux détaillés menés sur le terrain en écologie nutritionnelle sont indispensables pour évaluer cette hypothèse mais aussi pour déterminer pourquoi la femelle d'*Eulemur fulvus* ne montre pas de dominance. Idéalement de tels travaux devraient être complétés par des informations portant sur la physiologie du métabolisme, la reproduction et le comportement social de plusieurs espèces de lémuriens. Plus généralement, des travaux plus poussés sur les stratégies métaboliques chez les primates et les autres mammifères apporteraient une contribution aussi importante pour éclaircir les schémas de comportement adaptatif que l'a fait jusque là la recherche sur la sélection sexuelle et les stratégies de copulation.

INTRODUCTION

Research on lemurs is important for a variety of reasons. Above all, the time available to learn about many lemurs in natural environments is limited. The natural distribution of every lemur occupies only a fraction of Madagascar and most remaining primate habitat on the island is or soon will be under significant pressure from human exploitation (Mittermeier et al., 1992). Meanwhile, research on life history, physiology, genetics, infectious disease, and behavior can support efforts to preserve whatever small populations of lemurs might eventually be protected successfully. Because lemurs and other prosimians are "primitive" primates, sharing basic anatomical features with the world's original primates, they offer some unique perspectives on characteristics that may fundamentally distinguish primates among mammals. Furthermore, the lemurs comprise a diverse, monophyletic radiation of primates (Yoder et al., 1996) separate from others since the Eocene (Martin, 1990) and only lemurs among prosimians naturally form cohesive social groups. Consequently, alongside the world's other major primate groups, lemurs offer invaluable opportunities to evaluate hypotheses about mechanisms and constraints pertaining to primate development and evolution (e.g., Kappeler, 1996; Kappeler & Heymann, 1996). Of particular interest to many primatologists are hypotheses concerning social behavior (van Schaik & Kappeler, 1996).

Why Do Lemur Females Dominate Males?

One feature of lemur social behavior has garnered particular attention since the first quantitative studies: "female dominance" (Jolly, 1966; Evans & Goy, 1968; Richard & Heimbuch, 1975; Pollock, 1979). Unlike virtually all other primates, and perhaps any other major group of mammals (Ralls, 1976, Kappeler, 1993a), adult females in many lemur species appear to dominate males, individually and unconditionally (Jolly, 1984; Richard, 1987; but see Kappeler, 1993a). Social dominance garners priority of access to resources (e.g., Richard, 1978; Pollock, 1979; Rasamimanana & Rafidinarivo, 1993; Sauther, 1993), and, extending an argument begun by Hrdy (1981), Jolly (1984) proposed that unusually costly reproduction—due to sharp seasonality of food supply, effects of body size on the timing of reproduction, and rapid offspring growth rates—selected for female dominance in lemurs and also the few Anthropoids that show similar agonistic relations between the sexes. Richard and colleagues (Richard & Nicoll, 1987; Young et al., 1990) added an important wrinkle to this "female need" hypothesis when they suggested that, as generally hypometabolic mammals (Ross, 1992), female lemurs may characteristically elevate basal metabolic rate during reproduction, thereby exacerbating reproductive costs while energy supplies remain low throughout the Malagasy dry seasons.

The female need hypothesis emerged after a decade during which few western scientists were able to conduct research in Madagascar (1972–1982), and much of the explosion of behavioral research on lemurs since then has addressed this hypothesis, directly or indirectly. At present, it remains the favored explanation for the unusual evolution of female dominance among lemurs; but, opinions are divided on whether it is likely to be correct. Some authors see evidence accumulating in support of the hypothesis (e.g., Young et al., 1990; Sauther, 1993; Jolly, 1998), while others disagree (e.g., Tilden & Oftedal, 1995; Kappeler, 1996).

Unfortunately, attention to published details is sure to confuse many who refer to the relevant literature. Young et al. (1990), for example, presented comparative analyses supporting the female need hypothesis, documenting higher fetal growth rates in lemurs than in lorises, which lack female dominance (Bearder, 1987; see also Martin & MacLarnon, 1988). But, these authors also cited Pereira et al. (1990), who showed that female dominance is absent in *Eulemur fulvus rufus* (a taxon featured by Young et al., 1990) and warned that the trait had not yet been demonstrated for the vast majority of lemurs. Using a larger database, Kappeler (1992, 1996) confirmed lemurs' high fetal growth rates, but rejected Young et al.'s (1990) prediction of differential metabolic constraints on fetal development between lemurs and lorises, though his analysis could not test the idea. Since then, Tilden (1993, in review) has shown that relative rates of prenatal maternal energy transfer are greater in Lorisids (*Otolemur* spp.) than in Lemurids and that lorisid neonates contain significantly higher concentrations of fat and protein than do neonatal lemurs, highlighting just how little is potentially learned about maternal investment solely from data on offspring mass growth rates.

Nonetheless, Kappeler (1996) suggested that postnatal rates of maternal investment—as reflected by early infant mass growth rates—do not differ between lemurs and lorises, and Tilden and Oftedal (1995) found that mothers in both infra-orders provide infants comparable amounts of milk of uniformly low nutritive qualities. Authors of both works concluded that lemur females do not invest more energy in the most expensive phase of reproduction (lactation) than do lorisid females, and Kappeler (1996) went further. In a strongly-stated summary, surely delivered to provoke the additional research that will be necessary to confirm, qualify, or contradict his conclusions, Kappeler asserted that "the evolution of female dominance among Malagasy primates is not explainable as a direct result of an unusual set of physiological circumstances. There is neither empirical support for extraordinary high rates of reproductive investment in lemurs ... nor a compelling reason for a causal link between reproductive physiology and social behavior" [p. 884].

Of course, evolutionary biologists, including Kappeler (personal communication), generally believe compelling causal links between reproductive and social patterns to be the very essence of life. And, the important final section of Kappeler's (1996) discussion emphasizes that rates of reproductive investment can be judged properly only in relation to overall natural history contexts. These include species-typical socioecological conditions and entire arrays of physiological capacities. Gregarious lemur mothers, for example, might well be expected to engage greater costs of reproduction than lorisid mothers because those lemurs carry their active, rapidly-growing infants while foraging amidst competitive groupmates (Jolly 1984), rather than leaving them in a nest or "parking" them somewhere around which they forage solitarily, as do Lorisids (cf. van Schaik, 1989; Altmann & Samuels, 1992). In addition to these issues, the central element of Hrdy's (1981) original suggestion could not be accounted for in Kappeler's analyses–particular aspects of Madagascar's significant seasonality may have played crucial roles in the evolution of female social dominance among the gregarious lemurs.

Seasonality, Female Dominance, and the Myth of Reproductive Stress

With a modest proposal, we hope to forestall confusion that could obscure what we may be learning about the evolution of female dominance in lemurs. Like Kappeler (1996), we believe that female dominance does not function to help relieve peculiar levels of energetic stress experienced by female lemurs during reproduction. But, we do not suggest that "the evolution of female dominance is *unrelated* to the energetic costs of reproduction" (Kappeler 1993a; p. 154; emphasis ours). Rather, Jolly's female need hypothesis should be stood on its head: we suggest that female dominance is one of many tactics employed in lemur systems that enable females to *avoid* reproductive stress. As far as we know, today's female lemurs experience no more or less stress during efforts to reproduce than do females of other comparable mammals. What remains to be illuminated, of course, is the plethora of relevant details. Most important in needed efforts will be to accumulate detailed comparative data on metabolic physiology and nutritional ecology, as well as reproduction and social behavior.

Our argument takes its cue from an essay by King and Murphy (1985) published soon after Jolly's inspirational work. These authors argued that, despite widespread presumptions to the contrary, endotherms' attainment of adequate nutrition in nature is not necessarily difficult (see also Mrosovsky & Sherry, 1980). Even those that experience sharp annual cycles of ambient temperature and food availability do not necessarily experience periods of nutritional stress, which should be defined as nutrient demands exceeding rates of ingestion and extraction from storage such that reproduction or vital maintenance functions are impaired (Ankney, 1979; King & Murphy, 1985). Annual cycles of performance result from selection for life histories that minimize the potential for stress due to competing demands. Consequently, animals' nutritional status can be properly evaluated in relation to nutritional supply and demand only in the context of subjects' repertoires of behavioral and physiological tactics that function to minimize or avert discrepancies between the two.

Such compensatory adjustments generally entail building and drawing on nutrient reserves and reduction and reallocation of expenditures. Whereas dramatic physiological adjustments like those entailed in seasonal hibernation or fat storage prior to reproduction or migration readily attract attention, behavioral adjustments are also common, and while some are subtle, all are probably far more important than has been recognized to date. Regardless of empirical objectives, most animal behaviorists probably inwardly groan when subjects settle down for a nap during pre-scheduled observation time; but, such "laziness" (Herbers, 1981) or "loafing" (Ettinger & King, 1980) should be dissected and scrutinized, as it constitutes a crucial element of activity budgets for many animals and might represent an investment against unpredictable or environmentally forecasted episodes of cold weather, nutrient shortage, predation risk, or social challenge (cf. Lima, 1986; Lima & Dill, 1990; Lucas et al., 1993).

Many aspects of lemur anatomy and behavior reveal that these primates are comprehensively adapted to conserve energy. The "cuddliness" that attracts so much casual attention to lemurs, for example, derives primarily from the thick hair of their pelage, which presumably confers greater insulation than do the coats of many equatorial primates. Also, several species, such as Ruffed Lemurs (*Varecia variegata*), Ring-tailed Lemurs (*Lemur catta*), and sifakas (*Propithecus* spp.), characteristically engage in sunning behavior (Morland, 1993), almost reflexively on cold mornings when individuals move into patches of sun along a branch or when sunshine suddenly breaks through an otherwise overcast sky (Pereira, personal observation). That these taxa also "sun" in front of heat lamps in captivity further suggests that sunning helps to reduce the energy expenditures of thermogenesis. Most "large-bodied" lemurs also reduce heat loss in the coldest months of the year by resting in social huddles (Fig. 1; also Morland, 1993). Testimony to the value of this behavior to Ring-tailed Lemurs is provided by dominant individuals' routine commandeering of central huddle positions (Pereira,

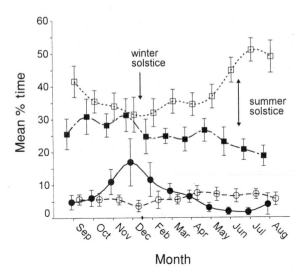

Figure 1. Effects of temperature and/or photoperiod on activity budgets of forest-living Ring-tailed Lemurs. Monthly mean percentages of time mature Ring-tailed lemurs in two forest-living groups at DUPC spent feeding, huddling, resting alone, and social grooming (dashed line, open circles) shown (± 2 S. E.; 1989–1992 data come from serial cohorts of young adult males and females; 1993 data come from older lactating adult females). Percentage of time resting alone (dotted line, open squares) declined along with mean daily temperatures following fall equinoxes (October–December) as time huddling in physical contact with groupmates increased (solid line, closed circles). N.B.: data were never collected during the month of January due to very low levels of activity. During the longest, warmest days each year (May–September), percentages of day-time spent huddling and feeding (dots and dashes, closed squares) reached minima while time resting alone reached maxima.

unpublished data). Conversely, Ring-tailed Lemurs increase daytime resting apart from groupmates in the warmest months of the year (Fig. 1) and pant, drool, and lick their wrists and palms during extremely hot, humid weather (Pereira, unpublished data). During warm periods, resting lemurs also give the impression of achieving unusually relaxed states; they may achieve energy savings unavailable to other mammals through an extreme ability to relax somatic musculature.

Less conspicuous seasonal adjustments made by lemurs are reported in this chapter, along with preliminary evidence of their endogenous regulation via a "biological clock." These comprise changes in rates of hair growth and levels of food intake, subcutaneous fatness, and two hormones known to help regulate nutrient partitioning in mammals, thyroxine and insulin-like growth factor 1. Endogenous control of these aspects of metabolic strategies is relevant to the current hypothesis to explain female dominance among lemurs in so far as such circannual rhythms bespeak the evolution of life histories that annually anticipate and prepare for changes in ambient conditions (Gwinner, 1986), thereby minimizing or even entirely averting physiological stress due to environmental seasonality. Focusing on Ring-tailed Lemurs (*Lemur catta*), the paragon of female dominance, and Red-fronted Brown Lemurs (*Eulemur fulvus rufus*), a closely-related species of similar size, shape, and group demography that entirely lacks formal dominance behavior (Pereira & Kappeler, 1997), we anticipated the possibility of finding species differences relevant to the female need hypothesis.

METHODS

Our interest in the seasonality of lemur life histories began with research on social behavior (Vick & Pereira, 1989; Pereira, 1993a), and growth rate and adult body mass (Pereira, 1993b). Two groups of Ring-tailed Lemurs maintained permanently in a 9-ha forest enclosure at the Duke University Primate Center (DUPC) were the primary subjects of this research. Near-daily behavioral data were collected from members of these two groups from 1986 until 1995, inclusive, during morning and late afternoon sessions of focal animal sampling (Altmann, 1974; Pereira, 1993a for details of subjects and sampling methods). Throughout this period, all group members were weighed monthly and beginning in 1992 other protocols were added to monthly measurements (see below). Two co-ranging groups of Red-fronted Brown Lemurs added to the monthly database and social behavior was studied in these groups in a series of separate investigations (e.g., Pereira et al., 1990; Kappeler, 1990, 1992, 1993b,c; Pereira & Kappeler, 1997; Pereira & McGlynn, 1997).

The formation and management of these semi-free-ranging groups of lemurs have been described elsewhere (Pereira & Izard, 1989; Pereira, 1993a,b); but, it is essential to mention that these lemurs foraged extensively from a variety of naturally occurring plants and were provisioned in different ways across the period of our research. From 1985 through 1988, large amounts of commercially-produced monkey chow were delivered daily to the lemurs ("high provisioning"). In 1989, an initial reduction of provisioning took place, and between 1990 and 1993, inclusive, amounts of chow were reduced still further, such that 5–6 l were delivered daily ("low provisioning") to each pair of separately-ranging groups of Ring-tailed and Red-fronted Brown Lemurs and a co-ranging group of Ruffed Lemurs (approximately 40 lemurs). In 1994 and 1995, levels of provisioning were not monitored closely and they may have exceeded those of the preceding period. Most important, however, is that sufficient provisions were delivered daily throughout all phases of our work so that, by design, dominant and subordinate lemurs all had access to adequate food and seasonal variation in nutrition must have been appreciably diminished in comparison to conditions that obtain in the wild (Pereira, 1993b).

Before proceeding, we must specify terminology used to delineate quarters of the annual cycle because lemurs occur naturally only south of the equator (Madagascar), whereas we studied lemurs north of the equator and seasonal changes in photoperiod, or "day-length," regulate aspects of physiology in almost all lemurs (van Horn, 1975; Rasmussen, 1985; Perret, 1992; Pereira, 1993b). All Lemuridae brought north of the equator, for example, shift reproduction such that most infants are born in March, April, or May, rather than in September, October, or November, as occurs in Madagascar (Rasmussen, 1985).

The longest day each year north of the equator, traditionally called the "summer" solstice (on or near 21 June), is the shortest day south of the equator, known as the "austral winter" solstice. The shortest day in the north, called the "winter" solstice (on or near 21 December), is the longest day in the south, the "austral summer" solstice. On days midway between the solstices, called equinoxes (about 21 March and 22 September), night and day are equal in length all over the earth. The equinox following the (austral) summer solstice is known as the (austral) "fall" or "autumnal" equinox and the equinox following the (austral) winter solstice is the (austral) "vernal" or "spring" equinox. Throughout the text, we use these traditional terms. We omit "austral" when referring either to particular times of year north of the equator or to particular photoperiodic times of year regardless of hemisphere. Readers will appreciate, however, that temperature changes in Madagascar and other subtropical and tropical sites are of secondary seasonal significance compared to annual or biannual changes in rainfall (but see Morland, 1993). At sites throughout Madagascar, most rain falls annually during the warmer months of the year (Richard & Dewar, 1991), increasing four to eight weeks after the austral spring equinox (in October or November) and waning around the time of the austral fall equinox (March).

Earlier analysis of body mass data for the Ring-tailed Lemurs (Pereira, 1993b) revealed three patterns suggesting adaptations that prepare individuals annually for the long dry season in southern Madagascar. First, under both high- and low-provisioning, adult male and female Ring-tailed Lemurs reliably increased body mass immediately before the fall equinox each year. Second, soon after the fall equinox, weanlings invariably reduced mass growth rates substantially for at least four to five months, despite our increasing of provisions slightly between the winter solstice and spring equinox to help the lemurs cope with winter temperatures (North Carolina, USA) colder than those normally experienced in Madagascar (Richard & Dewar, 1991). Finally, immatures who earliest attained body masses of approximately 1200g reduced growth rates earliest, and under high-provisioning conditions, when early growth rates were maximal, growth rate reductions following fall equinoxes tended to be greater than during the low-provisioning of subsequent years of research. Juveniles thus appeared to make compensatory adjustments in a normative seasonal scheduling of growth effort according to fatness or some other aspect of individual growth history.

These patterns led us to dissect several lemurid cadavers to describe the anatomy of white adipose tissue (Pereira & Pond, 1995), so that we could effectively add monthly skinfold measurements to our research protocol and try to document seasonal changes in subcutaneous fat storage. Measurements of several body lengths were also added to our monthly protocol (ulna, fibula, and trunk) as well as measurement of hair growth at a small patch shaved monthly on the underside of each lemur's tail. Serendipitously, shaving rings in fast-growing immatures' tails to identify them as individuals even in the trees had led us to discover seasonal hair growth in these animals, which do not exhibit conspicuous molts.

Monthly blood samples were collected from March 1994 through March 1995 to enable investigation of possible seasonal changes in circulating levels of insulin-like growth factor 1 (IGF-1) and thyroxine (T4), two hormones known to affect metabolism in humans and other mammals (e.g., Amato et al., 1993; Nagy et al., 1994, 1995; Suttie & Webster, 1995; Tomasi & Mitchell, 1996; Wolthers et al., 1996; Tomasi et al., 1998). IGF-1 (also known as somatomedin-C) is a small, single-chain polypeptide produced primarily by the liver. Long known as the mediator of effects of pituitary growth hormone, IGF-1 also helps regulate body composition. Low levels are associated with relatively high fatness in humans, laboratory rodents, and some wild mammals, and administration of IGF-1 to deficient humans reduces fat mass (e.g., de Boer et al., 1992; Marin et al., 1993; Webster et al., 1996). T4 is the main hormone produced by the thyroid gland and is enzymatically deiodinated in target tissues to triiodothyronine (T3), considered to be the "active" compound. Both of these hormones are transported in the blood by carrier proteins; but the small protein "free" fraction of each can leave capillaries and is thus available for deiodination or binding to intracellular receptors (Tomasi & Stribling, 1996), where they generally stimulate oxidative metabolism (Schmidt-Nielsen, 1990). Humans and other mammals deficient in thyroxine are generally obese.

Blood samples were refrigerated within an hour and centrifuged within 24 h of collection and serum aliquots were stored at −80°C until assays were performed. ELISA analyses of IGF-1 were performed using protocols and reagents obtained from Diagnostic Systems Laboratories, Inc. (Webster, Texas). After extracting IGF-1 from serum binding proteins, samples were incubated with anti-IGF-1 mouse monoclonal antibody labeled with the enzyme horseradish peroxidase in microtitration wells coated with another anti-IGF-1 mouse monoclonal antibody. Next, tetramethylbenzidine was added to wells for further incubation. Finally, stopping solution was applied and the degree of enzymatic turnover of the substrate was determined by dual wavelength absorbance measurement at 450 and 620 nm (ELISA plate reader, BioRad model 550). Diagnostic Systems' set of IGF-1 standards was used to establish a curve of absorbance values in relation to IGF-1 concentration (ng/ml).

Portions of blood serum samples were sent to the Yerkes Regional Primate Research Center Assay Services Laboratory (Lawrenceville, Georgia) for determinations of T4 levels via solid-phase ^{125}I radioimmunoassay, using the "Coat-A-Count" kit (Diagnostic Products Corporation; Los Angeles, California) for quantitative measurement of non-protein bound T4 levels. Free T4 antibody-coated tubes received 50 μl of calibrator standards, control solution, or subjects' serum, then received 1.0 ml of [^{125}I] Free T4. Tubes were vortexed and then incubated by waterbath for 60 min at 37°C. Finally, tubes were decanted, allowing 2–3 min of draining before performing 1 min of counting in a gamma counter.

Finally, in 1996, two mature males and three mature females of each species were transferred from DUPC groups to indoor housing at Bucknell University where daily feeding and room temperature were held constant while natural photoperiod was simulated electronically. Research on monthly changes in body mass, fatness, and hair growth continued and, on 1 July 1996, we began to quantify daily food intake. A mixed-species group of females and another of males each received one midday allotment of 6 chow biscuits per individual. Each morning, biscuits left over from the previous day were retrieved, counted, and discarded. Focal animal sampling (Altmann, 1974) was conducted thrice weekly during the first hour after provisioning to investigate whether species differences existed in seasonal patterns of food consumption.

RESULTS

Hair Growth and Fatness

Regardless of whether lemur subjects experienced seasonal changes in ambient temperature, males and females of both species grew hair on their tails only between spring and fall equinoxes, reliably terminating hair growth soon after fall equinoxes, real or simulated, and resuming hair growth only after subsequent spring equinoxes (Fig. 2). Likewise, regardless of whether seasonal changes in ambient temperature were experienced, males and females of both species reliably increased the size of subcutaneous depots of adipose tissue beginning after summer solstices, real or simulated, and reduced the sizes of these depots soon after winter solstices.

Some or all of five changes in living conditions may have helped cause unusual patterns of fatness to develop among the indoor-housed Bucknell lemurs in late 1997 and early 1998. First, beginning at the simulated fall equinox (September 1997), photoperiod was fixed to provide 12 h of light and 12 h daily for the subsequent two-year

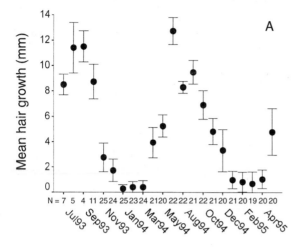

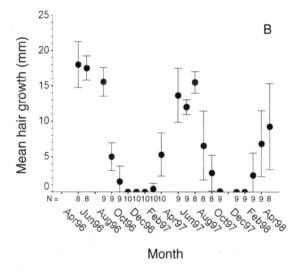

Figure 2. Seasonal changes in tail hair growth in Lemuridae. Panel A: average amounts of hair regrown on small patches shaved monthly on ventral surfaces of tails on adult Ring-tailed and Red-fronted Brown Lemurs in DUPC forest enclosures between summer solstices in 1993 and 1995 (± 2 S. E.). Annually, rates of hair growth declined around the fall equinox (beginning in September or October), ceased entirely in winter (January–March, inclusive), and began again after spring equinox (April). Panel B: subjects transferred to indoor conditions of simulated photoperiod and constant temperature and diet maintained this same seasonal pattern for two years. No species differences detected in either study; however, newly pregnant females grew hair for one or two months longer after the fall equinox than males and nonpregnant females.

period, to promote investigation of whether an endogenous "biological" clock (Gwinner, 1986) regulates phenomena under study. Next, in alternating weekly experiments throughout October and November, males or females were allowed to visit chambers normally occupied by the members of the opposite sex. Thereafter (beginning 1 December), cage tunnels between the males' and females' formerly separate rooms were permanently opened, combining our sub-colonies and providing members of both sexes much increased opportunity for physical exercise. From that point onward, the lemurs began consuming greater proportions of daily provisions and every one of the five females conceived and gestated a fetus to full term. To support continued study of seasonal changes in food intake, we increased the lemurs' daily provisions in late January 1998.

More data under the new conditions described above must accumulate before we can confidently interpret the departures from normative patterns seen between September 1997 and May 1998 (Fig. 3, panel B); however, it seems likely that the spontaneous opportunities for increased exercise and sexual interaction immediately before

Figure 3. Seasonal changes in subcutaneous fatness in Lemuridae. Panel A: mean sums of skinfold thicknesses measured monthly at abdominal and axillary sites on 20 to 35 mature semi-free-ranging DUPC Ring-tailed and Red-fronted Brown Lemurs between summer solstices in 1992 and 1995 (± 2 S. E.). Annually, fat depots at each site increased in size following the summer solstice and declined around or just after the winter solstice. In April or May each year, brief rises in subcutaneous adipose tissue coincided with spring flushes of natural forest foods. Panel B: subjects transferred to indoor conditions of simulated photoperiod and constant temperature and diet maintained this same seasonal pattern for nearly two years. With indoor animals, skinfold thicknesses measured at abdominal, axillary, and also ischial sites on mature individuals. Sample sizes shown beneath abscissa. No significant differences detected in either study between males and females or between the two study species. (See text for discussion of possible effects of procedures undertaken in late 1997 and early 1998.)

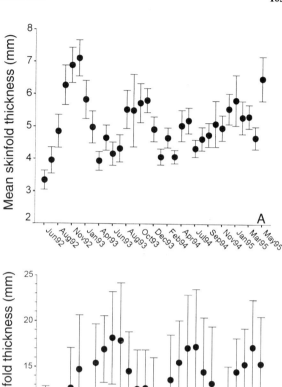
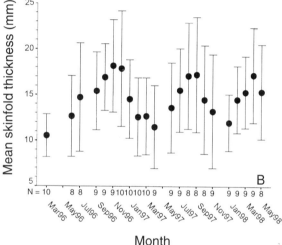

and during mating season 1997 led to reductions in fatness after which our increasing of daily provisions helped cause an appreciable compensatory rebound, leading to unusual fatness values throughout these periods.

Food Intake and Hormone Levels

Adult males and non-pregnant females of the two study species exhibited generally comparable seasonal changes in food intake in the Bucknell-based study, where food supply and temperature could be kept constant (Fig. 4). Focal animal sampling conducted during the first hour after provisioning (fall 1996-spring 1997) confirmed that, among males and among females, members of both species began leaving behind increasing proportions of daily provisions two to four weeks after the fall equinox, as hair growth began to decline.

After simulated fall equinoxes in 1997 and 1998, males left behind substantial proportions of chow significantly sooner than did females (Fig. 4), perhaps reflecting that attention to possible mating opportunities takes precedence over foraging for males

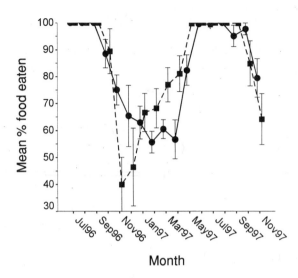

Figure 4. Seasonal changes in food intake. Mean daily percentages of chow consumed shown for each month (± 2 S. E.). Under simulated photoperiod, constant temperature, and constant food supply, adult males (dashed line, closed squares) and females (solid line, closed circles), housed separately throughout 1996 and 1997, consumed 100% of provisions almost daily throughout each summer (July–September, inclusive), left behind increasing proportions of provisions following fall equinoxes (Septembers), reached appetite minima in late fall or winter months (November–March), and again increased food intake beginning around or just after simulated spring equinoxes (March).

but not females, as in some other mammals (nearly all conceptions occur between late October and mid-December in our study populations; Rasmussen, 1985). For non-pregnant females, chow consumption reached its nadir from January through April, when less than 60% of available chow was consumed daily, on average. Along with hair growth, food intake began increasing around the time of the spring equinox, significantly sooner in males than in the non-pregnant females (Fig. 4). By the time of the summer solstice, members of both sexes again consumed 100% of provisions virtually every day.

In 1994, the semi-free-ranging adult male and female Ring-tailed Lemurs at the DUPC exhibited indistinguishable seasonal patterns of circulating IGF-1 and T4 (Fig. 5). Changes in levels of these hormones in the Red-fronted Brown Lemurs were similar, but may have been delayed by several weeks relative to the schedule shown by the Ring-tailed Lemurs. Small numbers of subjects precluded unequivocal evaluation of possible sex differences in Red-fronted Brown Lemurs or possible differences between the two species (Fig. 5).

Circulating levels of IGF-1 and T4 diminished just before the summer solstice and relatively low levels were maintained by all adults until the fall equinox. Immediately after the fall equinox, the more typical higher levels of these two hormones suddenly resumed (note small S.E. values, Fig. 5). Average monthly levels of circulating hormones in 1994 correlated negatively with monthly levels of adult food intake during the subsequent Bucknell-based study, and also with representative adult fatness and body weight values from research on DUPC lemurs lagged two months for the analysis (Fig. 6).

DISCUSSION

Adaptations to Prevent Stress during Dry Season

A cornerstone of life-history strategy for both Ring-tailed and Red-fronted Brown Lemurs comprises inter-related performance adjustments in response to envi-

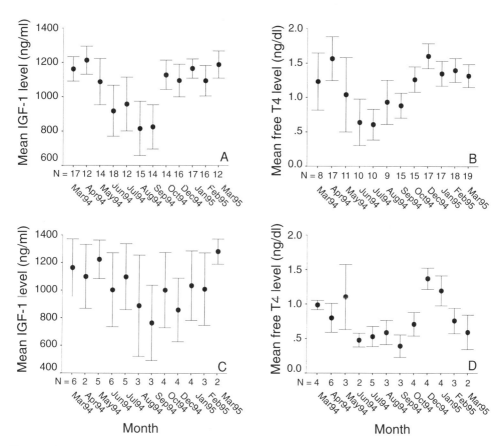

Figure 5. Seasonal changes in circulating levels of IGF-1 (Panels A and C) and free T4 (Panels B and D). Mean hormone levels shown for adult and adolescent subjects captured monthly (± 2 S. E.; sample sizes beneath abscissa). Circulating levels of both metabolically-active hormones declined in males and females of both species around time of the summer solstice (late June), remained low through late summer (July-September, inclusive), and thereafter reverted to the relatively high levels characteristic of fall, winter, and spring (October-early June, inclusive). Smaller numbers of mature Red-fronted Brown Lemurs captured monthly precluded satisfactory determination of normative patterns for this species, whereas larger numbers of Ring-tailed Lemur subjects rendered clear patterns.

ronmental cues that reliably forecast seasonal changes in ambient temperature and food supply. Like many birds and mammals (Wirtshafter & Davis, 1977; Mrosovsky, 1986, 1990), lemurs respond to exogenous stimuli by changing the level at which they regulate particular physiological states (rheostasis), determining set-points for growth rate (Pereira, 1993b), hair growth rate, appetite, fatness, and circulating levels of hormones that help to control these and other metabolic, anatomical, and behavioral adjustments. Most likely, these are responses chiefly to photoperiod, as they occurred in enclosure-living lemurs for which rich provisions were amply supplied throughout the year and persisted for two years after subjects were transferred to indoor housing where naturalistic photoperiod and ample diet continued while thermoneutral ambient temperature remained fixed.

These life history features clearly represent lemurid adaptation to predictably seasonal climatic and foraging conditions in Madagascar, though the particular array of

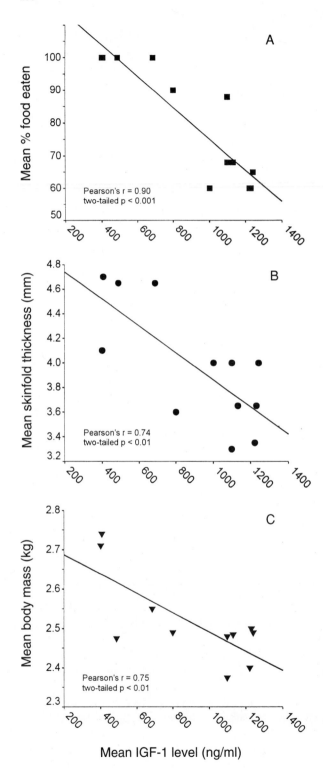

Figure 6. Mean monthly levels of circulating hormones in 1994 correlate negatively with mean monthly levels of adult food intake found during subsequent Bucknell-based study (Panel A) and also representative mean monthly adult body weights (Panel B: female Ring-tailed Lemurs) and fatness values (Panel C: male Ring-tailed Lemurs) that are lagged by two months for the analysis.

most important factors could only be identified through detailed new fieldwork. Both Ring-tailed and Red-fronted Brown Lemurs range in the south of Madagascar. The extreme south, the natural range of Ring-tailed Lemurs, is dry, experiencing as little as 40–60 cm annually; but, still more significant is that virtually all this rain typically falls from just before the austral summer solstice until the austral fall equinox (Richard & Dewar, 1991). Consequently, an annual "boom" period of foraging occurs from just before until just after this time of year (Sauther, 1992, 1993; Rasamimanana & Rafidinarivo, 1993; see also below).

The DUPC's Red-fronted Brown Lemurs derive from animals originally imported from both Madagascar's southeastern rainforest and its dry, southwestern deciduous forests and the animals we studied included a mix of them and hybrid descendants (D. Haring, personal communication). There is ample reason to expect, however, that Red-fronted Brown Lemurs from both regions show comparable seasonal patterns. First, our general pattern of findings extended not only between two divergent Lemurids, but also to yet a third, strikingly different member of this family, Ruffed Lemurs. The DUPC Ruffed Lemurs living in the same forest enclosure showed the same seasonal pattern of hair growth (Pereira, unpublished data), and researchers of wild Ruffed Lemurs report seasonal changes in activity designed to conserve energy during the austral winter months (Morland, 1993; Britt, personal communication; see also Rasamimanana & Rafidinarivo, 1993).

Second, and particularly important, striking convergences have begun to emerge among the limited datasets yet available describing phenology in Madagascar's major ecosystems. Work in the eastern rainforest (Overdorff, 1992, 1993; Hemingway & Overdorff, in press), in the extreme south (Sauther, 1992), and in the western deciduous forest (Ganzhorn, 1995; Sorg & Rohner, 1996) all reveal production of few new leaves or flowers between austral fall and spring equinoxes, increasing production of leaves and flowers around austral spring equinoxes, and high availability of important leaves and flowers sometimes accompanied by high fruit availability between austral summer solstices and fall equinoxes. Consideration of these data together with our results suggests that lemur life histories may be less linked to patterns of fruit production than previously thought (cf. Overdorff, 1993; Morland, 1993), but importantly tied to seasonal patterns of leaf and flower production. New research focused on the seasonality, nutrient content, and lemurs' styles of exploitation of leaves and flowers (cf. Overdorff, 1992), especially combined with information on metabolic physiology, reproduction, and social behavior, would probably advance our understanding of lemur natural histories significantly.

We suggest that success in foraging is most crucial to large-bodied lemurs just before, during, and just after the wet season, rather than throughout the dry season, as is often thought (e.g., Jolly, 1984; Overdorff, 1992), and that this is true especially for juveniles and adult females (Pereira, 1993b). Around the austral spring equinox, new infants are born, appetite begins to grow, and hair growth resumes in Ring-tailed and Red-fronted Brown Lemurs, just as new leaves, buds, flowers, and unripe fruit emerge in anticipation of the annual increase in rainfall (citations above). Three months later, when the period of peak rains has begun, both IGF-1 and T4 decline sharply, leading to a change in "nutrient partitioning," diverting excess nutrition daily into storage as adipose tissue. Females lactate for fast-growing infants at least two months longer (Sussman, 1977), gradually wean them, and then conceive again, just as the year's heaviest rains end. As food supplies dwindle and temperatures decline, IGF-1 and T4 again circulate at higher levels, appetite and activity levels decline, growth of hair and of juveniles' bodies is virtually suspended, portions of each day's energy needs are extracted from adipose tissue, and small, lean fetuses develop (Leutenegger, 1979; Tilden, in review).

One year of data collected since the Bucknell lemurs' light cycle was fixed at 12 h light: 12 h dark (September 1997-October 1998) provides preliminary evidence that lemurs' cycles of reproduction, subcutaneous fatness, hair growth, and appetite are endogenously controlled via a "biological clock." Appropriate further research will probably show that mechanisms for these changes relate, directly or indirectly, to correlated adjustments in circulating levels of IGF-1, free T4, and also leptin, a peptide secreted by adipocytes in proportion to level of lipid mass (Schwartz & Seeley, 1997; Widmaier et al., 1997). In passing, note that we documented hormonal changes in animals experiencing seasonal changes in temperature and forest foods, whereas discovery of seasonal changes in appetite and, perhaps, metabolic rate will often require controlled environments such as those we arranged at Bucknell. Cold temperatures and more time-intensive naturalistic foraging tasks in winter, for example, masked appetite changes in our lemurs while they were still living in the DUPC forest enclosures (Fig. 1). Likewise, in the context of seasonal environmental changes in Madagascar, the functional significance of the appetite changes documented in our laboratory will not necessarily be reflected in monthly percentages of time spent feeding. Work such as ours well illustrates the importance of interleaving results from nature, naturalistic environments, and laboratory settings.

"Hard-wired" circannual rhythms (Gwinner, 1986) in metabolic strategies reveals Ring-tailed and Red-fronted Brown Lemurs to have evolved as organisms that prepare for upcoming seasonal changes in environmental conditions and also inevitably make adjustments during cool-dry seasons that reduce costs of living. Lemurs' physiological and behavioral adjustments presumably minimize the difficulty of attaining adequate nutrition annually, despite significant seasonal changes in temperature and food supply. Nutrient demands, in other words, may rarely exceed levels of ingestion and extraction from storage to an extent that reproduction or other physiological functioning becomes impaired for many individuals ("stress;" Ankney, 1979; King & Murphy, 1985). Consequently, we agree with Kappeler (1996) that evidence is lacking to indicate that female lemurs experience unusually high costs of reproduction. It remains unknown how many female lemurs, if any, elevate basal metabolic rate during reproduction (cf. Young et al., 1990); but, the data presented here suggest that other winter adjustments might render such an expense unnecessary.

Metabolic Strategy and Social Behavior in Lemurs

Research in socioecology has yet focused almost exclusively on mating systems and proximate measures of male and female reproductive success (e.g., Rubenstein & Wrangham, 1986; Standen & Foley, 1989; Sterck et al., 1997). Endotherms spend most of their energy on thermoregulation, however (Schmidt-Nielsen, 1990), and adaptations that promote individual longevity should be paramount for iteroparous organisms. Indeed, variation in reproductive success seems best explained by lifespan across a variety of long-lived species (Clutton-Brock, 1988). This reasoning suggests that research on metabolic strategies would contribute as much toward illuminating animal adaptedness, including patterns of social behavior, as has research on sexual selection and mating strategies to date. Especially for seasonally-organized taxa, like lemurs, the significance of major features of social behavior is likely to be clarified by research on nutritional ecology and metabolic strategy. We explore this possibility below, in a review of the elements of female dominance, targeted aggression, and male-committed infanticide in large-bodied lemurs.

Female Dominance. Whether female dominance or other social patterns have evolved among Malagasy primates due to "unusual physiological circumstances" (Kappeler, 1996, p. 884) and precisely what causal links exist between reproductive physiology and social behavior in lemurs remain complex and fascinating questions requiring considerable new research. To begin, major environmental variables (e.g., photoperiod, temperature, nutrient supplies, local demography, predator and intruder pressures) can be expected to help regulate physiology and behavior, and social adjustments that enhance effects of tactics promoting success in migration, reproduction, and periods of high or low food supply should be particularly important for seasonal mammals that use neither burrows, tree holes, nor hibernation. In this context, our most intriguing result is certainly to have been unable to discern significant differences in several elements of basic metabolic strategy between Ring-tailed Lemurs, which show unconditional female dominance, and Red-fronted Brown Lemurs, which lack formal dominance behavior (definition in de Waal & Luttrell, 1985; also Pereira & Kappeler, 1997). Clearly, much remains to be learned about these primates.

If, as we suggest, foraging success is pivotal among lemurs when the foods become plentiful that determine relative success in nutrient storage (e.g., fat), the hormonal changes we documented may also help regulate adaptive changes in aggressivity over food, enhancing powerful animals' dominance over groupmates in species like Ring-tailed Lemurs, including all females' dominance over males. Kappeler (1990) emphasized that Ring-tailed females use their dominance over males in all social contexts; but, in the wild, Sauther (1993) documented a sharp increase in female aggressivity during feeding that began directly after the austral summer solstice, was maintained as infants were weaned, and declined around the austral fall equinox. Female dominance seems to have been augmented, as non-aggressive and non-feeding agonistic interaction between the sexes also peaked during the summer.

The males Sauther (1993) studied exhibited their own, but more modest peak in aggressivity during summer, supporting that feeding success is generally important to lemurs during the rains and that it is especially important to females (see Pereira, 1993a,b, 1995 on conflict among weanlings). In addition, sex differences in appetite change that we found suggest the possibility that males add fat in austral summers primarily to facilitate devoting most of their time and energy toward mating competition immediately thereafter. Males may routinely compensate during later stages of dry seasons by committing more time and energy to foraging than do non-pregnant females (Fig. 4).

In sum, female Ring-tailed Lemurs, and perhaps females in other gregarious lemur species, use dominance most crucially to maintain priority of access to food during their pivotal months of plentiful forage. Female dominance among lemurs may be an evolutionary legacy of former pair-bonding (Jolly, 1998; also Fietz, in press), where male subordination to females constitutes paternal investment (Pollock, 1979). But, as lemurs evolved greater gregariousness, perhaps quite recently (van Schaik & Kappeler, 1996), group-living should have exacerbated feeding competition (van Schaik, 1989) and the trait's function should have shifted, or been reinforced, such that female dominance became one of several life history traits that collectively enable females to avoid stress due to reproduction across months of scant nutrition and relatively cold weather. While seasonality has likely influenced the physiological and behavioral functioning of all large-bodied lemurs, only females undertake gestation and early lactation during annual long dry seasons (Jolly, 1984).

If this reasoning is correct, why do Red-fronted Brown Lemurs fail to exhibit female dominance? First, behavioral differences between Ring-tailed and Red-fronted Brown Lemurs extend far beyond the matter of female dominance. Rather, these Lemurids exhibit comprehensively different social systems that may represent equivalent alternatives in terms of female reproductive success (Pereira & McGlynn, 1997; Pereira & Kappeler, 1997); i.e., the environmental factors lemurs encountered, together with the physiological capacities they bore, may have been equally likely to have promoted the evolution of either of social system. Red-fronted Brown Lemurs lack formal dominance behavior, and all recent researchers have cited the possibility that a strong inclination to form dyadic partnerships may limit the manifestation and significance of pairwise competitive behavior, including dominance, in this species (Pereira & Kappeler, 1997; Pereira & McGlynn, 1997; Overdorff, 1998).

But, the natural ranges of *Lemur catta* and *Eulemur fulvus* do not overlap extensively, and differences in current or antecedent selection pressures, or both, may have promoted divergent social evolution. One possibility sees *Lemur*, *Varecia*, *Propithecus*, *Indri*, and *Eulemur* species having shifted to different extents in recent transitions from prior nocturnal lifestyles toward more cathemeral or virtually diurnal living (van Schaik & Kappeler, 1996; cf. Pereira, 1995). Small amounts of change along this evolutionary pathway should preclude the emergence of formal dominance behavior, perhaps explaining the agonistic behavior of *E. fulvus* (Pereira & Kappeler, 1997). Once equipped with formal dominance, females of other taxa might be permitted to abandon egalitarian partnerships with males and relatively small foraging parties (van Schaik & Kappeler, 1996, 1997) in favor of nepotistic relations with females and larger social groups (see Sterck et al., 1997; Jolly, 1998). Alternatively or additionally, relative omnivory may minimize feeding competition among *E. fulvus* (Pereira & Kappeler, 1997), thereby permitting female dispersal (Sterck et al., 1997) and promoting partnerships with males.

Much more detailed data than is available at present for *Eulemur macaco*, *E. coronatus*, *E. mongoz*, other *E. fulvus* subspecies, and *Hapalemur* could help clarify possible relationships among timing of activity, group size, and patterns of social dominance and bondedness in lemurs. And, new fieldwork comparing *Lemur catta*, other female-dominant species, and *E. fulvus* with regard to seasonal patterns in phenology, especially food preferences and nutrient profiles, and also female social relationships will be necessary to test whether differences in nutritional ecology and/or intensity of food competition play roles in these lemurs' divergent social evolution (cf. Overdorff, 1992, 1993, 1998). Additional and more detailed information on metabolic tactics in these and other species would be of essential complementary value.

Targeted Aggression. Long-term research on DUPC study groups and shorter projects with Ring-tailed Lemurs at the Berenty Reserve have shown that female Ring-tailed Lemurs living in large groups maintain adversarial relationships with particular peers (Pereira & Kappeler, 1997) whom they target for intense aggression and potential eviction on a distinctly seasonal basis (Vick & Pereira, 1989; Pereira, 1993a, 1995). At the DUPC, episodes of targeted aggression have begun almost exclusively around the time of the equinoxes (Fig. 7).

In the context of basic needs to conserve energy while also limiting female group

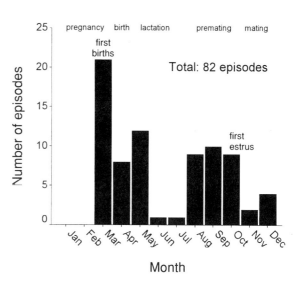

Figure 7. Seasonal onset of episodes of targeted aggression between forest-living female Ring-tailed Lemurs at DUPC. Numbers of episodes that began each month shown from daily records for Lc1 Group between 1980 and 1991, inclusive. Normative seasonal reproductive status of females indicated at top. Large majority of adult females (ca. 80%) conceived in early November and gave birth in late March. Most females failing to conceive during first estrus conceived in late December and gave birth in early May.

size in expanding populations (see Pereira & Kappeler, 1997), the seasonality of targeted aggression in Ring-tailed Lemurs may derive from reversals in relative costs and benefits of aggression and amicability. First, aggressors should maximize their efficiency by timing attacks to occur when subordinates can least afford to sustain them. Females' best interests are particularly vulnerable during birth seasons, especially because the earliest phase of lactation occurs at the end of the long dry season, and at least two months before annual rains restore ample food supplies. But, targeted aggression around the austral fall equinox may jeopardize victims' reproductive efforts nearly as effectively. Spatial peripheralization and the constant vigilance required to avoid intense attack by stealthy aggressors conjointly reduce foraging efficiency, diminishing victims' capacities to accumulate adipose tissue prior to conception. Moreover, estrous cycling can be disrupted in victims (Pereira, unpublished data), leading to late-born infants ultimately doomed by small size to bottom ranks in their birth cohorts (Pereira, 1993a, 1995) and, therefore, least chances to survive the long dry season (Pereira, 1993b).

Second, targeted aggression may rarely begin during winter because the tactic's energetic costs become outweighed by the benefits of social tolerance and gregariousness as temperatures decline and food and fat wane. To help minimize costs of winter, lemurs increase time spent resting in social huddles (Fig. 1; also Morland, 1993; Rasamimanana & Rafidinarivo, 1993). Future research may find that female pairs, friendly and adversarial alike, begin spending more time in proximity without exchange of aggression, using milder forms of aggression and more frequent reconciliation when conflicts do occur, and increasing time spent huddling even before the cold of winter becomes fully established. Changes in day-length may cue physiological changes that cause lemurs to suspend aggression and increase amicability. Research on birds and other mammals has revealed photoperiodic influences on self-maintenance behaviors, habitat preferences, selection of social partners, and rates of other social behaviors (e.g., Gorman et al., 1993; Moffatt & Nelson, 1994; Clayton & Cristol, 1996; Ferkin et al., 1996; Gahr & Kosar, 1996).

The different social structure of Red-fronted Brown Lemurs is reflected in the different seasonality of targeted aggression and different involvement of the two sexes in this species. Unlike Ring-tailed Lemurs, male Red-fronted Brown Lemurs both target and become targeted (Colquhoun, 1987; Vick & Pereira, 1989). Moreover, males sometimes assist female partners to attack other females and males are sometimes attacked by females assisted by males. Enhancing the impression that targeted aggression in Red-fronted Brown Lemurs functions mainly as an element of competition over participation in special male-female relationships (Pereira & Kappeler, 1997; Pereira & McGlynn, 1997) is that most episodes have begun during premating and mating seasons (Vick & Pereira, 1989). As in Ring-tailed Lemurs, however, targeted aggression in Red-fronted Brown Lemurs has begun relatively rarely during mid-winter or mid-summer.

Infanticide. Since Pereira and Weiss (1991) first published observations implicating male infanticide as an element of Ring-tailed Lemur natural history, many additional cases of infanticide in *Lemur* and *Eulemur* have accumulated, including several in Red-fronted Brown Lemurs (Pereira & McGlynn, 1997; Walker et al., in review). In search of a reasonable adaptive explanation, Pereira and Weiss (1991) gathered preliminary evidence from fieldwork with a variety of large-bodied lemurs suggesting that females rarely reproduced successfully in successive years. They proposed that a full course of lactation in a given year may limit females' abilities to store sufficient fat to succeed in reproduction the following year. If so, the classical advantage of infanticide exists for males in even these rigidly seasonal breeders: acceleration of female reproduction, so that it occurs during infanticidal males' brief reproductive tenures. Male infanticide has also been documented among wild Diademed Sifakas (*Propithecus diadema diadema*; Wright, 1995) and data from Verreaux's Sifakas (*P. verreauxi verreauxi*) corroborate that successful reproduction diminishes adult female body weight and decreases chances of reproduction in succeeding years (Richard et al., in review).

The coincidence of lactation, plentiful food, hair growth, and fat deposition, in fact, sets the stage for a number of possible male influences on female reproductive success and behavior in gregarious lemurs. Resident male Ring-tailed Lemurs, for example, may resist male immigration during birth seasons to enhance their chances for subsequent mating success in several ways (Pereira & Weiss, 1991). Of course, males may benefit from minimizing the number of competitors that co-reside in their groups. But also, males that actively deter birth-season immigration may be differentially attractive to females because of genetic benefits of mating with such vigorous males and because those males help to minimize both risk of infanticide and competition over precious spring or wet-season foods. Particularly interesting in this regard is Overdorff's (1998) report of male Red-fronted Brown Lemurs dispersing food competitors from the locations of their female partners and also being favored by their partners as mates. More work so carefully detailing polyadic social interactions in the contexts of different seasons and time spent foraging for different foods would be invaluable toward deepening our understanding of lemur behavioral biology.

Looking toward the Future

We have tried to show how research on metabolic strategies is at least as important as research on reproductive strategies to illuminating life histories and the adaptedness of behavior patterns in seasonal mammals such as lemurs. Clearly, though, our work has raised many more questions than it has answered, and we conclude by highlighting a few issues of particular interest for future investigations in each of four inter-related domains.

First, we must learn more about metabolic rates in a variety of lemurs, and it will be important to amass information on both basal and active rates (see discussion in Kappeler, 1996). While some information is available for prosimian primates (e.g., Müller, 1985; Martin & MacLarnon, 1988; Young et al., 1990; Ross, 1992; Kappeler, 1996; Schmid & Ganzhorn, 1996; Drack et al., this volume), all authors dealing with the available data for lemurs agree that they are distressingly thin. Investigation of effects of age, sex, social status, season, and reproductive status will be required to illuminate overall metabolic strategies and conditions that render growth, dominance contest, dispersal, or reproduction potentially stressful for lemurs. Beyond food preferences and other foraging elements, the behaviors central to understanding metabolic rates will be those most relevant to thermogenesis, including several that have gone surprisingly neglected in lemur research to date, namely huddling, sunning, and various modes of resting and "laziness" (Herbers, 1981; cf. Morland, 1993).

Second, other metabolic tactics, like those we explored, have yet to be investigated in any large-bodied lemur other than Ring-tailed and Red-fronted Brown Lemurs, and causal relations among such elements have yet to be demonstrated for any Lemurid or Indriid (but see van Horn, 1975; Rasmussen, 1985; Pereira, 1993b). What roles might leptin and other hormones play in lemur fatness and reproduction (cf. Schwartz & Seeley, 1997; Widmaier et al., 1997)? Do metabolically-active hormones directly mediate seasonal changes in female aggressivity? Do all large-bodied lemurs accumulate fat and grow winter pelage in summer, and do those tactics require some species to shift some activity into the night, thereby precluding the evolution of entirely diurnal lifestyles (Pereira, 1995)?

A crucial third objective will be to accumulate data on the phenologies and seasonally-changing nutrient contents and attractiveness of lemur foods. Until we learn more about which foods are most important to each species, or population, and how age- and sex-typical foraging preferences relate to specific metabolic tactics, we will be able to conclude little about how seasonality has guided the evolution of lemur natural histories. Is access to new leaves and flowers between the austral spring and fall equinoxes, rather than fruits, the factor that most importantly influences female reproductive success? Is energy not the whole story? Rather, do some foods contain certain nutrients, such as essential fatty acids, that pivotally influence females' chances for success, by modulating the effectiveness of physiological adjustments (cf. Florant, 1998)? Note that research questions in nutritional ecology absolutely require preservation of lemurs' natural environments and also that results in this domain should augment, with particular impact, our abilities to promote health and reproduction in captive colonies and protected natural populations of lemurs.

Finally, detailed data on social behavior are needed from the widest possible range of lemur species. We still lack an unequivocal inventory of which Lemuridae and Indriidae actually exhibit female dominance (cf. Pereira et al., 1990), for example, and few studies have identified the seasons or contexts that most reliably induce females to assert their dominance (cf. Pollock, 1979; Vick & Pereira, 1989; Sauther, 1993; also Fig. 7). What seasonal social adjustments are exhibited by females and males in species lacking female dominance? Is formal dominance behavior generally lacking among cathemeral species? Relatively omnivorous species? Finally, research should investigate how males influence female success with metabolic strategies. Do females guard against infanticide (cf. Pereira & Weiss, 1991) in all species where fat deposition and lactation co-occur? Do likely fathers? Are particular seasons or foods associated with male aggression that distances conspecifics from adult females (cf. Overdorff, 1998)?

Are males that are particularly efficient with aggression on behalf of females also preferred as mates?

ACKNOWLEDGMENTS

We are grateful to the editors for their invitation to participate in this volume, and to S. Goodman, P. Kappeler, and A. Jolly for helpful criticism of an earlier draft. We thank K. Glander, DUPC Director, for permission to conduct monthly measurements and blood draws with Duke's enclosure-living lemurs, D. Brewer and P. Feeser for expert care of those animals and assistance with monthly procedures, and M. Ravosa for special assistance with collection of the first year of data on fatness. We thank M. Gavitt for expert care of Bucknell's animal colonies, daily collection of left-over chow, and assistance with monthly measurements, A. Gleason, E. Racz, and K. Leonard collected data on post-provisioning feeding behavior and R. Chester, E. Crowley, A. Fincke, L. Selvaggi, and J. Smith assisted with monthly measurements at Bucknell. Portions of this work were supported by NICHD (R29-HD23243), the S. F. Hughes Endowment Fund, and Bucknell University. This is DUPC Publication 553.

REFERENCES

ALTMANN, J. 1974. Observational study of behavior: sampling methods. Behaviour, **49**: 227–267.
——, AND A. SAMUELS. 1992. Costs of maternal care: infant-carrying in baboons. Behavioral Ecology and Sociobiology, **29**: 391–398.
AMATO, G., C. CARELLA, S. FAZIO, G. LA MONTAGNA, A. CITTADINI, D. SABATINI, C. MARCIANO-MONE, L. SACCA, AND A. BELLASTELLA. 1993. Body composition, bone metabolism, and heart structure and function in growth hormone (GH)-deficient adults before and after GH replacement therapy at low doses. Journal of Clinical Endocrinology and Metabolism, **77**: 1671–1676.
ANKNEY, C. D. 1979. Does wing molt cause nutritional stress in Lesser Snow Geese? Auk, **96**: 68–72.
BEARDER, S. 1987. Lorises, bushbabies, and tarsiers: diverse societies in solitary foragers, pp. 11–24. *In* Smuts, B. B., D. L. Cheney, R. M. Seyfarth, R. W. Wrangham, and T. T. Struhsaker, eds., Primate Societies. The University of Chicago Press, Chicago.
CLAYTON, N. S., AND D. A. CRISTOL. 1996. Effects of photoperiod on memory and food storing in captive marsh tits, *Parus palustris*. Animal Behaviour, **52**: 715–726.
CLUTTON-BROCK, T. H. 1988. Reproductive Success. The University of Chicago Press, Chicago.
COLQUHOUN, I. C. 1987. Dominance and "fall-fever:" the reproductive behaviour of male brown lemurs (*Lemur fulvus*). Canadian Review of Physical Anthropology, **6**: 10–19.
DE BOER, H., BLOK, G. J., VOERMAN, H. J., P. M. DE VRIES, AND E. A. VAN DER VEEN. 1992. Body composition in adult growth hormone-deficient men, assessed by anthropometry and bioimpedance analysis. Journal of Clinical Endocrinology and Metabolism, **75**: 833–837.
DE WAAL, F. B. M., AND L. M. LUTTRELL. 1985. The formal hierarchy of rhesus monkeys: an investigation of the bared-teeth display. American Journal of Primatology, **9**: 73–85.
ETTINGER, A. O., AND J. R. KING. 1980. Time and energy budgets of the Willow Flycatcher (*Empidonax traillii*) during the breeding season. Auk, **97**: 533–546.
EVANS, C. S., AND R. W. GOY. 1968. Social behaviour and reproductive cycles in captive Ringtailed lemurs (*Lemur catta*). Journal of Zoology, London, **156**: 181–197.
FIETZ, J. C. 1999. Monogamy as a rule rather than an exception in nocturnal lemurs: The case of *Cheirogaleus medius*. Ethology, **105**: 259–272.
FERKIN, M. H., E. S. SOROKIN, AND R. E. JOHNSTON. 1996. Self-grooming as a sexually dimorphic communicative behaviour in meadow voles, *Microtus pennsylvanicus*. Animal Behaviour, **51**: 801–810.
FLORANT, G. L. 1998. Lipid metabolism in hibernators: the importance of essential fatty acids. American Zoologist, **38**: 331–340.

GAHR, M., AND E. KOSAR. 1996. Identification, distribution, and developmental changes of a melatonin binding site in the song controls system of the zebra finch. Journal of Comparative Neurology, **367**: 308–318.

GANZHORN, J. U. 1995. Low-level forest disturbance effects on primary production, leaf chemistry, and lemur populations. Ecology, **76**: 2084–2096.

GORMAN, M. R., M. H. FERKIN, R. J. NELSON, AND I. ZUCKER. 1993. Reproductive status influences odor preferences of the meadow vole, *Microtus pennsylvanicus*, in winter day lengths. Canadian Journal of Zoology, **71**: 1748–1754.

GWINNER, E. 1986. Circannual Rhythms. Springer-Verlag, New York.

HEMINGWAY, C. A., AND D. J. OVERDORFF. In press. Sampling effects on estimates of food availability: phenological method, sample size, and species composition. Biotropica.

HERBERS, J. M. 1981. Time resources and laziness in animals. Oecologia, **49**: 252–262.

HRDY, S. 1981. The Woman That Never Evolved. Harvard University Press, Cambridge, MA.

JOLLY, A. 1966. Lemur Behavior: A Madagascar Field Study. The University of Chicago Press, Chicago.

———. 1984. The puzzle of female feeding priority, pp. 197–215. *In* Small, M. F., ed., Female Primates: Studies by Woman Primatologists. Alan R. Liss, New York.

———. 1998. Pair-bonding, female aggression and the evolution of lemur societies. Folia Primatologica, **69**, Supplement **1**: 1–13.

KAPPELER, P. M. 1990. Female dominance in *Lemur catta*: more than just female feeding priority. Folia Primatologica, **55**: 92–95.

———. 1992. Female Dominance in Malagasy Primates. Ph D. thesis. Duke University, Durham, North Carolina.

———. 1993a. Female dominance in primates and other mammals, pp. 143–158. *In* Bateson, P. P. G., P. H. Klopfer, and N. S. Thompson, eds., Perspectives in Ethology, vol. 10. Plenum Press, New York.

———. 1993b. Variation in social structure: the effects of sex and kinship on social interaction in three lemur species. Ethology, **93**: 125–145.

———. 1993c. Reconciliation and post-conflict behaviour in ringtailed lemurs, *Lemur catta*, and redfronted lemurs, *Eulemur fulvus rufus*. Animal Behaviour, **45**: 901–915.

———. 1996. Causes and consequences of life-history variation among strepsirhine primates. The American Naturalist, **148**: 868–891.

———, and E. W. Heymann. 1996. Nonconvergence in the evolution of primate life history and socio-ecology. Biological Journal of the Linnean Society, **59**: 297–326.

KING, J. R., AND M. E. MURPHY. 1985. Periods of nutritional stress in the annual cycles of endotherms: fact or fiction? American Zoologist, **25**: 955–964.

LEUTENEGGER, W. 1979. Evolution of litter size in primates. The American Naturalist, **114**: 525–531.

LIMA, S. L. 1986. Predation risk and unpredictable feeding conditions: determinants of body mass in birds. Ecology, **67**: 377–385.

———, AND L. M. DILL. 1990. Behavioral decisions made under the risk of predation: a review and prospectus. Canadian Journal of Zoology, **68**: 619–640.

LUCAS, J. R., L. J. PETERSON, AND R. L. BOUDINER. 1993. The effects of time constraints and changes in body mass and satiation on the simultaneous expression of caching and diet-choice decisions. Animal Behaviour, **45**: 639–658.

MARIN, P., H. KVIST, G. LINDSTEDT, L. SJOSTROM, AND P. BJORNTORP. 1993. Low concentrations of insulin-like growth factor-I in abdominal obesity. International Journal of Obesity, **17**: 83–89.

MARTIN, R. D. 1990. Primate origins and evolution. Chapman Hall, London.

———, and A. M. MacLarnon. 1988. Comparative quantitative studies of growth and reproduction. Symposia of the Zoological Society of London, **60**: 39–80.

MITTERMEIER, R. A., W. R. KONSTANT, M. E. NICOLL, AND O. LANGRAND. 1992. Lemurs of Madagascar: an action plan for their future conservation 1993–1999. IUCN, Gland.

MOFFATT, C.A., AND R. J. NELSON. 1994. Day length influences proceptive behavior of female prairie voles (*Microtus ochrogaster*). Physiology and Behavior, **55**: 1163–1165.

MORLAND, H. S. 1993. Seasonal behavioral variation and its relationship to thermoregulation in ruffed lemurs (*Varecia variegata variegata*), pp. 193–203. *In* Kappeler, P. M., and J. U. Ganzhorn, ed., Lemur Social Systems and their Ecological Basis. Plenum Press, New York.

MROSOVSKY, N. 1986. Body fat: what is regulated? Physiology and Behavior, **38**: 407–414.

———. 1990. Rheostatis: The Physiology of Change. Oxford University Press, New York.

———, and D. Sherry. 1980. Animal anorexias. Science, **207**: 837–842.

MÜLLER, E. F. 1985. Basal metabolic rates in primates: the possible role of phylogenetic and ecological factors. Comparative Biochemistry and Physiology A, Comparative Physiology, **81**: 707–711.

Nagy T. R., B. A. Gower, and M. H. Stetson. 1994. Photoperiod effects on body mass, body composition, growth hormone, and thyroid hormones in male collared lemmings (*Dicrostonyx groenlandicus*). Canadian Journal of Zoology, **72**: 1726–1734.

———. 1995. Endocrine correlates of seasonal body mass dynamics in the collared lemming (*Dicrostonyx groenlandicus*). American Zoologist, **35**: 246–258.

Overdorff, D. J. 1992. Differential patterns in flower feeding by *Eulemur fulvus rufus* and *Eulemur rubriventer* in Madagascar. American Journal of Primatology, **28**: 191–203.

———. 1993. Similarities, differences, and seasonal patterns in the diets of *Eulemur rubriventer* and *Eulemur fulvus rufus* in the Ranomafana National Park, Madagascar. International Journal of Primatology, **14**: 721–753.

———. 1998. Are *Eulemur* species pair-bonded? Social organization and mating strategies in *Eulemur fulvus rufus* from 1988–1995 in southeast Madagascar. American Journal of Physical Anthropology, **105**: 153–166.

Pereira, M. E. 1993a. Agonistic interaction, dominance relation, and ontogenetic trajectories in ringtailed lemurs, pp. 285–305. *In* Pereira, M. E., and L. A. Fairbanks, eds., Juvenile Primates: Life History, Development, and Behavior. Oxford University Press, New York.

———. 1993b. Seasonal adjustment of growth rate and adult body weight in ringtailed lemurs, pp. 205–221. *In* Kappeler, P. M., and J. U. Ganzhorn, ed., Lemur Social Systems and their Ecological Basis. Plenum Press, New York.

———. 1995. Development and social dominance among group-living primates. American Journal of Primatology, **37**: 143–175.

———, and M. K. Izard. 1989. Lactation and care for unrelated infants in forest-living ringtailed lemurs. American Journal of Primatology, **18**: 101–108.

———, and P. M. Kappeler. 1997. Divergent systems of agonistic behaviour in lemurid primates. Behaviour, **134**: 225–274.

———, and C. A. McGlynn. 1997. Special relationships instead of female dominance for Redfronted lemurs, *Eulemur fulvus rufus*. American Journal of Primatology, **43**: 239–258.

———, and C. M. Pond. 1995. The organization of white adipose tissue in Lemuridae. American Journal of Primatology, **35**: 1–13.

———, and M. L. Weiss. 1991. Female mate choice, male migration, and the threat of infanticide in ringtailed lemurs. Behavioral Ecology and Sociobiology, **28**: 141–152.

———, R. Kaufman, P. M. Kappeler, and D. J. Overdorff. 1990. Female dominance does not characterize all of the Lemuridae. Folia Primatologica, **55**: 96–103.

Perret, M. 1992. Environmental and social determinants of sexual function in the male lesser mouse lemur. Folia Primatologica, **59**: 1–25.

Pollock, J. I. 1979. Female dominance in *Indri indri*. Folia Primatologica, **31**: 143–164.

Ralls, K. 1976. Mammals in which females are larger than males. Quarterly Review of Biology, **51**: 245–276.

Rasamimanana, H., and E. Rafidinarivo. 1993. Feeding behavior of *Lemur catta* females in relation to their physiological state, pp. 123–133. *In* Kappeler, P. M., and J. U. Ganzhorn, ed., Lemur Social Systems and their Ecological Basis. Plenum Press, New York.

Rasmussen, D. T. 1985. A comparative study of breeding seasonality and litter size in eleven taxa of captive lemurs (*Lemur* and *Varecia*). International Journal of Primatology, **6**: 501–517.

Richard, A. F. 1978. Behavioral Variation: Case Study of a Malagasy Lemur. Bucknell University Press, Lewisburg, Pennsylvania.

———. 1987. Malagasy prosimians: female dominance, pp. 25–33. *In* Smuts, B. B., D. L. Cheney, R. M. Seyfarth, R. M. Wrangham, and T. T. Struhsaker, eds., Primate Societies. The University of Chicago Press, Chicago.

———, and R. E. Dewar. 1991. Lemur ecology. Annual Review of Ecology and Systematics, **22**: 145–175.

———, and R. Heimbuch. 1975. An analysis of the social behavior of three groups of *Propithecus verreauxi*, pp. 313–333. *In* Tattersall, I., and R.W. Sussman, eds., Lemur Biology. Plenum Press, New York.

———, and M. E. Nicoll. 1987. Female social dominance and basal metabolism in a Malagasy primate, *Propithecus verreauxi*. American Journal of Primatology, **12**: 309–314.

———, R. E. Dewar, M. Schwartz, and J. Ratsirarson. In review. Weight change and female fertility in wild *Propithecus verreauxi*. Journal of Human Evolution.

Ross, C. 1992. Basal metabolic rate, body weight and diet in primates: an evaluation of the evidence. Folia Primatologica, **58**: 7–23.

Rubenstein, D. I., and R. W. Wrangham, eds. 1986. Ecological Aspects of Social Evolution: Birds and Mammals. Princeton University Press, Princeton, New Jersey.

SAUTHER, M. L. 1992. The effect of reproductive state, social rank and group size on resource use among free-ranging ringtailed lemurs (*Lemur catta*) of Madagascar. Ph. D. thesis, Washington University, St. Louis, Missouri.

———. 1993. Resource competition in wild populations of ringtailed lemurs (*Lemur catta*): Implications for female dominance, pp. 135–152. *In* Kappeler, P. M., and J. U. Ganzhorn, eds., Lemur Social Systems and Their Ecological Basis. Plenum Press, New York.

SCHMID, J., AND J. U. GANZHORN. 1996. Resting metabolic rates of *Lepilemur ruficaudatus*. American Journal of Primatology, **38**: 169–174.

SCHMIDT-NIELSEN, K. 1990. Animal Physiology. Cambridge University Press, New York.

SCHWARTZ, M. W., AND R. J. SEELEY. 1997. The new biology of body weight regulation. Journal of American Dietary Association, **97**: 54–58.

SORG, J.-P., AND U. ROHNER. 1996. Climate and phenology of the dry deciduous forest at Kirindy, pp. 57–80. *In* Ganzhorn, J. U., and J.-P. Sorg, eds., Ecology and economy of a tropical dry forest in Madagascar. Primate Report 46–1, Goettingen.

STANDEN, V., AND R. A. FOLEY, eds. 1989. Comparative Socioecology: The Behavioural Ecology of Humans and other Mammals. Blackwell Scientific Publications, Oxford.

STERCK E. H. M., D. P. WATTS, AND C. P. VAN SCHAIK. 1997. The evolution of female social relationships in nonhuman primates. Behavioral Ecology and Sociobiology, **41**: 291–310.

SUSSMAN, R. 1977. Socialization, social structure, and ecology of two sympatric species of *Lemur*, pp. 515–528. *In* Chevalier-Skolnikoff, S., and F. E. Poirier, eds., Primate Bio-social Development: Biological, Social, and Ecological Determinants. Garland Publishing, New York.

SUTTIE, J.M., AND J. R. WEBSTER. 1995. Extreme seasonal growth in arctic deer: comparisons and control mechanisms. American Zoologist, **35**: 215–221.

TILDEN, C. 1993. Reproductive energetics of prosimian primates. Ph. D. thesis, Duke University, Durham, North Carolina.

———. In review. Fetal growth rate does not reflect maternal reproductive investment during gestation in lower primates. Proceedings of the National Academy of Sciences.

———, and O. Oftedal. 1995. The bioenergetics of reproduction in prosimian primates: Is it related to female dominance?, pp. 119–131. *In* Alterman, L, G. A. Doyle, and M.K. Izard, eds., Creatures of the Dark. Plenum Press, New York.

TOMASI, T.E., AND D. A. MITCHELL. 1996. Temperature and photoperiod effects on thyroid function and metabolism in cotton rats (*Sigmodon hispidus*). Comparative Biochemistry and Physiology, **113A**: 267–274.

TOMASI, T. E., AND A. M. STRIBLING. 1996. Thyroid function in the 13-lined ground squirrel, pp. 263–269. *In* Geiser, F., A. J. Hulbert, and S. C. Nicol, eds., Adaptations to the Cold. New England Press, Armidale, New South Wales.

TOMASI, T., E. C. HELLGREN, AND T. J. TUCKER. 1998. Thyroid hormone concentrations in black bears (*Ursus americanus*): hibernation and pregnancy effects. General Comparative Endocrinology, **109**: 192–199.

VAN HORN, R.N. 1975. Primate breeding season: photoperiod regulation in captive *Lemur catta*. Folia Primatologica, **24**: 203–220.

VAN SCHAIK, C. P. 1989. The ecology of social relationships amongst female primates, pp. 195–218. *In* Standen, V., and R. A. Foley, eds., Comparative Socioecology: The Behavioural Ecology of Humans and other Mammals. Blackwell Scientific Publ., Boston, Mass.

———, and P. M. Kappeler. 1996. The social systems of gregarious lemurs: lack of convergence with Anthropoids due to evolutionary disequilibrium? Ethology, **102**: 915–941.

———. 1997. Infanticide risk and the evolution of male-female association in primates. Proceedings of the Royal Society of London, Series B, **264**: 1687–1694.

VICK, L.G., AND M. E. PEREIRA. 1989. Episodic targeting aggression and the histories of *Lemur* social groups. Behavioral Ecology and Sociobiology, **25**: 3–12.

WALKER, J., A. PITTS, T. ZAFISON, E. PRIDE, H. D. RABENANDRASANA, A. CALESS, AND A. JOLLY. In review. Infant killing and wounding in *Eulemur* and *Lemur*. International Journal of Primatology.

WEBSTER, J. R., I. D. CORSON, R. P. LITTLEJOHN, S. K. STUART, AND J. M. SUTTIE. 1996. Effects of season and nutrition on growth hormone and insulin-like growth factor-I in male red deer. Endocrinology, **137**: 698–704.

WIDMAIER, E. P., J. LONG, B. CADIGAN, S. GURGEL, AND T. H. KUNZ. 1997. Leptin, corticotropin-releasing hormone (CRH), and neuropeptide Y (NPY) in free-ranging pregnant bats. Endocrinology, **7**: 145–150.

WIRTSHAFTER, D., AND J. D. DAVIS. 1977. Set points, settling points, and the control of body weight. Physiology and Behavior, **19**: 75–78.

WOLTHERS, T., T. GROFTNE, N. MOLLER, J. S. CHRISTIANSEN, H. ORSKOV, J. WEEKE, AND J. O. JORGENSEN. 1996. Calorigenic effects of growth hormone: the role of thyroid hormones. Journal of Clinical Endocrinology and Metabolism, **81**: 1416–1419.

WRIGHT, P.C. 1995. Demography and life history of free-ranging *Propithecus diadema edwardsi* in Ranomafana National Park, Madagascar. International Journal of Primatology, **16**: 835–854.

YODER, A., M. CARTMILL, M. RUVOLO, K. SMITH, AND R. VILGALYS. 1996. Ancient single origin for Malagasy primates. Proceedings of the National Academy of Sciences, **93**: 5122–5126.

YOUNG, A.L., A. F. RICHARD, AND L. C. AIELLO. 1990. Female dominance and maternal investment in strepsirhine primates. The American Naturalist, **135**: 473–488.

7

CATHEMERAL ACTIVITY OF RED-FRONTED BROWN LEMURS (*EULEMUR FULVUS RUFUS*) IN THE KIRINDY FOREST/CFPF

Giuseppe Donati,[1] Antonella Lunardini,[1] and Peter M. Kappeler[2]

[1] Dipartimento di Etologia, Ecologia ed Evoluzione, Unità di
Antropologia, via S. Maria 55, I-56126 Pisa, Italy
lunardini@discau.unipi.it (corresponding authors)
[2] Deutsches Primatenzentrum, Abteilung
Verhaltensforschung/Ökologie, Kellnerweg 4, 37077 Göttingen
Germany

ABSTRACT

Cathemeral activity, which is defined by sequences of activity bouts and resting phases throughout the 24 hour cycle, is rare among primates but common among group-living lemurs. The proximate and ultimate causes and mechanisms underlying this activity pattern are still obscure. One group of Red-fronted Brown Lemurs (*Eulemur fulvus rufus*) was therefore observed in Kirindy Forest/CFPF for a total of 384 hours between March and June 1996 to investigate potential causes and correlates of their activity. Observations were equally distributed between diurnal and nocturnal activity cycles and intensified around periods of full and new moon. We found that Kirindy Red-fronted Brown Lemurs exhibited cathemeral activity throughout the study period and documented a significant increase of nocturnal and a concomitant decrease of diurnal activity during the transition between the wet and dry season (i.e., in April–May). Intensity and duration of activity were dependent upon lunar phase, due to a significant increase in activity during the nights of full moon. Furthermore, a balance between diurnal and nocturnal activity levels was observed. The animals moved on average 989 m during the day and 749 m at night within their 20.5 ha home-range. Distance traveled at night increased significantly during the dry season and nocturnal activity was also negatively correlated with minimum temperature. Red-fronted Brown Lemurs and their feeding patches were located significantly higher above ground at night. Considering that during the dry season less cover is provided by vegetation, these results suggest that the animals avoided diurnal exposure in the canopy and compensated for it with an increase of nocturnal foraging, especially during a full moon. Minimizing predation risks and thermoregulation benefits may therefore be among the main determinants of this behavioral strategy.

RÉSUMÉ

L'activité cathémérale, définie par une alternance de périodes d'activités et de phases de repos sur un cycle de 24 heures, est rare chez les primates mais commune chez les lémuriens vivant en groupes alors que les raisons, finalités et mécanismes sousjacents du schéma de cette activité sont toujours incompris. Un groupe de Lémurs à front roux (*Eulemur fulvus rufus*) a ainsi été observé dans la Forêt de Kirindy/CFPF pendant 384 heures au total entre mars et juin 1996 pour rechercher des causes et conséquences potentielles de la nature de leur activité. Les observations ont été équitablement réalisées entre les cycles d'activités diurnes et nocturnes mais ont été renforcées aux alentours des périodes de pleine et de nouvelle lunes. Nous avons trouvé que le Lémur à front roux de Kirindy présente une activité cathémérale tout au long de la période d'étude et montré une augmentation significative de l'activité nocturne avec son corollaire, la diminution de l'activité diurne, au cours de la transition entre les saisons sùche et humide (i.e. en avril-mai). L'intensité et la durée de l'activité dépendaient des phases lunaires avec une augmentation significative de l'activité au cours des nuits de pleine lune en même temps qu'un équilibre entre les niveaux d'activités diurne et nocturne étaient observé. Les animaux se déplaçaient sur 989 m en moyenne le jour et sur 749 m en moyenne la nuit au sein de leur domaine vital d'une surface de 20,5 ha. Les distances parcourues la nuit augmentaient sensiblement au cours de la saison sùche et l'activité nocturne était en corrélation négative avec la température minimum. Pendant la nuit, les Lémurs à front roux et leurs zones de nourrissage se trouvaient à des niveaux plus élevés au-dessus du sol. En considérant que la végétation procure moins de couvert pendant la saison sùche, ces résultats suggùrent que les animaux évitaient de s'exposer le jour dans la canopée et compensaient en augmentant leur activité de nourrissage nocturne, en particulier au cours de la pleine lune. Minimiser les risques de la prédation et rechercher les avantages de thermorégulation pourraient ainsi être les principaux déterminants de cette stratégie comportementale.

INTRODUCTION

Cathemerality has been defined by Tattersall (1987) as an activity pattern in which significant amounts of activity occur during the two phases of the day-night cycle. Since Conley's (1975) description of such an activity pattern in captive *Eulemur fulvus albifrons*, cathemerality has been described for many other group-living Lemuridae (Tattersall & Sussman, 1975; Sussman & Tattersall, 1976; Tattersall, 1977, 1979; Overdorff, 1988; Wilson et al., 1989; Colquhoun, 1993, 1998; Santini-Palka, 1994; Overdorff & Rasmussen, 1995; Curtis, 1997; Andrews & Birkinshaw, 1998; Birkinshaw & Colquhoun, 1998; Rasmussen, 1998a,b) and two anthropoids: the Night Monkey, *Aotus trivirgatus* (Wright, 1989) and the Howler Monkey, *Alouatta palliata* (Dahl & Hemingway, 1988).

The complex of biotic and abiotic determinants of this unusual behavioral phenomenon are still poorly understood (van Schaik & Kappeler, 1996). As in other mammals, several ecological and physiological pressures, such as temporal distribution of food resources and diet, ambient temperature, thermoregulation, interspecific competition, and predation risk have been suggested as possible determinants of this activity pattern. We begin by discussing these factors in more detail.

First, *Eulemur mongoz* and *E. rubriventer* have been observed feeding at night on resources available only during the hours of darkness, such as *Ceiba pentandra* and *Eucalyptus* flowers (Sussman & Tattersall, 1976; Overdorff, 1988). However, most cathemeral primates were observed to visit and feed on the same food items during both day and night (Tattersall, 1979, 1982; Overdorff & Rasmussen, 1995; Birkinshaw & Colquhoun, 1998; Colquhoun, 1998), so that a possible effect of resource availability on activity is not clear.

Second, seasonal variation in food availability may result in changes in lemur diet (Overdorff, 1993) and consequently, as Engqvist and Richard (1991) proposed, a shift in activity pattern. In particular, small-bodied primates without digestive specializations should cope with a seasonal increase of fibrous food items by minimizing the time in which no food is being processed, i.e., become cathemeral. Field observations and tests of food passage rate in *Eulemur* species seemed to confirm some dietary influence on activity (Tattersall, 1979; Overdorff, 1988; Colquhoun, 1993). However, more recent long-term data on year-round dietary composition in *E. macaco* (Colquhoun, 1998), *E. fulvus rufus*, *E. rubriventer* (Overdorff & Rasmussen, 1995), and *E. mongoz* (Curtis, 1997; Rasmussen, 1998a) revealed no clear relationship with the organization of circadian activity. We will investigate this factor by focussing on the transitional period between the wet and dry season.

Third, recent studies revealed a significant negative correlation between nocturnal activity levels and ambient temperatures in *Eulemur fulvus* and *E. mongoz* (Overdorff & Rasmussen, 1995; Curtis, 1997). Even though these results contrast with those of Tattersall's (1976) study of *E. mongoz*, which indicated a decrease of activity levels during the coldest hours of the day, they suggest that cathemerality may represent an additional means of behavioral thermoregulation, allowing animals to conserve energy during the coldest hours of the night by increasing nocturnal activity (see Curtis, 1997). This physiological explanation is strengthened by documentations of low basal metabolic rates and thermoregulatory flexibility (Daniels, 1984; Müller, 1985; Richard & Nicoll, 1987; Morland, 1993; Schmid, 1996; Schmid & Ganzhorn, 1996; Curtis, 1997; Drack et al., this volume). We will also examine variation in activity in relation to ambient temperature to examine this factor in more detail.

Fourth, as observed in other mammals, cathemeral activity may by advantageous by reducing competition for food resources. This explanation may apply to *Hapalemur simus* and *H. griseus*, for which species-specific 24h-activity patterns and feeding strategies may allow niche separation where they are sympatric (Santini-Palka, 1994). Two different kinds of cathemerality were also observed at Ampijoroa in sympatric *Eulemur fulvus* and *E. mongoz* (Rasmussen, 1998b). Testing predictions of this hypothesis by considering the activity of other sympatric species is beyond the scope of the present study.

Fifth, predation risk is one of the most important ecological factors influencing primate activity patterns (van Schaik & van Hooff, 1983; Isbell, 1994). Nature and extent of predation risk have been documented for several lemur species (Sauther, 1989; Goodman et al., 1993; Macedonia, 1993). Support for the hypothesis that cathemerality may represent an efficient antipredator strategy is provided by studies of several cathemeral primates. For example, *Eulemur rubriventer* was seen to visit the exposed branches of *Eucalyptus* trees only during nocturnal activity (Overdorff, 1988), *E. mongoz* increased its nocturnal activity when least cover was provided by vegetation (Curtis, 1997; Rasmussen, 1998a) and the normally nocturnal *Aotus trivirgatus* exhibited diurnal bouts of activity in areas where large diurnal raptors are absent (Wright, 1983). We will examine several predictions for habitat use to assess the importance of predation risk for activity.

Finally, a non-adaptive hypothesis on the evolutionary origin of cathemerality has recently been suggested. It proposes that cathemerality represents a transitional stage between nocturnality and diurnality (van Schaik & Kappeler, 1996). Among lemurs, massive changes of community structure in connection with the Holocene extinctions of one third of then extant lemurs and several large predators, may have enabled previously nocturnal species to expand their activities into the daylight (Martin, 1972; Tattersall, 1979; van Schaik & Kappeler, 1996). Others considered cathemerality as a stable activity pattern, probably ancestral to the genus *Eulemur* (Tattersall, 1982; but see also Erkert, 1989; Kappeler & Heymann, 1996). These hypotheses are inherently difficult to test directly, however.

The aim of this study was to describe the pattern of activity of a wild group of Red-fronted Brown Lemurs for the first time throughout the entire 24 hours cycle in a strongly seasonal habitat in the deciduous dry forest in western Madagascar. It focused on the influence of different ecological parameters, such as climate, moon-phase, ranging pattern, diet, time spent feeding, and temporal use of the forest levels. Particular attention was devoted to the transitional period between the pronounced wet and the dry seasons. Results from such a preliminary study may help to improve our understanding of the flexibility of activity and its potential correlates and constraints.

METHODS

Study Site

This study was conducted at the field station of the German Primate Center (DPZ) in Kirindy Forest/CFPF, a deciduous dry forest located at 44°39′E and 20°03′S (Sorg & Rohner, 1996), about 60 km north of Morondava. This forest is characterized by a wet and hot season, from November-December to March-April, and a dry and cold season, from April-May to October-November, with mean temperatures during the year ranging from a maximum of 30.7°C to a minimum of 19°C (see also Schmid & Kappeler, 1998). Annual rainfall is concentrated during the brief wet season, with a mean value of 767 mm per year. During the study period daily minimum and maximum temperature, as well as humidity were recorded. In addition to *Eulemur fulvus rufus*, seven sympatric lemur species are found in the area (Ganzhorn & Kappeler, 1996), all of which are nocturnal, except for *Propithecus verreauxi verreauxi* (Carrai & Lunardini, 1996).

Behavioral Observations

One group of Red-fronted Brown Lemurs was followed four days and four nights per month from dawn to dusk (10–12 h) and from dusk to dawn (12–13 h) between March and June 1996. Data were pooled for 24 h periods or per day and night time.

One day, night, or 24 h period was considered one unit for statistical analyses. Almost 400 h of observations, equally distributed between the two phases of the day-night cycle, were collected on the main study group consisting of four males and two females. Individuals were captured and subsequently marked with unique collars and pendants just before the onset of these observations. Habituation of the study group posed no problem; after two weeks, it was possible to approach the animals within a few meters. Particular attention was paid to concentrate observations around periods of new and full moon for each month in order to examine potential differences in light-related activity.

The activity of animals was recorded using scan sampling (Altmann, 1974) on each member of the group at five minute intervals. During each scan bout a maximum of 30 seconds were used to record data on activity and habitat use (Overdorff, 1996). The data collected with this method included activity (foraging, resting, huddling, climbing, locomoting, playing, auto- and allogrooming, and other social activities), type of food consumed, proximity to other group members, height in the forest, location of animals and location of feeding and resting sites within the study area locally known as CS7. Despite the use of a head-lamp and infra-red binocular glasses to enhance luminosity, it was difficult to recognize individuals and distinguish their behavior during nocturnal observations. Noises invariably associated with the animal's activity facilitated determination of their location and activity, however. Nevertheless, sufficient details could be observed to classify the main behavioral units and to quantify general activity.

Using these data it was possible to quantify different aspects of the lemurs' diet, such as proportion of feeding observations accounted for by each food item (ripe and unripe fruits, mature and new leaves, flowers, invertebrates and others). All feeding trees used for more than 5 seconds (i.e., feeding patches) were marked during observation sessions and identified with the help of a field assistant on a subsequent day. Moreover, the height, diameter at breast height, crown height, crown diameter, and shape of each tree were measured or estimated. More than 500 patches were marked with this method and 34 different plant species used by these lemurs were recognized.

Ranging and Habitat Use

The study area is equipped with a system of trails, creating a grid system with 25 m × 25 m quadrats. The grid quadrat and the position of feeding or resting patches (resting patches were defined as the places where the animals rested for more than 10 minutes) occupied by each animal were recorded every 5 minutes. Daily and nightly path lengths (DPL and NPL, respectively) were then calculated by connecting patches successively visited during each diurnal or nocturnal sampling. These values probably represent underestimates because we assumed linear distances between patches. As the group was fairly cohesive and each member visited most patches used by others, inter-individual variation was negligible.

Home-range size was calculated as a minimum convex polygon, using the spatial information mapped as daily and nightly paths. The home-range area was divided in 50 m × 50 m quadrats in order to analyze possible effects of time of day on home-range use. The core area consisted of all quadrats used by the group during more than 2% of the observations (cf. Overdorff, 1993). Height data were recorded for the following five categories: 0 m (indicating ground level), between 0 and 2 m (shrub layer), between 2 and 6 m (lower canopy), between 7 and 10 m (upper canopy), and over 10 m (emergent trees). Moreover, 7 m were used to distinguish between "high" and "low" feeding patches.

Statistical Analyses

The Mann-Whitney U-test was used to examine differences in daily activity between males and females. Two methods were used for the seasonal analysis of activity. First, descriptive statistics were employed to visualize activity profiles during the day-night cycle for each month. For this, the scan samples with activity of each group member were determined from the total amount of scan samples for each hour interval. Second, the total activity of the group during each diurnal or nocturnal observation session was compared with a Wilcoxon Sign-ranked test between the two 2-month blocks, March-April and May-June, to investigate seasonal differences in activity between the end of the wet season and the first part of the dry season. This monthly separation was introduced to allow seasonal comparisons. It was largely based on phenological and climatological data (Sorg & Rohner, 1996). A Wilcoxon Sign-ranked test was also used to compare daily and nightly activity between the two opposite lunar phases, the seasonal differences in diet and habitat use, as well as the monthly differences in the use of feeding patches between diurnal and nocturnal phases. Finally, Spearman's rank correlation coefficient was used to test, on a daily basis, relationships between the extent of nocturnal activity and diurnal activity, DPL, NPL, diurnal foraging, diurnal time spent foraging on leaves and flowers, minimum and maximum temperatures, humidity and day-length.

RESULTS

Patterns of Cathemeral Activity

The Red-fronted Brown Lemurs at Kirindy exhibited cathemeral activity during the entire study period. There were no significant differences between the activity of females and males (Mann-Whitney U-test: $U = 92$, $n_1 = 16$, $n_2 = 16$, n.s.). Figure 1 shows the monthly 24 hours activity pattern. A general trend for an increase of nocturnal and a decrease of diurnal activity emerged from the comparison between the activity profiles of March-April and May-June (Wilcoxon Sign-ranked test diurnal: $T = 1$, $n = 8$, $p < 0.01$; nocturnal: $T = 0$, $n = 8$, $p < 0.005$). Three main peaks of activity were recorded at the end of the wet season (March-April): the first one around sunrise (0500/0900 h), the second around sunset (1600/1800 h) and a third one between 2100 h and 0200 h. A short period of reduced diurnal activity (1000/1300 h) was also evident during these two months. The beginning of the dry season (May-June) revealed a significant increase in nocturnal activity associated with a drop of activity during the middle of the day. Despite the activity changes observed in the two phases of the day-night cycle the overall percentage of activity during the 24 hours did not vary significantly among the two 2- month blocks (Wilcoxon Sign-ranked test: $T = 7$, $n = 8$, n.s.).

Figure 2 illustrates the relation between activity and lunar phases. A significant increase of nocturnal activity during full versus new moon nights was noted (Wilcoxon Sign-ranked test: $T = 0$, $n = 7$, $p < 0.01$). However, the overall activity during the 24 hour

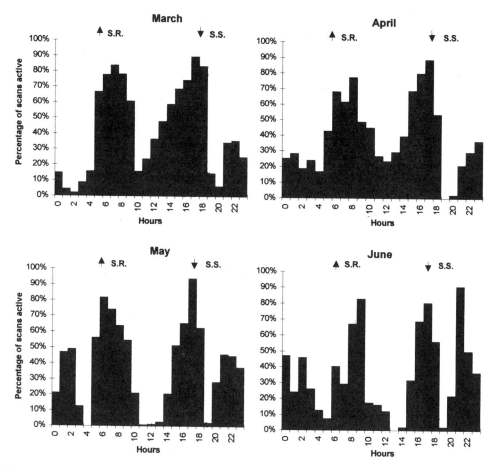

Figure 1. Patterns of cathemeral activity cycle of Red-fronted Brown Lemurs from March to June: bars represent the percentage of scans during which the animals were reported active throughout the 24 hours.

cycle did not vary between the two opposite moon phases (Wilcoxon Sign-ranked test: T = 5, n = 7, n.s.).

The associations between daily percentage of nocturnal activity and several variables of interest are summarized in Table 1. The balance between diurnal and nocturnal activity was confirmed by a significant negative correlation between nocturnal activity and both diurnal activity and diurnal foraging duration. At night the animals traveled more when they were more active, but both the distance traveled on the previous day and the percentage of fibrous food in the diet was not correlated with nightly activity. In addition, the animals were significantly more active during colder nights.

Ranging Patterns

During the study period the Red-fronted Brown Lemurs used a 20.5 ha home range (Fig. 3). The extent of the 24 hours core-area covered 20.7% of the total home range and the animals spent 57.8% of time in this area. The percentage of daily time spent within the diurnal core area (4.3 ha) was 49.4%, while about 80% of nightly time was spent within the nocturnal core area (4.0 ha).

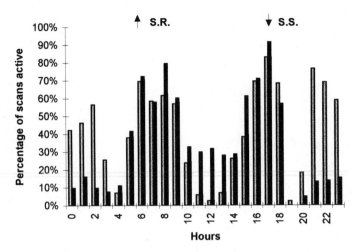

Figure 2. Patterns of cathemeral activity cycle of Red-fronted Brown Lemurs during the phases of new (in black) and full moon (in gray): bars rapresent the percentage of scans during which the animals were recorded active throughout the 24 hours.

Figure 4 shows variation in size between diurnal and nocturnal areas visited during each observation month. A trend of decreasing diurnal home ranges (from 15.8 ha in March to 10.3 ha in June) could be noted, together with a significant increase of the areas visited during the night (from 4.8 ha in March to 11.8 ha in June).

Daily path length did not vary significantly between the two 2-month periods (Wilcoxon Sign-ranked test: T = 6, n = 8, n.s.), while nightly path length showed a significant increase between April and May (Wilcoxon Sign-ranked test: T = 0, n = 8, $p < 0.005$). In May and June the distances traveled at night were almost as long as those traveled during the day (Table 2).

Habitat Use

Figure 5 depicts the percentages of time spent within the five main forest layers during diurnal and nocturnal activity. The animals remained in canopy layers most of the time during the 24 hours, but the low canopy was preferred during the day and the

Table 1. Spearman rank correlation coefficients between nocturnal activity of Red-fronted Brown Lemurs and potential ecological variables recorded during the study period; n = 16

	% of nocturnal activity
% of diurnal activity	−0.51*
Nightly path length	0.88**
% diurnal foraging	−0.44*
Daily path length	−0.12
% leaves and flowers in the diet	0.18
Min. temperature	0.66**
Max. temperature	−0.04
Rainfall	−0.39
Day-length	−0.24

*$p < 0.05$, **$p < 0.01$.

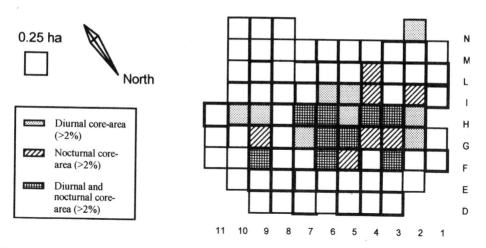

Figure 3. Home range used by Red-fronted Brown Lemurs during the entire study period. The core-area was considered to consist of all the quadrats covered in more than 2% of observations during the diurnal, the nocturnal, and the total cathemeral activity.

high canopy at night. Moreover, sporadic use of ground level and of the emergent layer showed opposite patterns during the day and night.

A dimensional analysis of the feeding patches used by our study animals is shown in Fig. 6a. A significant increase in the percentage of time spent foraging on high feeding patches between March-April and May-June was observed (Wilcoxon Sign-ranked test: $T = 0$, $n = 8$, $p < 0.005$). Moreover, high feeding patches were visited significantly more frequently at night than during the day throughout the study period (Fig. 6b, Wilcoxon Sign-ranked test: $T = 3$, $n = 9$, $p < 0.01$). Conversely, low feeding patches were visited significantly more frequently during daylight hours (Wilcoxon Sign-ranked test: $T = 31$, $n = 25$, $p < 0.005$).

Diet

Red-fronted Brown Lemurs ate food items from a total of 34 plant species. Three plant species made up more than 70% of the diet. During each observation month, one or two plant species were frequented for more than 50% of the foraging time (Table 3). Figure 7 shows the relative proportion of time spent by lemurs foraging on various

Table 2. Comparison between mean Daily Path Length (DPL; $n = 4$) and mean Nightly Path Length (NPL; $n = 4$) traveled by Red-fronted Brown Lemurs during each study month. Values are means ± standard deviations (S.D.)

	DPL	NPL
	Mean ± S. D. (m)	Mean ± S. D. (m)
March	1051 ± 108	476 ± 223
April	990 ± 69	528 ± 345
May	1003 ± 236	999 ± 444
June	914 ± 229	995 ± 380
Entire Study Period	989 ± 187	749 ± 435

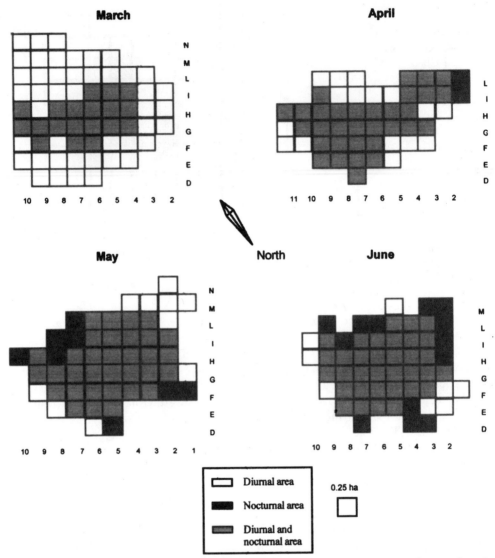

Figure 4. Comparison between diurnal and nocturnal areas used by Red-fronted Brown Lemurs throughout each observation month.

food items. The animals were mainly frugivorous, because on average 91.5% of the time spent feeding was devoted to fruit consumption. However, a significant increase in leaf intake was observed from April (3.5%) to May (12.2%) (Wilcoxon Sign-ranked test: $T = 0, n = 8, p < 0.005$) with a concomitant decrease in fruit consumption (from 95.8% to 87.8%). Throughout both seasons, ripe fruits were eaten more frequently than unripe ones, and mature leaves were eaten more often than new ones. Stems and bark were always eaten rarely.

The average percentage of time devoted to foraging over 24 hours was 21.3%, and did not vary significantly between seasons (Wilcoxon Sign-ranked test: $T = 7, n = 8$, n.s.). However, a significant decrease in diurnal and a significant increase in noctur-

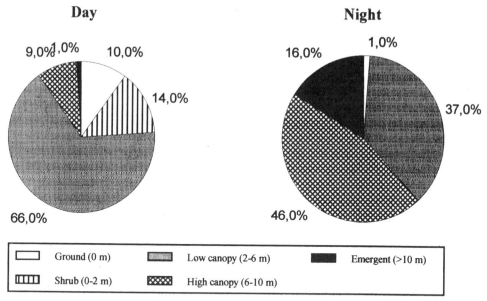

Figure 5. Percentage of scans spent by Red-fronted Brown Lemurs within the various height levels of the Kirindy Forest during the day and at night.

nal foraging times from March-April to May-June was observed (Wilcoxon Sign-ranked test diurnal: T = 1, n = 8, p < 0.01; nocturnal: T = 0, n = 8, p < 0.005).

DISCUSSION

This study of the activity and ecology of Red-fronted Brown Lemurs permits a preliminary assessment of potential mechanisms and functions of cathemerality. These lemurs were cathemeral during the entire transitional period between wet and dry season, but increased nocturnal activity and decreased diurnal activity as the dry season progressed. A more fine-grained analysis revealed negative correlations between diurnal and nocturnal activity and between percentage of nocturnal activity and diurnal time spent foraging, which may indicate an activity balance across the two parts of the day. These relationships also suggest that nocturnal activity may be regulated in response to previous diurnal activity, leading to behavioral compensation, both on a daily and on a monthly basis. A similar pattern has been reported for *Eulemur macaco* (Colquhoun, 1993), *E. mongoz* (Curtis, 1997), and for *E. rubriventer* (Overdorff & Rasmussen, 1995).

Ranging behavior of Red-fronted Brown Lemurs at Kirindy Forest was broadly similar to that of their conspecifics at Ranomafana National Park (Meyers, 1988; Overdorff, 1993). However, in the dry forest distances traveled at night were longer, distances traveled during the day shorter, and home ranges smaller than in the rain forest (cf. Sussman, 1975). Because ranging behavior is closely correlated with activity, day ranges decreased and night ranges increased with the transition to the dry season. At the end of the wet season the animals traveled almost exclusively during daylight hours. In contrast, in May and June ranges and distances covered at night were at least as large as those during the day.

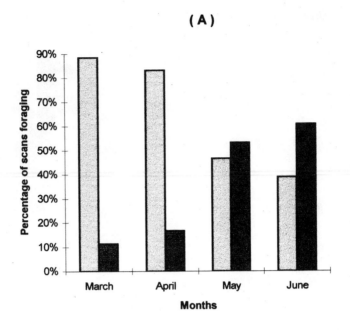

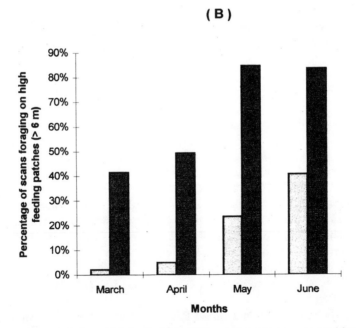

Figure 6. Percentage of scans during which the Red-fronted Brown Lemurs were reported foraging on high (<6 m, in darkish gray) and low (≤6 m, in light gray) feeding patches during the 24 hours (A) and percentage of scans spent exclusively on high feeding patches during the day (in light gray) and at night (in darkish gray) (B).

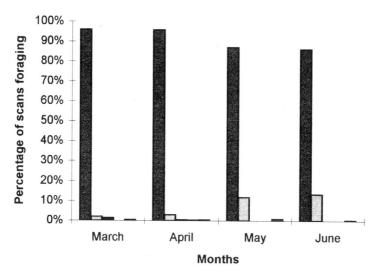

Figure 7. Percentage of scans during which the Red-fronted Brown Lemurs were reported foraging in various food category from March to June. Darkish gray bars = fruits, light gray bars = leaves, black bars = invertebrates, darkish white bars = flowers, white bars = miscellaneous.

Nocturnal activity of our animals increased during periods of full moon, suggesting that it partly depends on ambient light levels. As has been noted for captive and wild *Eulemur fulvus* ssp. (Erkert, 1989), wild *E. mongoz* (Harrington, 1978; Curtis, 1997), *E. macaco* (Colquhoun, 1993, 1998), and bushbabies (Nash, 1986), nocturnal activity may require a certain (moon) light threshold. On several nights, we noted that the onset of nocturnal activity was associated with the rise of the moon over the horizon. Furthermore, a sudden interruption of group activity, shortly after the beginning of a lunar eclipse occurred during one observation night with full moon.

The increase of nocturnal activity at the beginning of the dry season may be linked to the loss of foliage and the consequent increase of nocturnal light levels. In Kirindy Forest, for example, 44.6% of tree species shed their leaves in April and May (Sorg & Rohner, 1996). The lack of correlation between lunar phases and nocturnal activity recorded for two cathemeral *Eulemur* species in the eastern rain forest (Overdorff & Rasmussen, 1995) could be due to the covering effect of higher evergreen layers (Pariente, 1980; Curtis, 1997). However, our study animals were regularly active during periods of new moon, suggesting that light levels are not an absolute constraint on the nocturnal activity.

The results of this study provide no support for the hypothesis that cathemeral activity is controlled by fiber intake (Engqvist & Richard, 1991). Accordingly, species with a highly fibrous diet and a lack of gut specialization may need to extend their foraging time into the nocturnal phase in order to minimize the periods in which no food is being processed. In our study, nocturnal activity was not correlated with the amount of time spent foraging on leaves the previous day. In fact, a slight increase in leaf consumption was observed during the transition from the wet to the dry season. Because the animals were essentially frugivorous throughout the entire study period, however, it is even more unlikely that the observed changes in activity were related to the small amount of leaves in their diet. Moreover, these lemurs did not increase daily

Table 3. Vernacular and scientific names, mean height (M. H.), part eaten (P. E.; FR = fruit, LE = leaves) and percentages of diurnal (D) and nocturnal (N) foraging scans spent by Redfronted Brown Lemurs on the most visited (>2%) botanical species during each study month (scientific names from Sorg, 1996)

Vernacular	Genus and species	M. H. ± S. D. (m)	n	P. E.	% D	% N
March						
Amaninomby	*Terminalia boivinii*	5.1 ± 1.7	39	FR	81.5	51.3
Valotsy	*Breonia perrieri*	10.4 ± 2.1	13	FR	0.5	28.8
Tsivonino	*Noronhia* sp.	3.6 ± 1.4	18	FR, LE	6.2	1.8
Sakoambanditsy	*Poupartia sylvatica*	10.5 ± 2.8	4	FR	1.6	5.4
Maronono	?	4.5 ± 1.7	6	FR	1.8	3.6
April						
Amaninomby	*Terminalia boivinii*	4.7 ± 1.3	43	FR	72.8	4.6
Valotsy	*Breonia perrieri*	9.2 ± 2.7	9	FR	2.6	32.8
Fony	*Adansonia rubrostipa*	12.7 ± 2.5	10	FR	0.9	8.6
Taolankena	*Tarenna* sp.	2.9 ± 0.9	13	FR	3.7	0.0
Sely	*Grewia* sp.	4.5 ± 1.8	9	FR	3.8	0.0
Fiamy	*Ficus* sp.	14.0	1	FR	1.0	6.2
Tsivonino	*Noronhia* sp.	3.6 ± 1.4	4	FR, LE	3.2	0.0
May						
Latabarika	*Grewia cyclea*	7.9 ± 2.1	29	FR	16.7	51.2
Sely	*Grewia* sp.	4.2 ± 1.8	31	FR	36.6	6.2
Fiamy	*Ficus* sp.	14.0	1	FR	3.2	21.0
Fony	*Adansonia rubrostipa*	11.9 ± 3.0	14	FR	0.4	9.7
Taolankena	*Tarenna* sp.	3.4 ± 1.0	11	FR	8.2	0.0
Manary	*Dalbergia* sp.	5.1 ± 1.9	6	LE	3.7	4.4
Tikatikakala	?	4.1 ± 1.5	10	FR, LE	4.5	1.2
Mabiboanala	cf. *Anacardium*	5.7 ± 0.8	5	FR	5.5	0.0
Amaninomby	*Terminalia boivinii*	5.3 ± 1.9	2	FR	4.0	0.0
June						
Latabarika	*Grewia cyclea*	8.3 ± 2.5	41	FR	39.8	76.5
Sely	*Grewia* sp.	4.7 ± 1.3	36	FR	36.0	11.3
Tikatikatala	?	4.2 ± 1.5	11	FR, LE	7.5	1.2
Kily	*Tamarindus indica*	8.9 ± 2.3	3	FR, LE	1.0	5.6
Karimbolazo	?	3.6 ± 1.0	8	LE	4.8	1.4
Manary	*Dalbergia* sp.	4.6 ± 1.5	4	LE	2.9	1.4

foraging time with increasing leaf intake, but simply extended feeding bouts into the night.

Evidence for a dietary influence on nocturnal activity in other studies is equivocal. A correlation between nocturnal activity and unripe fruit consumption was described in *Eulemur fulvus rufus*, but not in sympatric *E. rubriventer* (Overdorff & Rasmussen, 1995). Activity and leaf intake were also uncorrelated in *E. macaco* (Colquhoun, 1998). Furthermore, in seasonally cathemeral *E. mongoz*, crude fiber intake was positively correlated with diurnal activity (Curtis, 1997). Finally, *Hapalemur* species are cathemeral (Santini-Palka, 1994; R. Martin, pers. comm.) in spite of their folivorous gut specialization. Thus, dietary habits of these and other lemurs seem to have little effect on cathemeral activity.

The combination of low basal metabolic rate and high body temperature characteristic of *Eulemur* species (Daniels, 1984; Müller, 1985), led Curtis (1997) to suggest that cathemerality may be a mechanism of behavioral thermoregulation. This hypothesis predicts that cathemeral lemurs remain active and thereby maintain their energy supply during the coldest part of the 24 h day. We found a positive correlation between nocturnal activity and minimum temperatures. Such a relationship was also described for *E. fulvus rufus* at Ranomafana (Overdorff & Rasmussen, 1995) and *E. mongoz* at Anjamena (Curtis, 1997). It should be noted, however, that the increase of nocturnal activity observed in our study took place during the first, warmer part of the night. Thus, more fine-grained data are necessary to further evaluate this hypothesis.

Finally, cathemerality in lemurs has also been related to antipredator behavior. Birds of prey may pose a particular risk for arboreal lemurs (e.g., Sauther, 1989; Langrand, 1990; Goodman et al., 1993), as indicated by their strong response towards raptors, which is difficult to explain without reinforcement (Csermely, 1996). Red-fronted Brown Lemurs at Kirindy Forest reacted to *Polyboroides radiatus*, *Buteo brachypterus* and various *Accipiter* species flying overhead with alarm calls and by moving quickly to lower forest layers, where they remained motionless.

Our analysis of habitat use suggested that cathemerality partly reflects an antipredatory strategy. For example, we noted a marked preference, particularly during feeding, for the upper canopy during nocturnal activity. Large trees such as *Ficus* sp. and *Breonia perrieri* were rarely visited during daylight hours, but were among the preferred nocturnal feeding patches. Also, at the end of the wet season the most frequently visited feeding patches comprised smaller trees, such as *Terminalia boivinii*, while at the beginning of the dry season the larger *Grewia cyclea* trees were preferred. In combination with the reduction of foliage, this switch ought to increase the exposure of lemurs during their diurnal feeding bouts. The exposure avoidance observed during the day and the increase of nocturnal foraging on higher feeding patches may reduce this risk, however. In addition, the frequency with which our Red-fronted Brown Lemurs joined groups of other conspecifics or sympatric Sifakas (*Propithecus v. verreauxi*) during diurnal foraging seemed to increase as the dry season progressed. This behavior should also decrease their predation risk.

Other cathemeral lemurs were also found to be more nocturnal when foraging on exposed patches (Overdorff, 1988), or when less cover is provided by vegetation (Curtis, 1997; Rasmussen, 1998a). In further support of this idea, in the eastern rain forest with less marked seasonal changes in leaf cover, cathemeral lemurs showed no seasonal variation in activity (Overdorff & Rasmussen, 1995). Moreover, *Hapalemur griseus* is mainly diurnal there (Wright, 1986; Grassi, 1997; Overdorff et al., 1997).

If cathemeral activity is partly a response to predation risk, it should also reflect predation risk at night. The Fossa, *Cryptoprocta ferox*, with its cathemeral activity and supreme climbing abilities presents the largest threat for lemurs at night (Goodman et al., 1993). By being also cathemeral, lemurs may have a greater chance at evading *Cryptoprocta* at night than strictly diurnal ones. At Kirindy, the diurnal *Propithecus verreauxi*, fall indeed prey to the *Cryptoprocta* more frequently than Red-fronted Brown Lemurs (Rasoloarison et al. 1995; C. Hawkins, pers. comm.). It is not clear whether this a cause or consequence of their activity pattern, however.

Determining causalities is further complicated by the fact that activity and predation risk are related to body size. For raptors, in particular, prey size clearly limits their prey spectrum. For example, in South America five different raptors are suspected to prey on the 800 g *Saimiri sciureus*, whereas only two of these prey on the 3 kg *Cebus apella* (Isbell, 1994). Among lemurs, cathemerality has been described for *Hapalemur* spp. and *Eulemur* spp., which are in size between some of the largest nocturnal species (i.e., *Lepilemur* sp.) and the smallest diurnal ones (*Lemur catta* and *Propithecus tatter-*

salli; Kappeler & Ganzhorn, 1993). Even though these limits are not clear-cut, because the nocturnal *Avahi* (but see Warren, 1994) and *Daubentonia* (Sterling, 1993) also fall in this size range, it has been suggested that these medium-sized lemurs retained some nocturnal activity as a response to the presence of diurnal raptors, which must have imposed an even great risk for them a few millennia ago (Goodman et al., 1993; Goodman, 1994a,b; van Schaik & Kappeler, 1996). Additional ecological field studies, as well as physiological studies of visual systems, may help to illuminate this aspect of lemur natural and evolutionary history eventually.

In conclusion, our preliminary study revealed interesting variation in activity of Red-fronted Brown Lemurs. Despite the limited nature of our data base, some potential determinants of cathemeral activity could be identified as unlikely or unimportant. Predation risk, thermoregulation and light availability seem to be more important in controlling this aspect of their behavior. Because of the small number and often preliminary nature of field studies of cathemeral activity, documenting variation in activity of many different taxa at different sites for entire annual cycles and relating it to ecological variables (cf. Curtis, 1997) remains an important goal for future lemur field studies.

ACKNOWLEDGMENTS

This work was carried out under the Accord de Collaboration between the Laboratoire de Primatologie et des Vertébrés de l'Université d'Antananarivo, the Abteilung Verhaltensphysiologie, Universität Tübingen, and the Deutsches Primatenzentrum. The authors thank the Ministùre des Affaires Etrangùres, the Direction des Eaux et Forêts, the Ministùre de l'Enseignement Supérieur, and especially B. Rakotosamimanana, for authorizing and supporting this research; the Centre de Formation Professionnelle Forestiùre in Morondava for their hospitality, S. Borgognini Tarli for her helpful criticisms of an earlier draft of this manuscript and for her continuous encouragement; I. Colquhoun, J. Ganzhorn, S. Goodman, and C. van Schaik for comments and discussion; V. Carrai for useful advices; C. Rakotondrasoa and R. Randriamarosoa for plant identification; G. Donati and E. Nappi for funding and support; C. Hawkins, R. Rasoloarison, J. Burkardt, D. Schwab, J. Fietz, S. Hussmann, S. Rümenap, S. Sommer, and S. Groos for their help and friendship in the field. A special thanks to F. Billi for her precious help and patience and to M. Donati, and F. Barbanera for their enthusiasm.

REFERENCES

ALTMANN, J. 1974. Observational study of behaviour: sampling methods. Behaviour, **49**: 227–267.
ANDREWS, J., AND C. BIRKINSHAW. 1998. A comparison between the daytime and night-time diet, activity and feeding height of the Black lemur, *Eulemur macaco* (Primates: Lemuridae), in Lokobe forest, Madagascar. Folia Primatologica, **69**: 175–182.
BIRKINSHAW, C., AND I. C. COLQUHOUN. 1998. Pollination of *Ravenala madagascariensis* and *Parkia madagascariensis* by *Eulemur macaco* in Madagascar. Folia Primatologica, **69**: 252–259.
CARRAI, V., AND A. LUNARDINI. 1996. Activity patterns and home range use of two groups of *Propithecus verreauxi verreauxi* in the Kirindy Forest, pp. 275–284. *In* Ganzhorn, J. U., and J. P. Sorg, eds., Ecology and Economy of a Tropical Dry Forest in Madagascar. Primate Report, **46–1**: 1–382.
COLQUHOUN, I. C. 1993. The socioecology of *Eulemur macaco*: A preliminary report, pp. 13–26. *In* Kappeler, P. M., and J. U. Ganzhorn, eds., Lemur Social Systems and their Ecological Basis. Plenum Press, New York.

———. 1998. Cathemeral behavior of *Eulemur macaco macaco* at Ambato Massif, Madagascar. Folia Primatologica, **69**: 22–34.
CONLEY, J. M. 1975. Notes on the activity pattern of *Lemur fulvus*. Journal of Mammalogy, **56**: 712–715.
CSERMELY, D. 1996. Antipredator behavior in lemurs: Evidence of an extinct eagle on Madagascar or something else? International Journal of Primatology, **17**: 349–354.
CURTIS, D. J. 1997. The Mongoose Lemur (*Eulemur mongoz*): A Study in Behaviour and Ecology. PhD Thesis, Zürich University, Zürich.
DAHL, J. F., AND C. A. HEMINGWAY. 1988. An unusual activity pattern for the mantled howler monkey of Belize. American Journal of Physical Anthropology, **75**: 201.
DANIELS, H. L. 1984. Oxygen consumption in *Lemur fulvus*: Deviation from the ideal model. Journal of Mammalogy, **65**: 584–592.
ENGQVIST, A., AND A. RICHARD. 1991. Diet as a possible determinant of cathemeral activity patterns in primates. Folia Primatologica, **57**: 169–172.
ERKERT, H. G. 1989. Lighting requirements of nocturnal primates in captivity: A chronobiological approach. Zoo Biology, **8**: 179–191.
GANZHORN, J. U., AND P. M. KAPPELER. 1996. Lemurs of the Kirindy Forest, pp. 257–254. *In* Ganzhorn, J. U., and J. P. Sorg, eds, Ecology and Economy of a Tropical Dry Forest in Madagascar. Primate Report, **46-1**: 1–382.
GOODMAN, S. M. 1994a. The enigma of antipredator behavior in lemurs: Evidence of a large extinct eagle on Madagascar. International Journal of Primatology, **15**: 129–134.
———. 1994b. Description of a new species of subfossil eagle from Madagascar: *Stephanoaetus* (Aves: Falconiformes) from the deposits of Ampasambazimba. Proceedings of the Biological Society of Washington, **107**: 421–426.
———, S. O'CONNOR, AND O. LANGRAND. 1993. A review of predation in lemurs: Implications for the evolution of social behavior in small, nocturnal primates, pp. 51–66. *In* Kappeler, P. M., and J. U. Ganzhorn, eds., Lemur Social Systems and their Ecological Basis. Plenum Press, New York.
GRASSI, C. G. 1997. A preliminary study of individual variation in diet of *Hapalemur griseus* at Ranomafana National Park, Madagascar. American Journal of Physical Anthropology, **24**: 119.
HARRINGTON, J. E. 1978. Diurnal behaviour of *Lemur mongoz* at Ampijoroa, Madagascar. Folia Primatologica, **29**: 291–302.
ISBELL, L. A. 1994. Predation on primates: Ecological patterns and evolutionary consequences. Evolutionary Anthropology, **3**: 61–71.
KAPPELER, P. M., AND J. U. GANZHORN. 1993. Evolution of primate communities and societies in Madagascar. Evolutionary Anthropology, **2**: 159–171.
KAPPELER, P. M., AND E. W. HEYMANN. 1996. Nonconvergence in the evolution of primate life history and socio-ecology. Biological Journal of the Linnean Society, **59**: 297–326.
LANGRAND, O. 1990. Guide to the Birds of Madagascar. Yale University Press, New Haven.
MACEDONIA, J. M. 1993. Adaptation and phylogenetic constrains in the antipredator behavior of ringtailed and ruffed lemurs, pp. 67–84. *In* Kappeler, P. M., and J. U. Ganzhorn, eds., Lemur Social Systems and their Ecological Basis. Plenum Press, New York.
MARTIN, R. D. 1972. Adaptive radiation and behaviour of the Malagasy lemurs. Proceedings Royal Society, London, **264**: 295–352.
MEYERS, D. M. 1988. Behavioral ecology of *Lemur fulvus rufus* in rain forest in Madagascar. American Journal of Physical Anthropology, **75**: 250.
MORLAND, H. S. 1993. Seasonal behavioral variation and its relationship to thermoregulation in ruffed lemurs (*Varecia variegata variegata*), pp. 193–203. *In* Kappeler, P. M., and J. U. Ganzhorn, eds., Lemur Social Systems and their Ecological Basis. Plenum Press, New York.
MÜLLER, E. F. 1985. Basal metabolic rates in primates—the possible role of phylogenetic and ecological factors. Comparative Biochemistry and Physiology A, Comparative Physiology, **81**: 707–711.
NASH, L. T. 1986. Influence of moonlight levels on travelling and calling patterns in two sympatric species of *Galago* in Kenya, pp. 357–367. *In* Taub, D. M., and F. A. King, eds., Current Perspectives in Primate Social Dynamics. Van Nostrand Reinhold Co., New York.
OVERDORFF, D. J. 1988. Preliminary report on the activity cycle and diet on the red-bellied lemur (*Lemur rubriventer*) in Madagascar. American Journal of Primatology, **16**: 143–153.
———. 1993. Ecological and reproductive correlates to range use in red-bellied lemurs (*Eulemur rubriventer*) and rufous lemurs (*Eulemur fulvus rufus*), pp. 167–177. *In* Kappeler, P. M., and J. U. Ganzhorn, eds., Lemur Social Systems and their Ecological Basis. Plenum Press, New York.
———. 1996. Ecological correlates to activity and habitat use of two prosimian primates: *Eulemur rubriventer* and *Eulemur fulvus rufus* in Madagascar. American Journal of Primatology, **40**: 327–342.

——, AND M. A. RASMUSSEN. 1995. Determinants of nighttime activity in "diurnal" lemurid primates, pp. 61–74. *In* Alterman, L., G. A. Doyle, and K. Izard, eds., Creatures of the Dark: The Nocturnal Prosimians. Plenum Press, New York.

OVERDORFF, D. J., S. G. STRAIT, AND A. TELO. 1997. Seasonal variation in activity and diet in a small-bodied folivorous primate, *Hapalemur griseus*, in southeastern Madagascar. American Journal of Primatology, **43**: 211–223.

PARIENTE, G. F. 1980. Quantitative and qualitative study of the light available in the natural biotope of Malagasy prosimians, pp. 117–134. *In* Charles-Dominique, P., H. M. Cooper, A. Hladik, C. M: Hladik, E. Pages, G. F. Pariente, A. Petter-Rousseaux, A. Schilling, and J.-J. Petter, eds., Nocturnal Malagasy Primates: Ecology, Physiology and Behavior. Academic Press, New York and London.

RASMUSSEN, M. A. 1998a. Ecological influences on cathemeral activity in the mongoose lemur (*Eulemur mongoz*) at Ampijoroa, northwest Madagascar. American Journal of Primatology, **45**: 202.

——. 1998b. Variability in the cathemeral activity cycle of two lemurid primates at Ampijoroa, northwest Madagascar. American Journal of Physical Anthropology, **40**: 183.

RASOLOARISON, R. M., B. P. N. RASOLONANDRASANA, J. U. GANZHORN, AND S. M. GOODMAN. 1995. Predation on vertebrates in the Kirindy Forest, western Madagascar. Ecotropica, **1**: 59–65.

RICHARD, A. F., AND M. E. NICOLL. 1987. Female social dominance and basal metabolism in a Malagasy primate, *Propithecus verreauxi*. American Journal of Primatology, **12**: 309–314.

SANTINI-PALKA, M. E. 1994. Feeding behaviour and activity patterns of two Malagasy bamboo lemurs *Hapalemur simus* and *Hapalemur griseus* in captivity. Folia Primatologica, **63**: 44–49.

SAUTHER, M. L. 1989. Antipredator behavior of free-ranging *Lemur catta* at Beza Mahafaly Special Reserve, Madagascar. International Journal of Primatology, **10**: 595–606.

SCHMID, J. 1996. Oxygen consumption and torpor in mouse lemurs (*Microcebus murinus* and *M.myoxinus*): Preliminary results of a study in western Madagascar, pp. 47–54. *In* Geiser, F., A. J. Hulbert, and S. C. Nicol, eds., Adaptations to the Cold: the Tenth International Hibernation Symposium. University of New England Press, Armidale.

SCHMID, J., AND J. U. GANZHORN. 1996. Resting metabolic rates of *Lepilemur ruficaudatus*. American Journal of Primatology, **38**: 169–174.

——, AND P. M. KAPPELER. 1998. Fluctuating sexual dimorphism and differential hibernation by sex in a primate, the gray mouse lemur (*Microcebus murinus*). Behavioral Ecology and Sociobiology, **43**: 124–132.

SORG, J. P. 1996. Vernacular and scientific names of plants of the Morondava region, pp. 339–346. *In* Ganzhorn, J. U., and J. P. Sorg, eds., Ecology and Economy of a Tropical Dry Forest in Madagascar. Primate Report, **46–1**: 1–382.

——, AND U. ROHENEr. 1996. Climate and tree phenology of the dry deciduous forest of Kirindy, pp. 57–80. *In* Ganzhorn, J. U., and J. P. Sorg, eds., Ecology and Economy of a Tropical Dry Forest in Madagascar. Primate Report, **46–1**: 1–382.

STERLING, E. 1993. Patterns of range use and social organization in aye-ayes (*Daubentonia madagascariensis*) on Nosy Mangabe, pp. 1–10. *In* Kappeler, P. M., and J. U. Ganzhorn, eds., Lemur Social Systems and their Ecological Basis. Plenum Press, New York.

SUSSMAN, R. W. 1975. A preliminary study of the behavior and ecology of *Lemur fulvus rufus* Audebert 1800, pp. 237–258. *In* Tattersall, I., and R. W. Sussman, eds., Lemur Biology. Plenum Press, New York.

——, AND I. TATTERSALL. 1976. Cycles of activity, group composition and diet of *Lemur mongoz mongoz* Linnaeus 1766 in Madagascar. Folia Primatologica, **26**: 270–283.

TATTERSALL, I. 1976. Group structure and activity rhythm in *Lemur mongoz* (Primates, Lemuriformes) on Anjouan and Moheli Islands, Comoro Archipelago. Anthropological Papers of the American Museum of Natural History, **53**: 367–380.

——. 1977. Ecology and behavior of *Lemur fulvus mayottensis*. Anthropological Papers of the American Museum of Natural History, **54**: 421–482.

——. 1979. Patterns of activity in the Mayotte lemur, *Lemur fulvus mayottensis*. Journal of Mammalogy, **60**: 314–323.

——. 1982. The Primates of Madagascar. Columbia University Press, New York.

——. 1987. Cathemeral activity in primates: A definition. Folia Primatologica, **49**: 200–202.

——, AND R. W. SUSSMAN. 1975. Observations on the ecology and behaviour of the Mongoose lemur *Lemur mongoz mongoz* Linnaeus (Primates, Lemuriformes) at Ampijoroa, Madagascar. Anthropological Papers of the American Museum of Natural History, **52**: 195–216.

VAN SCHAIK, C. P., AND J. A. R. A. M. VAN HOOFF. 1983. On the ultimate causes of primate social systems. Behaviour, **85**: 91–117.

VAN SCHAIK, C. P., AND P. M. KAPPELER. 1993. Life history, activity period, and lemur social systems, pp. 241–260. *In* Kappeler, P. M., and J. U Ganzhorn, eds., Lemur Social Systems and their Ecological Basis. Plenum Press, New York.

———. 1996. The social systems of gregarious lemurs: Lack of convergence with anthropoids due to evolutionary disequilibrium? Ethology, **102**: 915–941.

WARREN, R. D. 1994. Lazy leapers: a study of the locomotor ecology of two species of saltatory nocturnal lemur in sympatry at Ampijoroa, Madagascar. Ph.D. thesis, University of Liverpool, Liverpool.

WILSON, J. M., P. D. STEWART, G. S. RAMANGASON, A. M. DENNING, AND M. S. HUTCHINGS. 1989. Ecology and conservation of the crowned lemur *Lemur coronatus* at Ankarana, north Madagascar. Folia Primatolgica, **52**: 1–26.

WRIGHT, P. C. 1983. Day-active night monkeys (*Aotus trivirgatus*) in the Chaco of Paraguay. American Journal of Physical Anthropology, **60**: 272.

———. 1986. Diet ranging behaviour and activity pattern of the gentle lemur (*Hapalemur griseus*) in Madagascar. American Journal of Physical Anthropology, **69**: 283.

———. 1989. The nocturnal primate niche in the New World. Journal of Human Evolution, **18**: 635–658.

8

SOCIAL ORGANIZATION OF THE FAT-TAILED DWARF LEMUR (*CHEIROGALEUS MEDIUS*) IN NORTHWESTERN MADAGASCAR

Alexandra E. Müller[1]

[1] Anthropological Institute and Museum
University of Zurich
Winterthurerstrasse 190, 8057 Zurich, Switzerland
aem@aim.unizh.ch

ABSTRACT

The strictly nocturnal dwarf lemurs (*Cheirogaleus*) are extraordinary among primates in showing extensive torpor phases during the austral winter, which is thought to be an adaptation to seasonal variation in food availability. The Fat-tailed Dwarf Lemur (*C. medius*) is found throughout the west and south of Madagascar and enters torpor for six to eight months. Such a long period of dormancy may influence their social organization. A 20-month field study carried out at the Station Forestiùre d' Ampijoroa revealed a monogamous social pattern for *C. medius*.

Data from this study are presented on group composition, sleeping associations, and home range organization over two seasons in order to elucidate the basis of *C. medius*' unusual social organization. All studied groups contained one adult pair and their offspring from one or more birth seasons. Members of a group regularly sleep together, which might be an important means of social contact. In groups containing more than one offspring, brothers and sisters had the closest relationship. The adult pair remain together within the same home range for more than one season and presumably for life. Subsequently, some hypotheses about when monogamy should evolve are discussed.

RÉSUMÉ

Les cheirogales (*Cheirogaleus*) aux moeurs exclusivement nocturnes présente une caractéristique unique chez les primates avec ses longues phases de léthargie au cours de l'hiver austral qui correspondraient à une adaptation aux fluctuations saisonnières des disponibilités de nourriture. Le Petit Cheirogale (*C. medius*) qui est distribué dans tout l'ouest ainsi que dans le sud de l'île de Madagascar montre des phases de torpeur qui peuvent durer de six à huit mois. Une telle période d'inactivité pourrait avoir des conséquences sur l'organisation sociale de ces animaux. Une étude de 20 mois à la Station Forestière d'Ampijoroa a montré que le système social de *C. medius* était monogame.

Les données de cette étude, menée sur deux saisons, sont présentées sur la composition des groupes, les associations en dortoir et l'organisation des domaines vitaux pour élucider les bases de l'organisation sociale de *C. medius* qui est somme toute inhabituelle. Chaque groupe étudié contenait un couple adulte et leur progéniture d'une (ou plusieurs) saison(s) de reproduction antérieure(s). Un facteur qui pourrait être important dans les contacts sociaux a été observé dans le fait que les membres d'un groupe dorment régulièrement ensemble. Dans les groupes présentant plusieurs jeunes, les relations les plus étroites ont été observées entre les frères et les soeurs. Le couple adulte est resté uni sur le même domaine vital d'une saison à l'autre et le restera probablement pour la vie. Quelques hypothèses sont alors formulées et discutées sur les circonstances favorisant une évolution vers un système monogame.

INTRODUCTION

The strictly nocturnal dwarf lemurs (genus *Cheirogaleus*) are extraordinary among primates in showing extensive torpor phases during the austral winter (Petter et al., 1977). The Greater Dwarf Lemur (*C. major*) occurs in the eastern rainforest and is reported to spend up to three months per year in torpor (Petter et al., 1977, but see also Wright & Martin, 1995). The Fat-tailed Dwarf Lemur (*C. medius*) is found in western and southern dry forests, where seasonal variation in food availability is more pronounced. Individuals of this species may stay in torpor for six to eight months per year (Petter et al., 1977; Hladik et al., 1980).

Like most nocturnal prosimians, *Cheirogaleus medius* is usually seen alone during its nightly activities and is therefore designated as a solitary forager. This term was introduced by Bearder (1987) to avoid describing nocturnal prosimians as "solitary" in the sense that they lack a social network (Charles-Dominique, 1978). Several field studies on the social behavior of "solitary" nocturnal prosimians—focusing mainly on range overlap and sleeping associations (see Harcourt & Nash, 1986)—have shown that these species have different and complex patterns of social organization (e.g., Martin, 1972; Fodgen, 1974; Charles-Dominique, 1977; Charles-Dominique & Petter, 1980; MacKinnon & MacKinnon, 1980; Andrianarivo, 1981; Niemitz, 1984; Clark, 1985; Harcourt & Nash, 1986; Nash & Harcourt, 1986; Bearder, 1987; Crompton & Andau, 1987; Sterling, 1993; Gursky, 1995; Kappeler, 1997; Müller, 1998; Radespiel, 1998). These patterns show parallels to diurnal primates, with the difference that nocturnal prosimians (except *Avahi* spp.) are not seen in cohesive groups at night (Martin, 1981, 1995). It was accordingly proposed that the term "dispersed" be added when describing the social organization of these nocturnal prosimians (Eisenberg et al., 1972; Martin,

1981, 1995). Nevertheless, some authors still describe nocturnal prosimians as solitary while conceding that sleeping associations may occur. However, given that most nocturnal prosimians sleep with conspecifics (e.g. Martin, 1972; Andrianarivo, 1981; Harcourt & Nash, 1986; Bearder, 1987; Müller, 1998; Radespiel, 1998), such descriptions of their social organization are inadequate. "It is the age-sex composition of the social groups that differs, not the numbers" (Harcourt & Nash, 1986, p. 353). It is therefore essential to know the exact organization of these sleeping associations.

On the basis of range overlaps and sleeping associations, *Cheirogaleus medius* was recently described as monogamous (Müller, 1998; Fietz, in press). This is in contrast to many other prosimian species which are reported to be polygynous (Bearder, 1987; Kappeler, 1997). In addition to *C. medius*, other nocturnal, solitary foraging prosimians reported to live in family groups include *Galago zanzibaricus* (Harcourt & Nash, 1986) and *Phaner furcifer* (Charles-Dominique & Petter, 1980). In both species however, groups containing two females were also found (Charles-Dominique & Petter, 1980; Harcourt & Nash, 1986). *Tarsius* spp. are also often described as monogamous (e.g. MacKinnon & MacKinnon, 1980; Niemitz, 1984), but other researchers (e.g. Fodgen, 1974; Crompton & Andau, 1987) found no evidence of monogamy in their study populations. Based on her observations of some groups containing one adult male and two adult females, Gursky (1995) suggested that tarsiers are facultatively polygamous.

To date longer-term data on sleeping associations and group compositions in nocturnal prosimians have been lacking and it is therefore unknown whether relationships within groups are stable over years or for just one reproductive season. This is of special interest with respect to *Cheirogaleus medius*, whose yearly activity is interrupted by prolonged torpor. While the tendency for monogamous nocturnal prosimians to form sleeping associations is well known (Harcourt, 1986; Harcourt & Nash, 1986; Müller, 1998), the exact nature of family relationships remains unclear. In order to understand the basis of family group cohesion, we must first examine intragroup relationships between individuals (Chalmers, 1979).

In this paper, I present data on group composition, sleeping associations, and home range organization over two seasons in *Cheirogaleus medius* in order to elucidate the basis of their unusual social organization.

METHODS

Study Site

This study was conducted during two rainy seasons between November 1995 and June 1996 and between September 1996 and June 1997 at the Station Forestière d'Ampijoroa (16°19'S; 46°49'E) in northwestern Madagascar. Subsequently, in September and October 1997, the animals were recaptured in order to remove the radio collars. The study area is characterized as dry deciduous forest of western type (Ramangason, 1988) and has two distinct seasons with heavy rain from November to April. Annual rainfall was 1776mm in 1996 (U. Thalmann & A. Müller, unpublished data). Seven other sympatric lemurs occur in the area in addition to the study species, four of them being nocturnal. Data collection took place in "Jardin Botanique A" which is on a flat plateau with sandy soil. The forest is dense with undergrowth and lianas. The height of the canopy is 5–10m. Small trails were cut in order to establish a marked grid covering an area of approximately 5 ha with 10×10 m quadrats. Every quadrat had a code composed of a letter and a number. This grid facilitated following and precise location of study animals.

Trapping

Animals were captured with up to 20 traps during the first season (1995–1996) and up to 37 traps in the second season (1996–1997). The traps were made of wood and wire mesh, as described by Charles-Dominique and Bearder (1979) and had dimensions of $30 \times 50 \times 35$ cm. The traps were placed at a height of 1–1.5 m, wedged between two or more small trees or in a fork of a branch. Most traps were left at the same location during the entire study. As a rule, three traps (median and mode, range 1–5) were grouped at each capture location. The traps were baited with bananas several days before they were set in order to habituate lemurs to the presence and smell of these devices in their territories. During trapping, all traps were inspected regularly. Usually a trap night was completed within two hours because the animals would go to the traps immediately after leaving their sleeping holes.

Captured animals were marked with a radio collar (PD-2C transmitters from Holohil Systems Ltd., Ontario, Canada, weight 3.9 g) and/or a labeled collar; infants were marked by cutting hair on their tails. Age determination was based on body mass and is fully explained elsewhere (Müller, in press). Age class divisions are based on number of torpor phases passed: Infant = never in torpor ($0-^3/_4$ years), juvenile = once ($^3/_4-1^3/_4$ years), adolescent = twice ($1^3/_4-2^3/_4$ years), and adult = three or more times ($<2^3/_4$ years). Infants were born in January/February (pers. obs.).

Sleeping Site Determination and Observations

As far as possible, sleeping sites were determined every day using telemetry (TR-4 receiver with an RA-14K antenna from Telonics, Inc., Arizona, USA). All sleeping trees were marked with plastic flagging tape and a tag. The height of the sleeping hole (if known) was estimated to the nearest meter. The position of the sleep tree was determined using the grid system.

In the first season, radio-collared animals were followed for 42.5 h and sequentially localized for another 47 h. In the second season, animals were followed only (162.5 h), yielding an overall total of 252 h for both seasons. Both the sequential localizations and the follows were made between 1800 and 0600 h and were intended to last six hours but were sometimes shorter (e.g., because of heavy rainfall). During follows, an individual's position in the grid (determined by quadrat) and its activity were recorded every 2 min when possible. Other behavioral data were recorded *ad libitum* when the focal animal was visible. During localizations, positions of all collared individuals were determined sequentially.

Data Analysis

To test whether the animals slept in the same sleeping site by chance, the Kolmogorov-Smirnov goodness of fit procedure for continuous data was applied (Zar, 1984). To estimate expected frequencies, only sites used by both animals in the test period were considered as potential sites for individuals to meet. Expected frequencies (EF) were calculated as follows:

$$EF = N_{AB} \times \frac{n_{AB}}{n_A n_B}$$

where N_{AB} is the number of days both A and B were localized in their sleeping sites; n_{AB} is the number of sites used by A and B; n_A is the number of sites used by A; and n_B is the number of sites used by B. The significance level was set at $p < 0.05$, a p-value between 0.05 and 0.1 was interpreted as a tendency.

Home ranges were determined using a combination of observation data, trapping localizations, and sleeping site positions. The surface area was calculated as a minimum convex polygon. For groups whose home ranges were known for two seasons, I identified the quadrats that were located within the home range in order to determine which quadrats were used in both seasons and which quadrats were used in the first or second season alone. To compare home ranges over two seasons, an Overlap-Quotient (OQ)—based on the SÖrensen-Quotient—was calculated. This value can lie between 0% and 100%. The higher the value, the more similar are the two home ranges. OQ is calculated as follows (Mühlenberg, 1993):

$$OQ = \frac{2J}{n_A + n_B} \times 100$$

where J is the number of quadrats used in both seasons; n_A is the number of quadrats used in the first season; and n_B is the number of quadrats used in the second season. Range overlaps within groups were calculated with the same procedure.

To test whether the home range of a group was stable in size and location over both seasons, a procedure suggested by Fager (1957)—originally used for species associations—was used. This procedure tests whether the number of joint occurrences (in this case the number of quadrats used in both seasons), on the hypothesis of independent distribution, will have a hypergeometric distribution ($=H_0$). A one-tailed t-test is applied for "positive affinity" between quadrats used in the first season and quadrats used in the second season. The procedure operates as follows (Fager, 1957):

$$t = \left[\frac{(n_A + n_B)(2J - 1)}{2 n_A n_B}\right] \left[\sqrt{n_A + n_B - 1}\right]$$

where n_A is the number of quadrats used in the first season; n_B is the number of quadrats used in the second season; and J is the number of quadrats used in both seasons. If H_0 can be rejected, the quadrats used in the two seasons have a significant positive affinity, i.e. home ranges are the same over both seasons (Fager, 1957; for critical values see Zar, 1984, Table B.3).

RESULTS

Studied Animals and Group Compositions

In a total of 23 trap nights (= 437 trap sets) during the first season (1995–1996), ten different *Cheirogaleus medius* were caught: four females (#3, #4, #7, #8), three males (#1, #5, #6), and three infants (#9, #10, #11) born in 1996. During the second season (1996–1997), I caught 18 different individuals in 33 trap nights (= 1097 trap sets): seven

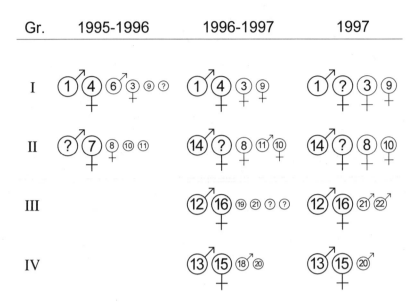

Figure 1. Group composition in *Cheirogaleus medius* groups between 1995 and 1997 at the Station Forestière d'Ampijoroa. Individuals with a question mark were not captured within this season but in the previous or following season. Different sizes of the circles show the different ages of the group members ranging from large circles = adults, medium-sized circles = adolescents, and small circles = juveniles. Unsexed circles are infants born in the same season.

females (#3, #4, #8, #9, #10, #15, #16), eight males (#1, #5, #11, #12, #13, #14, #17, #18), and three infants (#19, #20, #21) born in 1997. Both seasons yielded a total of 20 different captured individuals, of which eight were trapped in both seasons: three males (#1, #5, #11) and five females (#3, #4, #8, #9, #10). Between September and October 1997, one individual was caught that had not previously been trapped (#22, born in the same year). Due to their small size, the sex of infants could not be determined easily. Reliable sex determination was only possible once they were one year old.

In the first season (1995–1996) male #1 and females #4 and #7 were adult. Male #6 was classified as adolescent, and females #3 and #8 and male #5 were classified as juveniles. Three animals captured from March 1996 onwards (#9, #10, #11) were infants born in the same year. In the second season (1996–1997), males #1, #12, #13, #14, #17 and females #4, #15, and #16 were adult. Females #3 and #8 and males #5 were now classified as adolescents, whereas females #9 and #10 and males #11 and #18 were classified as juveniles. Three animals (#19, #20, and #21) were that year's infants (born in 1997).

With the exception of males #5 and #17, all of these animals belonged to one of four distinct groups (Fig. 1). These groups could easily be determined on the basis of trapping localities, sleeping sites, and coinciding home ranges (see also Müller, 1998). All four groups contained one adult pair. The young male #5 had a range that overlapped with three groups and might be a "floater" (see Fietz, this volume). The trapping locality of the adult male #17 indicated that he was a neighbor of group IV. In three groups (I, II and IV), pairs already had offspring when observations on them commenced (groups I and II in 1995, group IV in 1996) and one pair (group III) had offspring only in the coming birth season.

Home Ranges and Overlaps

As animals from the same family groups have coinciding home ranges (Müller, 1998, unpublished data), "group ranges" rather than individual's ranges are presented here. Group ranges were determined by pooling all localizations from all group members.

In the first season of the study (1995–1996), group ranges of two groups (group I and II) and of male #5 were determined: Group I's home range was 1.91 ha and group II's home range 1.74 ha (Table 1). In group I, all members were radio-collared; in group II, only two females were radio-collared (Table 1). Range overlap between group I and II was 4.6%. The range size of male #5 was 2.36 ha (Müller, 1998; Table 1). The group ranges of groups I and II and of male #5 in the 1995–1996 season are presented in Figure 2.

In the second season (1996–1997), group range size was between 2.00 ha (group IV) and 2.73 ha (group III) (Table 1). As in the first season, all members of group I were radio-collared; in group II, all individuals wore radio-collars except for the adult female; and in groups III and IV only the adult pairs were radio-collared (Table 1). Also the young male #5 was radio collared. Range overlaps between groups were between 5.2% and 21.9% (I–II: 11.7%; I–III: 21.9%; I–IV: 6.4%; II–III: 5.2%; II–IV: 7.6%). The ranges of groups I, II, III, IV and of male #5 in the 1996–1997 season are presented in Figure 2.

Ranges of groups in which animals were radio-collared for both years were more than 80% consistent in size and location over the two seasons (OQ; Table 1). In all

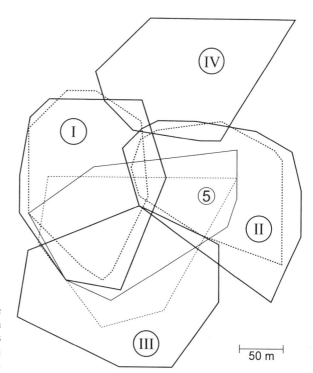

Figure 2. Group ranges of *Cheirogaleus medius* over two seasons at the Station Forestière d'Ampijoroa. Dashed line is group ranges from the 1995–1996 season and solid line from the 1996–1997 season.

Table 1. Home range sizes of *Cheirogaleus medius* family groups for the two seasons at the Station Forestière d'Ampijoroa. GR = group range size in ha; F = number of hours group members were followed; OQ = Overlap-Quotient; t = calculated value using the formula of Fager (1957)

Group	1995–1996 season Radio-collared animals	GR [ha]	F [h]	1996–1997 season Radio-collared animals	GR [ha]	F [h]	OQ	t
I	#1,#3,#4,#6	1.91	24	#1,#3,#4,#9	2.48	87	87.6%	16.444
II	#7,#8	1.74	8	#8,#10,#11,#14	2.48	25	83.5%	14.352
III	—	—	10.5	#12,#16	2.73	18.5	—	—
IV	—	—		#13,#15	2.00	10	—	—
#5	#5	2.36		#5	2.30	22	81.1%	13.685

groups, the range stayed significantly ($p < 0.001$) the same during both season (Table 1). The comparison between the ranges of the two seasons is shown in Figure 2. The finding that ranges of the second season are larger than ranges from the first season is most probably attributable to the larger data set from the second season. The number of hours group members were followed are presented in Table 1 (for localizations see Müller, 1998).

Sleeping Sites

Each animal used between four (#7 in 1995–1996 and #11 in 1996–1997, respectively) and 22 (#5 in 1996–1997) sleeping sites within one season (median 8.5 over both years) (Table 2). As female #7 and male #11 were collared for just a few weeks, a limited number of sleeping site localizations were made (Table 2). The actual number of sleeping sites utilized may be higher. Animals that were radio-collared over both seasons (#1, #3, #4, #5, and #8) generally used the same sleeping sites in the second season as in the first (Table 2). Most sleeping sites were used by more than one member of the same group; some were used by all group members. Members of one group never slept at sites used by other groups. Only the young male #5 used some of the sleeping sites of group I, II and III. While he was never seen sharing a hole with another animal in the first season (1995–1996), from April 1997 he slept regularly in the same tree as the young females of group I and II. Although he slept in the same tree as these young females, he did not always share the hole with them. In at least one sleeping site he slept in a hole separate from the other group.

The majority of the sleeping trees used were live trees, both in terms of the actual number of trees used as well as with respect to the number of days the animals slept in them (Table 3). The average height of sleeping site entrance was 5 m (median; range 1.5–12 m, $n_{trees} = 43$). The holes at 29 sleeping sites could not be described, either because the hole was hidden in dense foliage and the animals could never be seen leaving the hole or because the animals used these sites so rarely that there was no opportunity to find out where it was.

Seven sleeping site trees were also used by *Lepilemur edwardsi*, although they usually used either different holes in the same tree or the same hole at different times. In one case, however, a *Cheirogaleus* family group (group III) occupied a hole simultaneously with a *L. edwardsi* individual for several consecutive days.

Table 2. Sleeping sites used by *Cheirogaleus medius* individuals in the Station Forestière d'Ampijoroa. ID = identification number of animal; Gr. = group; L = number of localizations at sleeping sites; S = number of sleeping sites; used by others = number of sleeping sites used also by other animals; Sum = total number of sleeping sites used during the two seasons; in both seasons = number of sleeping sites used in both seasons; ad = adult; adol = adolescent; juv = juvenile; inf = infant

Study animals			1995–1996 season				1996–1997 season				1995–1997	
ID	Sex	Gr.	Age	L	S	used by others	Age	L	S	used by others	Sum	in both seasons
1	M	I	ad	203	15	12	ad	234	18	15	22	11
4	F	I	ad	148	12	11	ad	121	13	11	16	9
6	M	I	adol	53	7	7	—	—	—	—	—	—
3	F	I	juv	132	13	12	adol	192	17	15	18	11
9	F	I	inf	—	—	—	juv	196	13	12	—	—
14	M	II	—	—	—	—	ad	123	7	5	—	—
7	F	II	ad	35	4	2	—	—	—	—	—	—
8	F	II	juv	109	7	3	adol	253	7	6	9	5
10	F	II	inf	—	—	—	juv	123	5	5	—	—
11	M	II	inf	—	—	—	juv	24	4	4	—	—
12	M	III	—	—	—	—	ad	115	8	6	—	—
16	F	III	—	—	—	—	ad	222	12	6	—	—
13	M	IV	—	—	—	—	ad	110	6	5	—	—
15	F	IV	—	—	—	—	ad	120	5	5	—	—
5	M	—	juv	180	9	2[1]	adol	228	22	11[2]	25	6

[1] Both sleeping sites are also used by members of group I.
[2] Sleeping sites used by members of group I (n = 7), II (n = 2) and III (n = 2).

Sleeping Associations within Groups

Animals within groups did not sleep together by chance. For all combinations except one (adult female #4 and adolescent male #6: D = 0.051, ns.), the Kolmogoroff-Smirnov goodness of fit procedure gave highly significant results indicating that individuals sleep more often together than expected (father #14 and adolescent daughter #8: D = 0.148, $0.01 < p < 0.005$; all other combinations: $D \geq 0.210$, $p < 0.001$). Excluding the combination "mother and adolescent son", animals slept together 24–74% of the

Table 3. Number of dead trees used as sleeping sites and percentage of days slept in dead trees by four *Cheirogaleus medius* groups and one single male in the Station Forestière d'Ampijoroa. Gr. = group; No. of trees = total number of trees used as sleeping sites; No. of † trees = number of dead trees used as sleeping site; † trees used (%) = percentage of dead trees used as sleeping sites; slept in † trees (%) = percentage of days slept in dead trees

	1995–1996 season				1996–1997 season			
Gr.	No. of trees	No. of † trees	† trees used (%)	Slept in † trees (%)	No. of trees	No. of † trees	† trees used (%)	Slept in † trees (%)
I	19	5	26.3	27.0	24	5	20.8	19.0
II	10	1	10.0	13.9	10	2	20.0	20.5
III					14	3	21.4	15.6
IV					6	1	16.7	26.5
#5	9	2	22.2	37.8	22	6	27.3	27.9

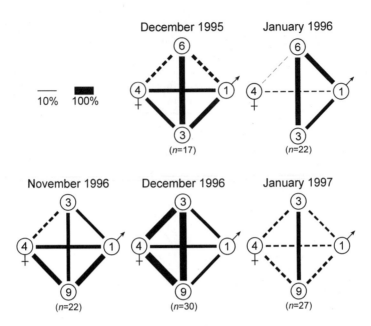

Figure 3. Sleeping associations of *Cheirogaleus medius* within group I with all members radio-collared. The thickness of the line shows the percentage of days in which an individual slept together with the other individual. Solid line = individuals slept together significantly more often than expected, dashed line = individuals did not sleep together significantly more often than expected.

time (median 47.6%). But it should be noted that the chance of meeting at the same site is not equal for all combinations, due to different numbers of sleeping sites used. The ratio between days slept together and the probability of meeting by chance lies between 2.58 and 11.77 (median 5.18, excluding the combination #4 + #6, where the ratio is 1.43). This means that, on average, animals slept together five times more often than would be expected by chance. These results give an overall picture of sleeping site associations, but a closer examination is warranted. The results could be biased because the data for the different combinations are not of equal duration nor do they cover exactly the same period. Therefore, additional analyses for each combination and month were conducted.

Associations within Group I. All members of group I were radio-collared between 15 December 1995 and 22 January 1996 and between 29 October 1996 and 1 February 1997. Their sleeping associations are shown in Figure 3. The thickness of the line shows the percentage of days in which an individual slept together with the other individual. In November 1996 and in December of both seasons, significant associations were found between all members of the family with only few exceptions (mother and adolescent daughter in November 1996; and parents and adolescent son in December 1995). In January of both seasons, the animals tended to be dispersed while sleeping. Whereas in January 1996 the two offspring slept more often than would expected together as well as with their father, in January 1997 they slept only with each other more often than expected. The entire group slept together on only a few occasions (Table 4).

Table 4 shows the percentage of days on which animals slept alone. In January 1996, the adult female was separated from her family 90% of the time. The actual per-

Table 4. Percentage of days individuals of group I slept alone. ID = identification number of individuals; adol. = adolescent; juv. = juvenile

ID	Sex	1995–1996 season			1996–1997 season			
		Age	Dec.	Jan.	Age	Nov.	Dec.	Jan.
1	M	adult	29.4	28.6	adult	27.3	53.3	51.9
4	F	adult	35.3	90.5	adult	40.9	6.7	48.2
6	M	adol.	29.4	14.3	—	—	—	—
3	F	juv.	5.9	28.6	adol.	59.1	6.7	33.3
9	F	infant	—	—	juv.	13.6	3.3	22.2
all alone			0.0	9.5		7.4	0.0	3.7
all together			17.7	0.0		22.2	30.0	3.7

centage of days she slept alone is certainly less than this, however, since she had infant offspring at this time. Remarkably, the adolescent daughter slept alone nearly 60% of the time in December 1996. Apart from this exception, offspring were less likely to sleep alone than adults. Only rarely did all group members sleep alone. For the remaining months, only isolated combinations could be analyzed. These are discussed below in comparison with combinations of other family groups.

Adult Pairs. Pair 1 + 4 slept together significantly more than expected by chance in November and December of both seasons and in March 1996 (no data for March 1997), but not in January and February (although there is a tendency in February 1996). Pair 12 + 16 slept more together than expected in December but not in January, and pair 13 + 15 slept together significantly more than expected by chance from December to February, but not in November (they shared the same sleeping site 50% of the time but they used only two sites during this month). All these data are presented in Table 5.

All adult pairs slept apart for several days during the 1996 and 1997 birth seasons. Female #4 (group I) used a particular tree which neither she nor other group members

Table 5. Sleeping associations of *Cheirogaleus medius* within adult pairs. ID = identification number of the individuals (male + female); s = significant ($p < 0.05$), i.e. the adult pair slept together significantly more often than expected; ns = not significant ($p > 0.1$), i.e. the adult pair did not sleep together significantly more often than expected. T = tendency ($0.05 \leq p \leq 0.1$), i.e the adult pair showed a tendency to sleep together more often than expected. Sample sizes are in parentheses

ID	Season	Month				
		N	D	J	F	M
1 + 4	1995–1996	s (15)	s (30)	ns (31)	T (29)	s (30)
1 + 4	1996–1997	s (22)	s (30)	ns (28)	ns (14)	—
12 + 16	1996–1997	—	s (25)	ns (27)	—	—
13 + 15	1996–1997	ns (20)	s (30)	s (26)	s (16)	—

Table 6. Sleeping associations of *Cheirogaleus medius* between mother and offspring; for details of codes see Table 5

		Offspring		Month		
ID	Season	Sex	Age	N	D	J
4 + 6	1995–1996	M	adolescent	—	ns (17)	ns (30)
4 + 3	1996–1997	F	adolescent	ns (25)	s (30)	ns (28)
4 + 3	1995–1996	F	juvenile	—	s (28)	ns (21)
4 + 9	1996–1997	F	juvenile	s (25)	s (30)	ns (28)

used at any other time. In 1996, she slept in this tree from 5–6 January and from 8–14 January, then she switched to another tree and stayed there alone between 15 and 24 January. In 1997, she slept again in that tree between 2 and 8 February. Female #16 (group III) slept apart from her mate between 13 January and 2 February 1997, but did not use a special tree. Female #15 (group IV) slept apart from her mate between 28 January and 5 February 1997, but during this period changed her sleeping site twice. She also did not use a special tree.

Mothers and Offspring. In November 1996 the mother #4 slept significantly more often than expected with her juvenile daughter #9 but not with her adolescent daughter #3. In December, she slept significantly more than expected with all her daughters but not with her son #6. In January, she slept with none of her offspring more than expected. All data are presented in Table 6.

Fathers and Offspring. A significant association was found between father #1 and his adolescent son #6 in January but not in December. Father #1 slept significantly more often than expected with his adolescent daughter #3 in November and December but not between January and May (no data for February). On the other hand, father #14 slept significantly more than expected with his adolescent daughter #8 in February but not between November and January. With his juvenile daughter #3, male #1 slept more than expected in December and April and displayed a tendency in January but not in March or May (no data for November and February). One year later (1996–1997 season), he slept significantly more often than expected with his other juvenile daughter #9 in September and displayed a tendency in December but not in October, January, April or March (no data for March and April). All data are presented in Table 7.

Brothers and Sisters. In all tested combinations and months, brothers and sisters slept together significantly more often than expected, regardless of the sex and age of the individuals. All data are presented in Table 8.

Young Male #5. In the first season (1995–1996), male #5 was never found sleeping with another individual. During the second season (1996–1997), he was always seen alone until 4 April 1997, when he was localized at the same site as juvenile female #9 from group I. They stayed together until 14 April, when male #5 switched to the adolescent female #3, also from group I. On 24 April, female #9 joined them; on 28 April, they were all joined by adult male #1 (assumed to be the father of the two females).

Table 7. Sleeping associations of *Cheirogaleus medius* between father and offspring; for details of codes see Table 5

		Offspring		Month							
ID	Season	Sex	Age	O	N	D	J	F	M	A	M
1 + 6	1995–1996	M	adolescent	—	—	ns (17)	s (30)	—	—	—	—
1 + 3	1996–1997	F	adolescent	—	s (22)	s (30)	ns (28)	—	ns (18)	ns (29)	ns (31)
1 + 3	1995–1996	F	juvenile	—	—	s (28)	T (21)	—	ns (14)	s (30)	ns (21)
1 + 9	1996–1997	F	juvenile	ns (19)	s (22)	T (30)	ns (28)	—	—	ns (29)	ns (31)
14 + 8	1996–1997	F	adolescent	—	ns (25)	ns (30)	ns (27)	s (27)	—	—	—

In this sleeping tree, male #5 did not always use the same hole as the family. On 10 May, the adult male changed sleeping sites but then returned on 20 May. On this day, male #5 changed his sleeping partner to juvenile female #10 from group II. When female #10's radio collar ceased functioning on 3 July, they were still together. Male #5 stayed in that site and on 18 July adolescent female #8 (group II) began using the site and stayed until 30 July.

Dispersal from Natal Areas

Only limited data could be collected on the dispersal of young individuals. All four young females stayed with their families until the study was terminated at the end of October 1997 (see also Fig. 1). At this time, females #9 and #10 were $1^{3}/_{4}$ years old and females #3 and #8 were $2^{3}/_{4}$ years old (Müller, in press). It remains unclear at which age females leave their natal group. During the entire study, neither female #3 nor female #8 had offspring although female #3 showed vaginal opening at the end of October in both 1996 and 1997.

Two of the three collared young males disappeared from their natal range during the study. Male #6 left when he was 2 years old; male #11 (the twin brother of female #10) disappeared by the age of $1^{1}/_{4}$ years. Both males disappeared in a fashion similar to that described for *Galago zanzibaricus* by Harcourt and Nash (1986): Male #6 could

Table 8. Sleeping associations of *Cheirogaleus medius* within brothers and sisters; for details of codes see Table 5

				Month						
ID	Season	Sex	Age	N	D	J	F	M	A	M
6 + 3	1995–1996	M + F	different	—	s (17)	s (21)	—	—	—	—
3 + 9	1996–1997	F + F	different	s (25)	s (30)	s (28)	—	—	s (29)	s (31)
8 + 10	1996–1997	F + F	different	—	—	—	—	s (26)	s (29)	—
8 + 11	1996–1997	F + M	different	—	—	—	—	s (14)	—	—
10 + 11	1996–1997	F + M	same	—	—	—	—	s (17)	—	—

not be located at a sleeping site for five successive days, then returned for one day and finally disappeared completely. Male #11 could not be located for a day on two occasions; he reappeared for one and three days, respectively, and then disappeared completely. Male #6 was located for the last time on 11 February 1996, male #11 on 21 March 1997. It might be, however, possible that both males died, but if so I would have found the transmitters.

The third young male (#5) stayed in his range for the duration of the study. But this male appeared to be unrelated to other animals, and his range overlapped up to 100% with groups I, II, and III. Between 13 and 15 February, he slept far away from his usual range (smallest distance to range border: 13 February = 300 m, 14 February = 80 m, 15 February = 280 m) but then returned and remained in his home range until the end of the study (last seen on 12 October 1997).

DISCUSSION

Preliminary results indicating that *Cheirogaleus medius* at Ampijoroa is monogamous (Müller, 1998) were confirmed. Additionally, it was shown that an adult pair stays together in the same home range for more than one season and that these ranges remained more-or-less the same size and at the same location over both seasons. All studied groups contained one adult pair and their offspring from one or more birth seasons. Because offspring remain in their natal range for more than one year and because adult pairs have new offspring every year, the adult pair might be forced to remain together in order to maximize the survivability of their offspring.

There was some evidence that males disperse at a younger age than females. Although none of the four young females left the natal range during the study, the absence of more than one reproducing female in all four groups indicates that young females do not remain in their natal ranges and matriarchies do not exist. Hence, females presumably disperse, but at an older age than males. As males disperse before reaching adult size and sexual maturity, they are presumably forced to remain solitary for a few years while attempting to acquire a range and a mate. Whether this applies to young male #5 is unclear. The fact that he had several interactions with other young females may support this hypothesis, but further investigations are needed. It might also be that he is a "floater" (see Fietz, this volume). Females do not leave their natal areas immediately after becoming sexually mature (Müller, in press). They may serve a role in helping to raise younger siblings (see below).

Like other small nocturnal prosimians, Fat-tailed Dwarf Lemurs often sleep together. In groups containing more than one offspring, brothers and sisters had the closest relationship. In groups III and IV, the adult pair slept more often together than the pair from group I. This may be because group I used more sleeping sites than groups III and IV, so the chance of meeting was smaller. Considering the different number of sleeping sites used by each group, the adult pair of group I slept together relatively more often than pairs from other groups. However, if the animals actively search for each other, a higher number of sleeping sites should not affect the absolute percentage of days spent sleeping together. Another variable to consider is that both groups III and IV consisted of only two and three animals, respectively. Groups containing less than four animals must sleep together—otherwise one animal will be alone. It is hence possible that these pairs did not have a closer relationship than those of group I, but less combination possibilities. A general "sleeping association pattern" could not be found, but two things were notable. First, the close relation between brothers and sisters: in all months they slept together significantly more often than expected. Second, females slept alone on several days in January/February, presumably after giving birth. Similar behavior by females has also been reported for *Galago zanzibaricus* (Harcourt, 1986).

But why do *Cheirogaleus medius* individuals sleep together? Since they do not forage as cohesive groups, sleeping associations might be the best way to socialize and the advantages of sociality may be substantial. Clark (1985) pointed out three benefits animals may gain from their sociality: 1) Ectoparasite-born diseases can be reduced by allogrooming. Indeed, as in *Galago crassicaudatus* (Clark, 1985), allogrooming in *C. medius* was directed to body regions that are difficult to reach (e.g. head, ears, neck). Allogrooming was observed after the animals left their sleeping site and when they met during the night. With the exception of single male #5, who was found to be carrying ticks on two occasions, none of the animals had noticeable ectoparasites. 2) Alloparenting by older offspring may be important. The fact that brothers and sisters very often slept together supports this suggestion. Siblings were also observed together at night. Older offspring may take care of younger siblings while parents are occupied with care of new-borns or range defense. 3) Information on food resources and sleeping sites may be shared. This certainly is a very important benefit, but difficult to prove. The fact that family groups did use the same sleeping sites in two consecutive seasons and my inference (see below) that *Cheirogaleus* feeds on predictable food sources would also support this hypothesis.

There seems to be no need for the dwarf lemurs to sleep together for thermoregulatory reasons as individuals do not sleep without exception in groups. Also individuals (especially males) of the much smaller mouse lemurs (*Microcebus murinus*) often sleep alone (Martin, 1972; Schmid, 1997; Radespiel, 1998). Interestingly, Schmid (1997) found that mouse lemurs sleep always in groups when in torpor. Sleeping together might be also advantageous for *Cheirogaleus* when in torpor (however male #5 did hibernate alone in 1996), but during their active period it is not indispensable for thermoregulatory reasons. Sleeping together could be a form of mate guarding. It is more convenient for a male to be already close to the female when she is leaving her sleeping site instead of looking for her while she is foraging and traveling. Radespiel (1998) reported a *Microcebus* male waiting at the entrance of a nest hole for estrous females.

Like other solitary foragers, *Cheirogaleus medius* lives in a social network. Surprisingly, they are organized in small family groups. This is puzzling because monogamy is rare among mammals (Kleiman, 1977) and not the prevailing social system for nocturnal solitary foragers (Martin, 1981; Bearder, 1987; Charles-Dominique, 1995). Monogamy is commonly thought to occur because females are too widely dispersed for males to defend more than one female, or because the female needs the help of the male to rear offspring (Kleiman, 1977). According to Wittenberger and Tilson (1980), paternal care is never indispensable in primates, in the sense that females cannot rear any offspring without the male's help, but might be important. Essential paternal care (Fietz, in press) and a form of indirect paternal investment (Müller, in press) were reported for *C. medius*, but whether those are really indispensable remains unknown. In species for which paternal care is "only" very important, monogamy should evolve when the benefits of pairing with an already mated male (e.g. superior male and habitat, cooperation with other females) do not compensate the costs of polygyny which is often loss of paternal help (Wittenberger & Tilson, 1980). It is also possible that the costs of polygyny are too high because of food competition between the females. It might then be that the females are too widely distributed and a male is unable to defend a range big enough for more than one female and her offspring (Kleiman, 1977).

Group range sizes in *Cheirogaleus medius* lie between 2.0 and 2.7 ha. Monogamous primates usually occupy small territories (e.g., Tenaza, 1975; Robinson et al., 1987) with a territory being defined as "a defended area" (Nice, 1941; Chalmers, 1979). Apart from the need to defend a home range, the issue of range defensibility must be addressed, i.e. is an animal able to successfully defend his range? According to the formula from Mitani and Rodman (1979), a *C. medius* male is able to defend its home range (A. Müller, unpublished data). Nevertheless, *C. medius* groups show range overlap of up to 20%, which would mean that they cannot keep their ranges exclusive. Other monogamous primates such as mongoose lemurs, gibbons, and titi monkeys all display territorial behavior, but also have overlapping ranges (Kinzey, 1981; Leighton, 1987; Curtis, 1997) despite their defensibility according to the criteria of Mitani and Rodman (1979). Obviously, defending a range or showing territorial behavior does not automatically mean that a group will be able to maintain exclusive use. The Mitani-Rodman formula was criticized for oversimplification and subsequently modified by several authors (Martin, 1981; Lowen & Dunbar, 1994). Martin (1981) proposed calculating the ratio of daily path length to boundary length, as territory defense involves patrolling the boundary rather than crossing it. But even if an animal is able to patrol its range it cannot be omnipresent and monitoring its entire area. Lowen and Dunbar (1994) took detection distance into account, and proposed a detection distance of 500 m. This distance is probably too great for most forest-dwelling species. However, in all formulas important devices for range defence such as vocalizations or scent marking are neglected.

Cheirogaleus medius appear to defend their ranges as the overlapping areas are not as big as in non-defended areas (see Cheney, 1987). They presumably defend patches (feeding trees, sleeping sites) rather than areas, as animals from one group never utilized other groups' sleeping or feeding trees. Direct confrontations were never observed (but see Fietz, in press), and other indications of intergroup aggression such as injuries and scars were not detected (only one female had a small scratch in her ear). As in many prosimians, scent marking may also play an important role in territory defense (Schilling, 1979, 1980; Fietz, 1999).

Dwarf lemurs undergo prolonged torpor phases in the dry season and show a remarkable decrease in body mass during this time of inactivity (Petter et al., 1977; see also Müller, in press). During their active time, however, they have not only to eat for daily survival, but they must also store fat for the next torpor phase and might be therefore more dependent on food than other lemurs. A sudden bottleneck in food supply requiring greater travel distances or temporary emigration (as reported by Overdorff [1993] for *Eulemur fulvus rufus*) could be fatal. It could be therefore hypothesized that *Cheirogaleus medius* is specialized on resources that are predictable over time and evenly spaced, and that exact knowledge and access to resources is essential for them. Thus, to guarantee a sufficient food supply, those resources have to be defended. Probably, a male can only defend a home range big enough for one female and her offspring and this might also force him into monogamy. Of course, much of what has been said in this paragraph is highly speculative, however, it highlights the need for more detailed studies on *Cheirogaleus* ecology.

ACKNOWLEDGMENTS

Most of all I am indebted to Urs Thalmann, Michele Rasmussen, and Don Reid for all their help and support in Madagascar during my field work. I thank the Karons for their hospitality and helpfulness during my stay in Madagascar. Thanks go to Benno Schoch for helping to prepare the traps and to Arsùne Vélo for field assistance. I am very grateful to Bob Martin and Gustl Anzenberger for their support during the study and for taking care of administrative matters during my absence in the field. Jörg Ganzhorn, Steve Goodman, Bob Martin, Michele Rasmussen, Christophe Soligo, Urs Thalmann, and three anonymous reviewers are thanked for helpful comments on the manuscript. Thanks go to Lorenz Gygax and Karin Isler for statistical advice. The research was conducted under an "Accord de Cooperation" between the Universities of Zurich (Switzerland) and Mahajanga (Madagascar) and the governmental institutions of Madagascar (Commission Tripartite) gave research permission. The project was financed by the A.H. Schultz Foundation, the Family Vontobel Foundation, the Goethe Foundation, the G. and A. Claraz Donation, and the Swiss Academy of Natural Sciences (SANW).

REFERENCES

ANDRIANARIVO, A. J. 1981. Etude comparée de l'organisation sociale chez *Microcebus coquereli*. Mémoire de Diplome d'Etudes Approfondies, Université de Madagascar.

BEARDER, S. K. 1987. Lorises, bushbabies, and tarsiers: diverse societies in solitary foragers, pp. 11–24. *In* Smuts, B. B., D. L. Cheney, R. M. Seyfarth, R. W. Wrangham, and T. T. Struhsaker, eds., Primate Societies. The University of Chicago Press, Chicago.

CHALMERS, N. 1979. Social Behaviour in Primates. Edward Arnold, London.

CHARLES-DOMINIQUE, P. 1977. Ecology and Behaviour of Nocturnal Primates. Duckworth, London.

——. 1978. Solitary and gregarious prosimians: evolution of social structures in primates, pp. 139–149. *In* Chivers, D. J., and K. A. Joysey, eds., Evolution. Recent Advances in Primatology, Vol. 3. Academic Press, London.

——. 1995. Food distribution and reproductive constraints in the evolution of social structure: nocturnal primates and other mammals, pp. 425–438. *In* Altermann, L., G. A. Doyle, and M. K. Izard, eds., Creatures of the Dark. The Nocturnal Prosimians. Plenum Press, New York.

——, AND S. K. BEARDER. 1979. Field studies of lorisid behavior: methodological aspects, pp. 567–629. *In* Doyle, G. A., and R. D. Martin, eds., The Study of Prosimian Behavior. Academic Press, London.

CHARLES-DOMINIQUE, P., AND J.-J. PETTER. 1980. Ecology and social life of *Phaner furcifer*, pp. 75–96. *In* Charles-Dominique, P., H. M. Cooper, A. Hladik, C. M. Hladik, E. Pagùs, G. F. Pariente, A. Petter-Rousseaux, J.-J. Petter, and A. Schilling, eds., Nocturnal Malagasy Primates. Ecology, Physiology, and Behavior. Academic Press, London.

CHENEY, D. L. 1987. Interactions and relationships between groups, pp. 267–281. *In* Smuts, B. B., D. L. Cheney, R. M. Seyfarth, R. W. Wrangham, and T. T. Struhsaker, eds., Primate Societies. The University of Chicago Press, Chicago.

CLARK, A. B. 1985. Sociality in a nocturnal "solitary" prosimian: *Galago crassicaudatus*. International Journal of Primatology, **6**: 581–600.

CROMPTON, R. H., AND P. M. ANDAU. 1987. Ranging, activity rhythms, and sociality in free-ranging *Tarsius bancanus*: A preliminary report. International Journal of Primatology, **8**: 43–71.

CURTIS, D. J. 1997. The Mongoose Lemur (*Eulemur mongoz*): A Study in Behaviour and Ecology. Ph.D. thesis, University of Zürich, Zürich.

EISENBERG, J. F., N. A. MUCKENHIRN, AND R. RUDRAN. 1972. The relation between ecology and social structure in primates. Science, **17**: 863–874.

FAGER, E. W. 1957. Determination and analysis of recurrent groups. Ecology, **38**: 586–595.

FIETZ, J. 1999. Monogamy as a rule rather than a exception in nocturnal lemurs: The case of the Fat-tailed Dwarf Lemur (*Cheirogaleus medius*). Ethology, **105**: 259–272.

FODGEN, M. P. L. 1974. A preliminary study of the western tarsier, *Tarsius bancanus* Horsefield, pp. 151–165. *In* Martin, R. D., G. A. Doyle, and A. C. Walker, eds., Prosimian Biology. Duckworth, London.
GURSKY, S. L. 1995. Group size and composition in the spectral tarsier, *Tarsius spectrum*: implications for social organization. Tropical Biodiversity, **3**: 57–62.
HARCOURT, C. S. 1986. *Galago zanzibaricus*: Birth seasonality, litter size and perinatal behaviour of females. Journal of Zoology, London, **210**: 451–457.
HARCOURT, C. S., AND L. T. NASH. 1986. Social organization of galagos in Kenyan coastal forests: I. *Galago zanzibaricus*. American Journal of Primatology, **10**: 339–355.
HLADIK, C. M., P. CHARLES-DOMINIQUE, AND J.-J. PETTER. 1980. Feeding strategies of five nocturnal prosimians in the dry forest of the west coast of Madagascar, pp. 41–74. *In* Charles-Dominique, P., H. M. Cooper, A. Hladik, C. M. Hladik, E. Pagùs, G. F. Pariente, A. Petter-Rousseaux, J.-J. Petter, and A. Schilling, eds., Nocturnal Malagasy Primates. Ecology, Physiology, and Behavior. Academic Press, London.
KAPPELER, P. M. 1997. Intrasexual selection in *Mirza coquereli*: Evidence for scramble competition polygyny in a solitary primate. Behavioral Ecology and Sociobiology, **45**: 115–127.
KINZEY, W. G. 1981. The titi monkeys, genus *Callicebus*, pp. 241–276. *In* Coimbra-Filho, A. F., and R. A. Mittermeier, eds., Ecology and Behaviour of Neotropical Primates. vol. 1. Academia Brasileira de Ciências, Rio de Janeiro.
KLEIMAN, D. G. 1977. Monogamy in mammals. The Quarterly Review of Biology, **52**: 39–69.
LEIGHTON, D. R. 1987. Gibbons: Territoriality and monogamy, pp. 135–145. *In* Smuts, B. B., D. L. Cheney, R. M. Seyfarth, R. W. Wrangham, and T. T. Struhsaker, eds., Primate Societies. The University of Chicago Press, Chicago.
LOWEN, C., AND R. I. M. DUNBAR. 1994. Territory size and defendability in primates. Behavioral Ecology and Sociobiology, **35**: 347–354.
MACKINNON, J. R., AND K. S. MACKINNON. 1980. The behavior of wild spectral tarsiers. International Journal of Primatology, **1**: 361–379.
MARTIN, R. D. 1972. A preliminary field-study of the lesser mouse lemur (*Microcebus murinus* J. F. Miller 1777). Zeitschrift für Tierpsychologie, Supplement **9**: 43–89.
———. 1981. Field studies of primate behaviour. Symposia of the Zoological Society of London, **4**: 287–336.
———. 1995. Prosimians: from obscurity to extinction?, pp. 535–563. *In* Altermann, L., G. A. Doyle, and M. K. Izard, eds., Creatures of the Dark. The Nocturnal Prosimians. Plenum Press, New York.
MITANI, J. C., AND P. S. RODMAN. 1979. Territoriality: the relation of ranging pattern and home range size to defendability, with an analysis of territoriality among primate species. Behavioral Ecology and Sociobiology, **5**: 241–251.
MÜHLENBERG, M. 1993. Freilandökologie. Quelle & Meyer Heidelberg, Wiesbaden.
MÜLLER, A. E. 1998. A preliminary report on the social organisation of *Cheirogaleus medius* (Cheirogaleidae; Primates) in north-west Madagascar. Folia Primatologica, **69**: 160–166.
———. In press. Aspects of social life in the fat-tailed dwarf lemur (*Cheirogaleus medius*): inferences from body weights and trapping data. American Journal of Primatology.
NASH, L. T., AND C. S. HARCOURT. 1986. Social organization of galagos in Kenyan coastal forests: II: *Galago garnettii*. American Journal of Primatology, **10**: 357–369.
NICE, M. M. 1941. The role of territory in bird life. American Midland Naturalist, **26**: 441–487.
NIEMITZ, C. 1984. An investigation and review of the territorial behaviour and social organisation of the genus *Tarsius*, pp. 117–127. *In* Niemitz, C., ed., Biology of Tarsiers. Gustav Fischer, Stuttgart.
OVERDORFF, D. J. 1993. Similarities, differences, and seasonal patterns in the diets of *Eulemur rubriventer* and *Eulemur fulvus rufus* in the Ranomafana National Park, Madagascar. International Journal of Primatology, **14**: 721–753.
PETTER, J.-J., R. ALBIGNAC, AND Y. RUMPLER. 1977. Mammifères Lémuriens (Primates Prosimiens). Faune de Madagascar. ORSTOM, Paris.
RADESPIEL, U. 1998. Die soziale Organisation des grauen Mausmakis (*Microcebus murinus*, J. F. Miller 1777)—Eine freilandökologische und laborexperimentelle Studie. Ph.D. thesis, Hannover. Dr. Köster, Berlin.
RAMANGASON, G. S. 1988. Flore et végétation de la forêt d'Ampijoroa, pp. 130–137. *In* Rakotovao, L., V. Barre, and J. Sayer, eds., L'Equilibre des Ecosystèmes Forestiers à Madagascar: Actes d'un Séminaire International. IUCN, Gland.
ROBINSON, J. G., P. C. WRIGHT, AND W. G. KINZEY. 1987. Monogamous cebids and their relatives: Intergroup calls and spacing, pp. 44–53. *In* Smuts, B. B., D. L. Cheney, R. M. Seyfarth, R. W. Wrangham, and T. T. Struhsaker, eds., Primate Societies. The University of Chicago Press, Chicago.
SCHILLING, A. 1979. Olfactory communication in prosimians, pp. 461–542. *In* Doyle, G. A., and R. D. Martin, eds., The Study of Prosimian Behavior. Academic Press, London.

———. 1980. Seasonal variations in the fecal marking of *Cheirogaleus medius* in simulated climatic conditions, pp. 181–190. *In* Charles-Dominique, P., H. M. Cooper, A. Hladik, C. M. Hladik, E. Pagùs, G. F. Pariente, A. Petter-Rousseaux, J.-J. Petter, and A. Schilling, eds., Nocturnal Malagasy Primates. Ecology, Physiology, and Behavior. Academic Press, London.

SCHMID, J. 1997. Torpor beim Grauen Mausmaki (*Microcebus murinus*) in Madagaskar: Energetische Konsequenzen und ökologische Bedeutung. Ph.D. thesis, University of Tübingen, Tübingen.

STERLING, E. J. 1993. Patterns of range use and social organization in aye-ayes (*Daubentonia madagascariensis*) on Nosy Mangabe, pp. 1–10. *In* Kappeler, P. M., and J. U. Ganzhorn, eds., Lemur Social Systems and their Ecological Basis. Plenum Press, New York.

TENAZA, R. R. 1975. Territory and monogamy among Kloss' gibbons (*Hylobates klossii*) in Siberut Island, Indonesia. Folia Primatologica, **24**: 60–80.

WITTENBERGER, J. F., AND R. L. TILSON. 1980. The evolution of monogamy: hypotheses and evidence. Annual Review of Ecology and Systematics, **11**: 197–232.

WRIGHT, P. C., AND L. B. MARTIN. 1995. Predation, pollination, and torpor in two nocturnal prosimians: *Cheirogaleus major* and *Microcebus rufus* in the rain forest of Madagascar, pp. 45–60. *In* Altermann, L., G. A. Doyle, and M. K. Izard, eds., Creatures of the Dark. The Nocturnal Prosimians. Plenum Press, New York.

ZAR, J. H. 1984. Biostatistical Analysis, 2nd edition. Prentice-Hall, Inc., Englewood Cliffs, N.J.

9

DEMOGRAPHY AND FLOATING MALES IN A POPULATION OF *CHEIROGALEUS MEDIUS*

Joanna Fietz[1]

[1] Deutsches Primatenzentrum
Abteilung Verhaltensforschung / Ökologie
Kellnerweg 4, 37077 Göttingen, Germany
joanna.fietz@t-online.de

ABSTRACT

According to recent field studies, the Fat-tailed Dwarf Lemur (*Cheirogaleus medius*) lives in permanent pairs together with the subadult offspring from the previous breeding season, within defended territories. Obligate paternal care was suggested to be responsible for the evolution of pair-living in this species. Aims of this study were to describe the population dynamics in *C. medius* and to draw attention to the occurrence of floating males as possible alternative to monogamy in these nocturnal lemurs.

A mark-recapture study in the Kirindy Forest/CFPF, western Madagascar, was combined with observations and radio-tracking of 34 individuals during the rainy seasons from 1995 to 1998. The study area was saturated with territories during the whole study period. Nevertheless, demographic variables such as population density, sex ratio, and age structure varied strongly between years. In 1996 and 1998 population density was highest (280 and 360 individuals/km^2) and up to 21% of the complete study population were subadults, while in 1995 and 1997 only about 4% of the captured animals were not adult and population density was relatively low. Consistent with this bi-annual fluctuation, no infants were recorded in 1998. In two of the four study periods, the sex ratio of adults was male biased. These surplus males (= floaters) were not associated with a given female and differed significantly from males with pair-bonds (= territorial males) concerning movement patterns and sleeping habits. The large home ranges of these floaters overlapped mutually as well as with the territories of territorial pairs. Apart from testis size, that was significantly smaller in floaters during the mating season, morphometric data were not different between male of those two categories. Floaters were between two and at least four years old and sexually mature.

Disappearance rate was not different between floaters and territorial males. The results suggest that floaters represent a pool of potential breeders, that may be

New Directions in Lemur Studies, edited by Rakotosamimanana et al.
Kluwer Academic / Plenum Publishers, New York, 1999.

prevented from pair-living by skewed sex ratios, high population densities and a subsequent lack of vacant territories. But since they have the chance to sneak copulations with females living in pairs, the possibility that floaters represent an alternative mating strategy in *Cheirogaleus medius*, cannot be ruled out. Future genetic studies about paternity and reproductive success of floaters and pair-bonded males, combined with demographic data, will show whether or not floaters represent an evolutionary stable strategy.

RÉSUMÉ

D'après des études menées récemment, le Petit Cheirogale *Cheirogaleus medius* vit en couples permanents avec sa progéniture subadulte de la période reproductive précédente dans des territoires défendus. Les soins paternels obligatoires avaient été suggérés comme étant responsables de l'évolution de la vie en couple pour cette espèce. L'objet de la présente étude était de décrire les dynamiques de la population chez *C. medius* et d'attirer l'attention sur l'existence de males «vagabonds» comme une alternative possible à la monogamie chez ces lémuriens nocturnes.

Une étude de marquage/recapture dans la Forêt Kirindy/CFPF dans l'ouest de Madagascar a été complétée par des onservations et un suivi effectué avec un système de radio-tracking sur 34 individus pendant les saisons des pluies de 1995 à 1998. Cette zone d'études était saturée de territoires au cours de l'ensemble de la période d'étude. Cependant, les variables démographiques comme la densité de la population, le sexe ratio et la structure de l'âge variaient considérablement d'une année à l'autre. En 1996 et 1998 les densités de population étaient les plus grandes (280 and 360 individus/km^2) et les individus sub-adultes représentaient près de 21% de la population étudiée alors qu'en 1995 et 1997 4% seulement des animaux capturés n'étaient pas adultes et les densités de population étaient relativement faibles.

En accord avec cette fluctuation bisannuelle, aucun enfant n'a été enregistré en 1998. Dans deux des quatre périodes d'études le sexe-ratio des adultes penchait vers les mâles. Ces mâles surnuméraires (mâles vagabonds) n'étaient pas associés à une femelle donnée et se différenciaient sensiblement des mâles d'un couple formé (mâles territoriaux) par ses mouvements et ses habitudes de sommeil. Les grandes zones vitales de ces satellites se chevauchaient et débordaient sur les territoires des couples. En dehors de la taille des testicules qui étaient sensiblement plus petits chez les vagabonds au cours de la période de reproduction, les données morphométriques n'étaient pas différentes entre les mâles des deux catégories. Les vagabonds avaient entre 2 et 4 ans au moins et étaient à maturité sexuelle.

Le taux de disparition n'était pas différent entre les vagabonds et les mâles territoriaux. Les résultats suggèrent que les vagabonds représentent un réservoir de reproducteurs potentiels qui sont empêchés à la vie en couple par des taux de sex-ratio défavorables, des densités de population élevées et donc un manque de territoires disponibles. Mais dans la mesure où ils ont une chance de chaparder une copulation à des femelles vivant en couple, la possibilité que ces vagabonds représentent une stratégie de reproduction alternative chez *Cheirogaleus medius* ne peut être éliminée. Des études génétiques futures sur la paternité et les succès de reproduction des satellites et des mâles en couple, combinées avec des données démographiques montreront si les vagabonds représentent ou non une stratégie stable dans l'évolution.

INTRODUCTION

Alternative mating tactics occur in a variety of species and may depend on age, size, condition, or phenotype of the respective individuals. Either alternative tactics yield equal reproductive success or they follow unavoidable constraints (Andersson, 1994). Sexually mature individuals, that are not permanently associated with other conspecifics, could represent a possible alternative mating strategy to permanent bisexual groups. Floaters are defined as sexually mature individuals, that are not associated with conspecifics and differ from territorial males in their movement pattern. While for mammals examples for floating males are rare: Elephant Seals, *Mirounga angustirostris* (Le Boeuf, 1974), Red Deer, *Cervus elaphus* (Clutton-Brock et al., 1979), and Fallow Deer, *Dama dama* (Thirgood, 1990), the occurrence of floaters is well known in birds (Smith, 1978, 1984, 1987; Matthysen, 1989; Smith & Acrese, 1989; Rohner, 1996). Here floaters generally differ from other members of the population by having a lower rank and larger home ranges (Smith, 1987). The existence of such floaters can be interpreted mainly in two different ways: either territory holders are the "better" males, and the occupation of a territory depends on male quality. In this case, territory holders could be expected to be larger or in a better body condition. In addition they should have a notably higher reproductive success than floaters. This has been shown for the Song Sparrow (*Melospiza melodia*), where territoriality was greatly superior to floating as a social and reproductive strategy for yearling males (Smith & Acrese, 1989). Or alternatively, the mixture of floaters and territorial males in a given population could represent different evolutionary stable strategies. In this case lifetime reproductive success of each male category should be similar. Either way, a pool of non-breeding, sexually mature individuals could be essential for the flexibility of a population. This was shown in the Rufous-collared Song Sparrow (*Zonotrichia capensis*) and the Black-capped Chickadee (*Parus atricapillus*). Here floaters of both sexes occur and can quickly be recruited once, if a territory holder or a regular flock member of either sex dies (Smith, 1978, 1987).

For a variety of different primate species it is known that solitary ranging individuals occur: for example, Black Lemurs, *Eulemur macaco* (Petter, 1962), Alaotran Gentle Lemur, *Hapalemur griseus alaotrensis* (Mutschler et al., 1998), Japanese Macaques, *Macaca fuscata* (Nishida, 1966; Kawai & Yoshiba, 1968), Hanuman Langurs, *Presbytis entellus* (Laws & Vonder Haar Laws, 1984), Yellow Baboons, *Papio cynocephalus* (Slatkin & Hausfater, 1976), Redtailed Guenons, *Cercopithecus ascanius* (Jones & Bush, 1988), Kloss's Gibbons, *Hylobates klossii* (Tilson, 1981; Mitani, 1984; Brockelmann et al., 1998), Gorillas, *Gorilla gorilla* (Schaller, 1963), Howler Monkeys, *Alouatta palliata* (Carpenter, 1934), Spider Monkeys, *Ateles geoffroyi* (Carpenter, 1935), Saddleback Tamarin, *Saguinus fuscicollis*, Cotton-top Tamarin, *S. oedipus*, and Pygmy Marmoset, *Cebuella pygmaea* (Goldizen, 1987). In spite of the wide occurrence of this behavior, detailed information about the biology of these species, particularly with regard to if individuals remain floaters on the long-term, is still lacking. It is therefore unclear, if solitarily ranging individuals are in a transient state, migrating to a new territory, or real floaters.

Fat-tailed Dwarf Lemur (*Cheirogaleus medius*) occur in the woodlands of western and southern Madagascar, and belongs to a group of small (130 g) nocturnal primates (Cheirogaleidae). They are exceptional among primates, because they spend up to seven months hibernating within tree holes (Petter, 1978; Hladik et al., 1980; Petter-Rousseaux, 1980). Animals emerge from hibernation in November and mate at the beginning of the rainy season during December (Hladik et al., 1980; Fietz, 1999). After a gestation period of 61–64 days (Foerg, 1982), females usually give birth to two young. In the middle of April, adults have started hibernating after accumulating a great quantity of fat (Petter et al., 1977; Petter, 1978; Hladik et al., 1980; Petter-Rousseaux, 1980). Their diet consists of flowers, nectar, fruits with a seasonally varying proportion of insect prey (Petter et al., 1977; Hladik et al., 1980; Petter-Rousseaux & Hladik, 1980). While in captivity *C. medius* are sexually mature within the first year (Foerg, 1982), wild *C. medius* reached sexual maturity within their second year (Fietz, 1999). According to recent field studies, *C. medius* lives in permanent pairs, within defended territories, together with the subadult offspring from the previous year (Müller, 1998, this volume; Fietz, in press). Obligate paternal care was suggested to be the selective factor driving pair-living in this species (Fietz, 1999).

The aims of this paper are to describe population dynamics of *Cheirogaleus medius* over a period of four years and to describe the occurrence of non-pair-living males (= floaters) in some years. Body measurements, age, mortality rate, disappearance rate, and territorial behavior are compared between territory holders and floaters, to arrive at a better understanding whether or not the existence of floaters represent an alternative mating strategy.

MATERIALS AND METHODS

Study Area and Study Period

This research was carried out in the Kirindy Forest/CFPF. The study site is part of a dry deciduous forest of western Madagascar (60 km northeast of Morondava), a forestry concession held by the Centre de Formation Professionelle Forestière de Morondava (CFPF). The Morondava region is characterized by a dry season of eight months. Most rain falls between December and February, with an annual mean of 800 mm; for detailed description of the forest see Ganzhorn and Sorg (1996), The study area is about 25 ha (500 × 500 m). Observations were made from 1995 to 1998, during four study periods (March 1995; November 1995–May 1996; October 1996–April 1997; and December 1997–March 1998).

Trapping. Sherman traps (7.7 × 7.7 × 30.5 cm) were set each month for four consecutive nights (1995: March, November, December; 1996: January–May, November, December; 1997: January, February, December; 1998: January.). Traps were placed every 50 m at the intersections of a grid system of trails. During the second study period two traps were placed at each intersection, while in the following study periods only one trap was placed per intersection, because of a limited number of traps in the following years. Traps were baited with bananas and fixed on shrubs or trees at 40–200 cm height. Each day the traps were in operation they were opened and baited in the late afternoon and checked early the following morning. Animals were released in the late afternoon at the site of capture.

Body Measurements. Body measurements follow Schmid and Kappeler (1994) and were taken of each animal the first time they were captured. Body mass (to the nearest gram with a 300 g scale) and testis size (length and width of one testis) were measured each time an animal was trapped. Testis volume was calculated according to the volume of a spherical ellipsoid (Kappeler, 1997) as:

Volume = $\pi*W^2/6$; L = testis length and W = width of one testis.

Animals were generally not anaesthetized but simply restrained, with the assumption that any resultant error in body measurements and mass were not gender-biased. Captured animals were sexed and individually marked with passive transponders, that were injected subdermally (Trovan, EURO I.D. Usling GmbH, Weilerswist, Germany).

Radio Tracking. A TRX-1000 receiver (Wildlife Materials Inc., Carbondale, Illinois, USA) with a three element yagi antenna and a TR-4 receiver (Telonics, Inc., Mesa, Arizona, USA) protected by an environmental housing (Sexton Photographics, Salem, Oregon, USA) with a flexible two element yagi antenna were used. Animals were radio-collared with 4g TW-4 button cell tags (Biotrack Ltd., Wareham, Dorset, UK). A total of 34 individuals (21 males and 13 females) were equipped with radio collars. Each animal was followed and observed for up to six hours per night (18.00–24.00 or 24.00–6.00). Positions of the tagged animal were taken continuously and defined by cross-bearings obtained from two different intersections of the defined trail system. Locations of the animals were triangulated with the computer software Track3 (A. Ganzhorn, unpublished). Home range data were analyzed with Tracker (1994) using the minimal convex polygon technique (Kenward, 1991; Camponotus, 1994). A total of 521 radio tracking hours were accrued during this study.

Nest Hole Occupation. During the day, radio-collared animals were located in their nest sites. The transponder reading system has a range of about 10cm through wood. Since all captured individuals were individually marked with transponders, it was possible to check each reachable nest hole (below 2.5m) additionally with the transponder reading system. With this method is was possible to verify whether or not the radio-collared animal was sleeping with a marked conspecific.

Definitions and Data Sets

Infants: Individuals born within the respective study period.

Subadults: Individuals one year old (passed one torpor phase), not sexually mature, and differ in body size and body mass from adult individuals (Fietz, 1999).

Adults: Individuals at least two years old (passed two torpor phases). Females were considered to be adult if they showed any signs of estrous, pregnancy or lactation, or when they were observed to take care of their infants. To determine pregnancy, females were palpated. Males were considered as adult if they had visibly developed testes.

Floaters: Adult males not associated with a given female and differed from territorial males with regards to movement patterns and sleeping habits.

Age: Minimum age of trapped individuals could be determined, due to subsequent retrapping in different study years. If an individual was captured first as an infant or subadult, exact age could be determined. Four individuals were not measured, sexed or aged.

Relative number of sleeping holes: Number of different sleeping holes used by a given individual, divided by the total number of nest hole checks for this individual. This denominator was used because the period of time that individual animals were radio-collared differed.

Rate of communal sleeping: Percentage of nest hole checks in which individuals were found to sleep with conspecifics. Only nest hole checks with the transponder reading system were considered.

Exclusivity: Percentage of nest holes used by a given individual that belonged to another sleeping group.

Absence from study area: Percentage of complete nest hole checks that a given individual was not recorded within the study area but known to be alive because it was recorded afterwards.

Disappearance: Individual no longer recorded within the study area.

Mortality: Radio-collar of individual was found with signs of predation.

For comparisons of home range size and sleeping habits between the two male categories, only data derived from the same study period (1996/97) were considered. For comparisons of body measurements between territorial males and floaters only data of radio-collared males were included. This ensured that males were correctly attributed to one of the two categories (floater or territorial male). For the analyses of testis volume and body mass, only data collected during the mating season (December) were analyzed.

Statistical Analyses

The Mann-Whitney test (MWU) was used to examine differences in body measurements and home range size (SAS, 1987; Lorenz, 1988). The Fisher's Exact Test (SAS, 1987) was performed to test gender differences in mortality and the rate of animals that were no longer recorded, and whether sex ratio of subadults differed significantly from an expected sex ratio of 1:1.

RESULTS

Demographic Structure

Population Density and Age. During this study a total of 154 individuals were trapped (63 males; 52 females; 39 subadults and infants; Table 1), and with recaptures 637 animals were captured. Population density and age structure of the population showed bi-annual fluctuations during the study period (Fig. 1). In 1995/96 and 1997/98, population density was relatively high with 280 and 360 individuals/km^2, respectively. In these two blocks of years subadults accounted for 21% of the whole study population. In 1994/95 and 1996/97 the population density was much lower, 188 and 212 individuals/km^2, respectively, and only 4.3% and 3.8%, respectively, of the captured animals were subadult (Table 1). Consistent with this bi-annual fluctuation, no infants were recorded to be born in 1998.

Sex Ratio. Sex ratios of trapped individuals varied between the four study periods, but were not significantly different from an expected 1:1 ratio (Fig. 2). In the

Table 1. Number of captured adults, subadults, and individuals of unknown status of *Cheirogaleus medius* during the respective study periods in the Kirindy Forest/CFPF

Study periods	Adult males	Adult females	Subadults	Unknown status	Total
1994/95	19	24	1	2	46
1995/96	32	23	14	1	70
1996/97	30	21	2	0	53
1997/98	35	35	19	1	90

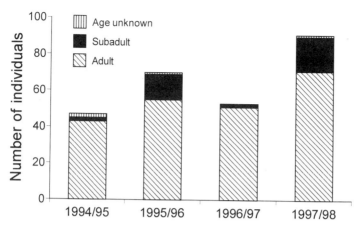

Figure 1. Numbers of subadult and adult individuals of *Cheirogaleus medius* captured during the respective study periods in the Kirindy Forest/CFPF.

first study period the number of captured females exceeded that of males. In 1995/96 and 1996/97 the sex ratio was male-biased, while in 1997/98 the same number of males and females were captured. In 1995/96 a total of 12 individual subadults were trapped. Seven of these were recaptured and sexed during subsequent study periods; six of them were males. The observed sex ratio of these recaptured individuals differed significantly from a ratio of 1:1 (n = 7, Fisher's Exact Test, p = 0.05).

Mortality and Disappearance Rate. Mortality of radio-collared animals was 23% and not gender biased (Table 2; Fisher's Exact Test, p = 1.0). About 44% of the radio-collared animals disappeared, there was no significant difference between males and females (Table 2; Fisher's Exact Test, p = 0.3). In 1996/97 four floaters and seven territorial males were radio-collared (Table 3). Two of the territorial males were known to be predated upon. The difference in mortality between the two male categories was not significant (Fisher's Exact Test, p = 0.5). Four territorial males and three floaters disappeared between 1996/97 and 1997/98. Concerning the disappearance rate, there

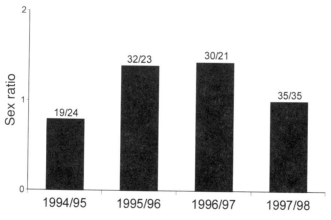

Figure 2. Sex ratio of captured *Cheirogaleus medius* during the respective study periods in the Kirindy Forest/CFPF (number of males/number of females).

Table 2. Number of radio-collared male and female *Cheirogaleus medius* and number of those that were known to be predated upon or disappeared

	Radio-collared	Dead	Disappeared
Males	21	5	11
Females	13	3	4

Table 3. Number of radio-collared territorial male and floater *Cheirogaleus medius* and number of those that were known to be predated upon or disappeared

	Radio-collared	Dead	Disappeared
Territorial Male	7	2	4
Floater	4	0	3

was no significant difference between the territorial males and floaters (Fisher's Exact Test, p = 1.0).

Comparison between Floaters and Territorial Males

In 1996/97 and 1997/98 five males were captured that differed from territorial males with regard to home range use and sleeping habits. For these individuals, which are considered floaters home ranges (n = 4) were significantly larger than those of territorial males (n = 8; Table 4). They overlapped mutually as well as with the territories of territorial pairs (Fig. 3). Floaters used a significantly greater number of sleeping holes than territorial males. Floaters were never found to sleep with conspecifics, while territorial males were often found to sleep with their partners or/and their offspring. Despite systematic searching with transponder and radio-tracking devices, floaters were, in contrast to territorial males, not always present in the study area. Numerous parameters associated with the home range size and sleeping habits of territorial and floater males were significant (Table 4). While body mass and body measurements did not differ between territorial males and floaters, testis volume was significantly smaller in floaters during the mating season (Table 5). The age of the floaters ranged between

Table 4. Comparison of home range size and sleeping habits between floaters and territorial males of *Cheirogaleus medius*. Values are medians and 25/75% quartiles; MWU = Mann-Whitney U-test

Variable	Floater	n	Territorial male	n	MWU test z	p
Home range size (m^2)	39,302 25,425/115,889	4	14,007 9421/19,519	8	2.7	0.007
Relative number of sleeping holes	0.2 0.1/0.3	4	0.09 0.03/0.1	6	2.1	0.03
Rate of communal sleeping (%)	0 0/0	4	56.3 23.5/89.7	5	2.1	0.03
Exclusivity of sleeping hole use (%)	14.8 13.6/26.3	4	0 0/0	7	2.3	0.02
Absence from study area (%)	9.2 7.6/24.7	4	0 0/0.2	6	2.8	0.006

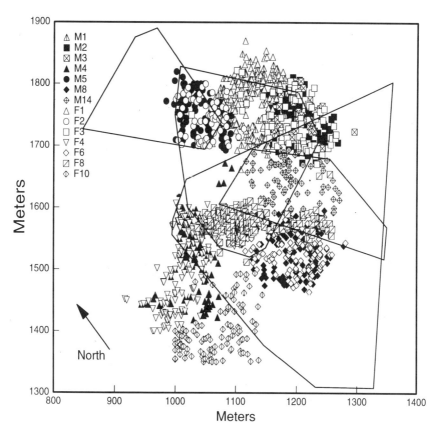

Figure 3. Home ranges of males (= M) and females (= F) *Cheirogaleus medius* that were on territories (open or closed symbols) and floaters (large black lined polygons) during the 1996/97 field season in the Kirindy Forest/CFPF.

two years, which is the first year of reproductive activity, to at least four years. Territorial males included individuals, that were between two and at least five years old.

Combats. Altogether four fights between floaters and conspecifics were observed, all of them during the mating season. Three fights took place within one night between a floater (at least three years old) and three different conspecifics. First the floater was

Table 5. Comparison of body measurements between floaters and territorial males of *Cheirogaleus medius*. Values are medians and 25/75% quartiles; MWU = Mann-Whitney U-test

Variable	Floater	n	Territorial male	n	MWU test z	p
Testis volume (mm^3)	110.6 83.2/154.8	4	201.0 139.7/363.5	11	2.1	0.04
Body mass (g)	116 106.5/123	5	128 112/160	15	0.7	n.s.
Head width (mm)	26.9 25.6/27.5	5	27 26.5/27.6	15	0.7	n.s.
Head length (mm)	42.7 40.7/43.8	5	43.9 42.6/45.2	15	1.3	n.s.
Hindfoot length (mm)	35.1 32.7/36.6	4	36.4 34.4/39.5	15	1.3	n.s.

encountered in a tree and pursued by an unmarked individual, and both individuals fell out of the tree. The floater was then chased away by another adult radio-collared male at the border of the radio-collared male's territory. In the third combat a marked subadult individual chased the floater twice from a feeding tree, and then the floater disappeared without further aggressive encounters. The fourth fight took place within the territory of an radio-collared couple. A two year old floater was approached by an unmarked individual, which, due its small body size was probably a subadult. After a short chase the floater fell from the tree.

Mate-Guarding

Copulations were observed in two pairs. These pairs were not in each others company for at least 27% and 30% of the observed time (18.00–24.00). Distance between partners and the presence of floaters, during the time when pairs were not together, could not be estimated, since it was not possible to focus on more than one animal during the night.

DISCUSSION

During the four different study periods the study site was highly saturated with densely packed territories that abutted one another. Given the 25 ha study area and territories with an average size of about 1.5 ± 0.5 ha (Fietz, 1999), the expected number of pairs within the study area should have been about 16. The observed number of adult individuals within the study site ranged between 43 and 70, with varying sex ratios. But even taken only the number of adult females (21 and 35) into account, as a measure for the maximal number of pairs, this density exceeds the expected number of pairs. Since *Cheirogaleus medius* is pair-living with both sexes defending exclusive territories (Fietz, 1999), the extremely high number of adult individual, could be explained by variation in territory size and edge effects at the border of the study site. In a population were the local density has reached its carrying capacity and no vacant territories are available, animals can either remain within the territories of their parents or try to compete for a territory of their own as shown for the Florida Scrub Jay (Woolfenden & Fitzpatrick, 1984, 1990; Fitzpatrick & Woolfenden, 1992). In fact, in *C. medius* offspring remained with their parents until the second year, when they reached sexual maturity. They slept with the parents and the new set of offspring in their sleeping holes, but were never observed to provide help with the newborn (Fietz, 1999). In captivity *C. medius* reached sexual maturity within their first year (Foerg, 1982), while wild *C. medius* are sexually mature within their second year. It could be possible that young *C. medius* are sexually suppressed as long as they are living in a family group and that sexual maturity was therefore delayed as seen in species of tamarins (Ziegler et al., 1987; Savage et al., 1988; Abbott, 1984) and in *Microcebus murinus* (Perret, 1985, 1986, 1995, 1996; Perret & Schilling, 1987; Schilling & Perret, 1987).

In one instance, two adult females and one adult male were observed to be members of the same sleeping group. These two females reproduced and raised their offspring together, as soon as a neighboring home range became vacant, one of the two females moved to the adjoining territory and mated there with a different male (Fietz, 1999). This would suggest that in *Cheirogaleus medius* females are the philopatric sex. This supposition is further supported by the observations that those two females and another neighboring female interacted socially, including such activities as grooming, playing, and spent part of the hibernation period together. Family groups or pairs are found to hibernate together within the same nest holes, but contradictory to former assumptions, they emerge during the dry season, change sleeping holes, and may split up or form new sleeping associations (Fietz, unpublished data). Unlike males, females were able to hold their territories, even after their partners had died (Fietz, 1999). Furthermore, the observation of two females living on the same territory could explain the occurrence of surplus females in 1994/95, since floating females were never observed in this study.

Even though population density can be regarded as saturated during the whole study period, the absolute number of animals trapped at the site varied strongly. Differences in population densities between the years can be mainly attributed to a bi-annually occurring high numbers of subadults. After periods with high population densities, only a few subadults were recorded the following year. New-born litters within their nest holes, can be detected acoustically and by the nest-guarding behavior of both parents. In 1998 no litter was found, despite the fact that some females were known to have been pregnant that season. Explanations for this reproductive failure could be, that either pregnant females aborted or reabsorbed their embryos, or the new-born died immediately after birth. However, the reasons for this lack of successful reproduction such as resource abundance or climatic factors were not investigated in this study.

In two of the four study years, the number of captured *Cheirogaleus medius* males exceeded the number of females. Mortality and disappearance rate were not gender biased and can not be the cause for a skewed sex ratio. Significant bias in the sex ratio of subadults suggests that in some years more males than females are born. Captive female *Microcebus murinus* overproduced sons when grouped prior to mating (Perret, 1990, 1995, 1996), which is in accordance with the local resource hypotheses (Clark, 1978). Infant sex ratio of captive *C. medius*, that were caged in family groups, showed the same tendency (Foerg, 1982). Whether or not female *C. medius* are able to influence the sex ratio of their offspring, like shown for *M. murinus* (Perret, 1990, 1995, 1996) remains unknown.

Surplus adult males (= floaters) in this study used large home ranges that they did not seem to mark or defend against conspecifics. While territorial individuals marked their territories quite frequently, floaters were observed to fecal mark only on two occasions. In fights with conspecifics, floaters were generally defeated. If territory holders are the "more fit" males, one would expect that they might be larger than floating males with regard to body size or condition. However, no morphometric difference was found between these two groups of males. Floaters were sexually mature individuals, but even during the breeding season they were not reproductively active, perhaps due to the fact that no territories were available. Saturation was likely because vacant territories were occupied immediately during the next active period. In this study no floater was observed to establish itself on a territory. In three cases, males were observed to settle within territories where a territorial female was still present. Two of these cases, males were unmarked individuals. The third individual was an already marked two year old male which had been living with his family group, and he then took over a vacant territory, two territories distant from that of his family (Fietz, 1999). The fact that floaters have significantly smaller testes could be explained by sexual suppression of floaters by territorial males. In captive *Microcebus murinus*, lower ranking males have reduced testis size. Perret and Schilling (1987) and Perret (1977,

1985, 1992, 1995) showed that subordinate *M. murinus* are sexually suppressed by volatile chemosignals present in the urine of dominant males. Both sexual suppression of floaters by territorial males and mate guarding of estrus females are possibilities to prevent extra-pair copulations between females and strange males. Nevertheless, observations have shown that females of *Cheirogaleus medius* in estrus were not strictly mate guarded by their males and should theoretically have the opportunity to perform extra pair copulations with floaters or neighboring males.

The occurrence of floaters does not seem to be restricted to the Kirindy population. In a population of *Cheirogaleus medius* in Ampijoroa (Müller, 1998, this volume), one male showed the typical characteristics of a floater. He moved within a big home range, that overlapped with territories of other groups, used a high number of different sleeping sites that belonged to other sleeping groups, and showed no clear associations to conspecifics.

The results of this field project suggest that floaters represent a pool of potential male breeders, that are probably prevented from pair-living by high population densities and an associated lack of vacant territories. Since these floater males did not differ morphologically in body condition or age from territorial males, they presumably have the chance to sneak copulations with females on territories. The possibility that floaters represent an alternative mating strategy in *Cheirogaleus medius*, can not be discounted. Ongoing genetic studies associated with paternity and reproductive success of floaters and pair-bonded males will elucidate the question if floaters are making the best out of a bad job or represent an evolutionary stable strategy.

ACKNOWLEDGMENTS

I am grateful to the Commission Tripartite of the Malagasy Government, the Laboratoire de Primatologie et des Vértebrés, l'Université d'Antananarivo, and the Ministère pour la Production Animale and the Département des Eaux et Forêts for permits to work in Madagascar. I also thank the Centre de Formation Professionelle Forestière de Morondava for their hospitality and permission to work on their concession. Joachim Burkhardt, Eva Bienert, Jutta Schmid, Dorothea Schwab, Sonja Schaff, Simone Sommer, Simone Teelen, and Barbara König helped in numerous ways with this field project. I thank Jörg Ganzhorn, Joachim Burkhardt, Peter Kappeler, and Klaus Schmidt-Koenig for their valuable advice, support, and help. Peter Kappeler provided trapping data from March 1995. Jörg Ganzhorn, Steve Goodman, and two anonymous referees provided valuable comments on earlier drafts of this paper. Financial aid by the Deutsches Primatenzentrum, the HSPII, and the Deutsche Forschungsgemeinschaft to Graduiertenkolleg "Primatologie" made this study possible.

REFERENCES

ABBOTT, D. H. 1984. Behavioral and physiological suppression of fertility in subordinate marmoset monkeys. American Journal of Primatology, **6**: 169–186.

ANDERSSON, M. 1994. Sexual Selection. Princeton University Press. Princeton, New Jersey.

BROCKELMAN, W. Y., U. REICHARD, U. TREESUCON, AND J. RAEMAEKERS. 1998. Dispersal, pair formation and social structure in gibbons (*Hylobates lar*). Behavioral Ecology and Sociobiology, **42**: 329–339.

CAMPONOTUS, A. B., AND RADIO LOCATIONS SYSTEMS, A. B. 1994. Tracker: Wildlife Tracking and Analysis Software User Manual. Version 1.1. Solna and Huddinge, Sweden.
CARPENTER, C. R. 1934. A field study of the behavior and social relations of howling monkeys. Comparative Psychology Monographs, **10**: 1–168.
———. 1935. Behavior of Red Spider Monkeys in Panama. Journal of Mammalogy, **16**: 171–180.
CLARK, A. B. 1978. Sex ratio an local resource competition in a prosimian primate. Science, **210**: 163–165.
CLUTTON-BROCK, T. H., S. D. ALBON, S. D. GIBSON, AND F. E. GUINESS. 1979. The logical stag: adaptive aspects of fighting in red deer (*Cervus elaphus* L.). Animal Behaviour, **27**: 211–225.
FIETZ, J. 1999. Monogamy as a rule rather than exception in nocturnal lemurs: The case of the Fat-tailed Dwarf Lemur *Cheirogaleus medius*. Ethology, **105**: 259–272.
FITZPATRICK, J. W., AND G. E. WOOLFENDEN. 1992. Florida Scrub Jay, pp. 201–218. *In* Newton, I., ed., Lifetime Reproduction in Birds. Academic Press, London.
FOERG, R. 1982. Reproduction in *Cheirogaleus medius*. Folia Primatologica, **39**: 49–62.
GANZHORN, J. U., AND J. P. SORG, eds. 1996. Ecology and Economy of a Tropical Dry Forest in Madagascar. Primate Report, **46-1**: 1–382.
GOLDIZEN, A. W. 1987. Tamarins and marmosets, pp. 34–43. *In* Smuts, B. B., D. L. Cheney, R. M. Seyfarth, R. W. Wrangham, and T. T. Struhsaker, eds., Primate Societies. The University of Chicago Press, Chicago.
HLADIK, C. M., P. CHARLES-DOMINIQUE, AND J. J. PETTER. 1980. Feeding strategies of five nocturnal prosimians in the dry forest of the west coast of Madagascar, pp. 41–73. *In* Charles-Dominique, P., H. M. Cooper, A. Hladik, C. M. Hladik, E. Pages, G. F. Pariente, A. Petter-Rousseaux, J.-J. Petter, and A. Schilling, eds., Nocturnal Malagasy Primates: Ecology, Physiology and Behavior. Academic Press, New York.
JONES, W. T., AND B. B. BUSH. 1988. Movement and reproductive behavior of solitary male Redtail Guenons (*Cercopithecus ascanius*). American Journal of Primatology, **14**: 203–222.
KAPPELER, P. M. 1997. Intrasexual selection and testis size in strepsirhine primates. Behavioral Ecology, **8**: 10–19.
KAWAI, M., AND K. YOSHIBA. 1968. Some observations on the solitary males among Japanese Monkeys. Primates, **9**: 1–12.
KENWARD, R. E. 1991. Ranges IV, Version 1.5. Software for analyzing animal location data. Institute of Terrestrial Ecology, Wareham.
LAWS, J. W., AND J. VONDER HAAR LAWS. 1984. Social interactions among adult male langurs (*Presbytis entellus*) at Rajaji Wildlife Sanctuary. International Journal of Primatology, **5**: 31–50.
LE BOEUF, B. J. 1974. Male-male competition and reproductive success in elephant seals. American Zoologist, **14**: 163–176.
LORENZ, R. J. 1988. Biometrie. Gustav Fischer Verlag, Stuttgart.
MATTHYSEN, E. 1989. Territorial and nonterritorial settlings in juvenile Eurasian Nuthatches (*Sitta europaea* L.) in summer. Auk, **106**: 560–567.
MITANI, J. C. 1984. The behavioral regulation of monogamy in gibbons (*Hylobates muelleri*). Behavioral Ecology and Sociobiology, **15**: 225–229.
MÜLLER, A. E. 1998. A preliminary report on the social organization of *Cheirogaleus medius* (Cheirogaleidae; Primates) in north-west Madagascar. Folia Primatologica, **69**: 160–166.
MUTSCHLER, T., A. T. C. FEISTNER, AND C. M. NIEVERGELT. 1998. Preliminary field data on group size, diet and activity in the Alaotran Gentle Lemur *Hapalemur griseus alaotrensis*. Folia Primatologica, **69**: 325–330.
NISHIDA, T. 1966. A sociological study of solitary males monkeys. Primates, **7**: 141–204.
PERRET, M. 1977. Influence du groupement social sur l'activation sexuelle saisonniére chez le male de *Microcebus murinus* (Miller, 1777). Zeitschrift für Tierpsychologie, **43**: 159–179.
———. 1985. Influence of social factors on seasonal variations in plasma testosterone levels of *Microcebus murinus*. Zeitschrift für Tierpsychologie, **69**: 265–280.
———. 1986. Social influences on oestrus cycle length and plasma progesterone concentrations in the female Lesser Mouse Lemur (*Microcebus murinus*). Journal of Reproductive Fertility, **77**: 303–311.
———. 1990. Influence of social factors on sex ratio at birth, maternal investment, and young survival in a prosimian primate. Behavioral Ecology and Sociobiology, **27**: 447–454.
———. 1992. Environmental and social determinants of sexual function in the male Lesser Mouse Lemur (*Microcebus murinus*). Folia Primatologica, **59**: 1–25.
———. 1995. Chemocommunication in the reproductive function of mouse lemurs, pp. 377–392. *In* Alterman, L., G. A. Doyle, and M. K. Izard, eds., Creatures of the Dark: The Nocturnal Prosimians. Plenum Press, New York.
———. 1996. Manipulation of sex ratio at birth by urinary cues in a prosimian primate. Behavioral Ecology and Sociobiology, **38**: 259–266.

——, AND A. SCHILLING. 1987. Intermale sexual effect elicited by volatile urinary ether extract in *Microcebus murinus* (Prosimians, Primates). Journal of Chemical Ecology, **13**: 495–507.
PETTER, J.-J. 1962. Ecological and behavioral studies of Madagascars lemur in the field, pp. 267–281. *In* Buettner-Janusch, J., ed., The Relatives of Man. Annals of the New York Academy of Sciences, **102**.
——. 1978. Ecological and physiological adaptations of five sympatric nocturnal lemurs to seasonal variations in food production, pp. 211–223. *In* Chivers, D. J., and J. Herbert, eds., Recent Advances in Primatology, Vol. 1, Academic Press, New York.
——, R. ALBIGNAC, AND R. RUMPLER. 1977. Faune de Madagascar: Mammifères Lémuriens. Vol. 44. ORSTOM/CNRS, Paris.
PETTER-ROUSSEAUX, A. 1980. Seasonal activity rhythms, reproduction, and body weight variations in five sympatric nocturnal prosimians, simulated light and climatic conditions, pp. 211–223. *In* Charles-Dominique, P., H. M. Cooper, A. Hladik, C. M. Hladik, E. Pages, G. F. Pariente, A. Petter-Rousseaux, J.-J. Petter, and A. Schilling, eds., Nocturnal Malagasy Primates: Ecology, Physiology and Behavior. Academic Press, New York.
——, AND C. M. HLADIK. 1980. A comparative study of food intake in five nocturnal prosimians in simulated climatic conditions, pp. 211–223. *In* Charles-Dominique, P., H. M. Cooper, A. Hladik, C. M. Hladik, E. Pages, G. F. Pariente, A. Petter-Rousseaux, J.-J. Petter, and A. Schilling, eds., Nocturnal Malagasy Primates: Ecology, Physiology and Behavior. Academic Press, New York.
ROHNER, C. 1996. The numerical response of great horned owls to the snowshoe hare cycle: consequences of non-territorial "floaters" on demography. Journal of Animal Ecology, **65**: 359–370.
SAS/STAT. 1987. Guide for personal computers. SAS Institute Inc., Cary, NC.
SAVAGE, A., T. E. ZIEGLER, AND C. T. SNOWDON. 1988. Sociosexual development, pair bond formation, and mechanisms of fertility suppression in female Cotton-Top Tamarins (*Saguinus oedipus oedipus*). American Journal of Primatology, **14**: 345–359.
SCHALLER, G. B. 1963. The Mountain Gorilla: Ecology and Behavior. The University of Chicago Press, Chicago.
SCHILLING, A., AND M. PERRET. 1987. Chemical signals and reproductive capacity in a male prosimian primate (*Microcebus murinus*). Chemical Senses, **12**: 143–158.
SCHMID, J., AND P. M. KAPPELER. 1994. Sympatric mouse lemurs (*Microcebus* spp.) in western Madagascar. Folia Primatologica, **63**: 162–170.
SMITH, S. M. 1978. The "underworld" in a territorial sparrow: adaptive strategy for floaters. American Naturalist, **112**: 571–582.
——. 1984. Flock switching in chickadees: Why be a winter floater? American Naturalist, **123**: 81–98.
——. 1987. Responses of floaters to removal experiments on wintering chickadees. Behavioral Ecology and Sociobiology, **20**: 363–367.
SMITH, J. N. M., AND P. ACRESE. 1989. How fit are floaters? Consequences of alternative territorial behaviors in a migratory sparrow. American Naturalist, **133**: 830–845.
SLATKIN, M., AND G. HAUSFATER. 1976. A note on the activities of a solitary male baboon. Primates, **17**: 311–322.
THIRGOOD, S. J. 1990. Alternative mating strategies and reproductive success in Fallow Deer. Behaviour, **116**: 1–9.
TILSON, R. L. 1981. Family formation strategies of Kloss's Gibbons. Folia Primatologica, **35**: 259–287.
WOOLFENDEN, E., AND J. W. FITZPATRICK. 1984. The Florida Scrub Jay: Demography of a Cooperatively Breeding Bird. Princeton University Press, Princeton, New Jersey.
——. 1990. Florida Scrub Jays: a synopsis after 18 years of study, pp. 211–223. *In* Stacey, P. B., and W. D. Koenig, eds., Cooperative Breeding in Birds: Long Term Studies of Ecology and Behavior. Cambridge University Press, Cambridge.
ZIEGLER, T. E., A. SAVAGE, G. SCHEFFLER, AND C. T. SNOWDON. 1987. The endocrinology of puberty and reproductive functioning in female Cotton-Top Tamarins (*Saguinus oedipus*) under varying social conditions. Biology of Reproduction, **37**: 618–627.

10

INFLUENCE OF SOCIAL ORGANIZATION PATTERNS ON FOOD INTAKE OF *LEMUR CATTA* IN THE BERENTY RESERVE

Hantanirina Rasamimanana[1]

[1] Ecole Normale Supérieure, Université d'Antananarivo, BP 881
Antananarivo (101), Madagascar
hrasamim@syfed.refer.mg

ABSTRACT

On the basis of previous work on the food preferences of Ring-tailed Lemurs (*Lemur catta*) with respect to habitat composition, there is good evidence that the distribution of plant resources affects the social dispersion pattern of this species. In addition, females have feeding priority. In order to understand the relationships between social organization of *L. catta* and food intake, the feeding habits of males and females were studied by means of focal animal sampling at the Berenty Reserve. The number of troop members feeding in the same food patch as the focal animal at the beginning and at the end of each feeding session was recorded. There seems to be no detrimental effect on the general condition of males associated with female dominance. The male strategy appears to benefit from these social organization patterns, which allows them to eat more efficiently than the females despite of the fact they were always displaced wherever they ate. Females show more variation in numerous feeding parameters than males. Within *L. catta*, males and females seemed to have evolved different feeding strategies to cope with differences in their energy requirements, particularly with regards to reproduction.

RÉSUMÉ

Beaucoup de travaux sur les comportements alimentaires ont démontré l'effet de la répartition des plantes consommées sur le mode de dispersion en groupes sociaux chez les primates. L'étude de l'influence de la taille de chaque groupe social sur le mode d'alimentation du *Lemur catta* a été menée dans la réserve de Berenty par l'observation focale de chacun des 13 adultes mâles et 13 adultes femelles de deux troupes de maki. Ainsi tous les paramètres définissant le mode d'alimentation de l'animal focalisé ont été notés, aussi bien que le nombre d'individus mangeant la même essence au début et à la fin de chaque session d'alimentation. *L. catta* se caractérise par la priorité des femelles aux ressources alimentaires, traduite par la dominance des femelles sur les mâles. Pourtant les mâles ne semblent en aucun cas souffrir de cette dominance, si bien qu'ils tirent toujours profit de n'importe quelle dispersion sociale pour manger plus efficacement que les femelles et cela malgré le fait qu'ils soient constamment déplacés par ces dernières oú qu'ils se nourrissent. Les femelles quant à elles, ont différents modes d'alimentation selon qu'elles se dispersent socialement ou individuellement pour se nourrir. Chez *L. catta*, mâle et femelle semblent avoir développé chacun différentes stratégies d'alimentation pour faire face aux diverses demandes d'énergie requise par leur état physiologique respectif, en particulier pour la reproduction.

INTRODUCTION

Feeding practices are important parameters used to understand the behavioral and ecological differences between living primates (Fleagle, 1988). It has been proposed that food choice is the result of the evolutionary history of each species which have been specifically molded to the conditions of its environment (Sussman, 1974). Food competition has been argued to have an important influence on social organization patterns (Watts, 1985). In this paper, I use data on food intake of male and female Ring-tailed Lemurs (*Lemur catta*), associated with social organization patterns, to examine the influence of these variables on differences in the consumption rates between the sexes.

Ring-tailed Lemurs live in social groups containing up to 25 individuals of both sexes, excluding infants of the year (Jolly et al., 1993). The sex ratio varies, although for adults it tends to be close to 1.0 (Sussman, 1974; Jolly, 1985). Females and their female offspring remain in the natal group. Most of the males migrate from their natal group at the age of between three to five years, generally during the period of the year between September and November, falling between the early and late births seasons (Sussman, 1992; Jolly et al., 1993; Sauther, 1993). No pronounced body size or weight dimorphism exists in *Lemur catta*.

Jolly et al. (1993) consider Ring-tailed Lemurs to be territorial because they have an aggressively defended core area. Sauther (1993) does not classify them as territorial, because of significant home range overlap between groups, and different groups share feeding, sleeping, and resting sites. However, there are core areas which are intensively used and defended; these can change seasonally and can be used by other groups. Ring-tailed Lemurs are characterized by female dominance (Jolly, 1966; Budnitz & Dainis, 1975; Taylor, 1986).

Dominance hierarchies are based on the outcome of agonistic encounters (Pereira et al., 1990; Sauther, 1993). For example, a dominant approaches an individual, the individual responds with submissive signals or direct retreat. In other cases, when a subordinate approaches the dominant individual, the latter does not retreat or show any submission. Female Ring-tailed Lemurs virtually never retreat on approach of males nor signal submissively towards them, and males almost never act aggressively towards females (Pereira et al., 1990). In certain contexts, females approach the males aggressively, and at other times not aggressively, but in both contexts males retreat showing submissive behavior. In many cases when females displace a male, they do not occupy the spot vacated by the male, but rather return to their original place before the encounter (Rasamimanana & Rafidinarivo, 1993). Kappeler (1990) noted that female aggressiveness towards males is not only displayed during feeding but also in other social contexts.

Many authors have suggested that female dominance can be explained by the high energy requirements needed for reproduction. Hrdy (1981) noted that the general pattern for seasonally breeding primates is that the male feeding strategy is different from female feeding patterns. In general, food intake by the male is lower than by the female, but during the reproductive season he eats considerably more, when he can gain weight rapidly. Hrdy hypothesized that males use this strategy to compensate for female feeding priority, in particular exploiting differences in social organization patterns during feeding.

Food distribution, availability, and palatability affect the social organization patterns of *Lemur catta* (Sauther, 1992). This species feeds in the various vertical strata of the forest (Sussman, 1974). About 25% of the feeding time is spent on the ground eating leaves of herbaceous species, soil, and the ripe and fallen fruits of *Tamarindus indica* and *Crateva excelsa* (Sauther, 1992). The remaining 75% of feeding time is off the ground, either in large trees or in bushes. During feeding periods the troop members spread out for short distances, similar to other species of primates (e.g., baboons, capuchins, vervets, and macaques). While moving from one resource to another, dominant female *L. catta* tend to lead the group, which follows in the order: dominant males and subordinate females, followed in turn by the most subordinate males (Sussman, 1974). Even given this female feeding dominance, the physical condition of the different troop members seems to be similar within each season.

I hypothesize that when foraging with the entire troop, male *Lemur catta* eat less efficiently than females, especially when feeding on fruit; males spend more time feeding alone than females; and males will spend less time feeding per bout with the entire troop than in subgroups or alone. Further, within the social organization patterns found in *L. catta*, males have adopted strategies that allow them to maximize food intake without competing for resources with females. Such strategies might include high average consumption rates when they forage alone and eating a greater percentage of nutrient rich food, such as fruits. The purpose of this chapter is to investigate various parameters of differences in the feeding techniques of male and female *L. catta* in the context of different social organizational patterns and test the above hypothesis.

METHODS

I compared 13 adult males and 13 adult females living in two different troops of *Lemur catta* in the Berenty Private Reserve, southern Madagascar (see Budnitz & Dainis [1975] and Nicoll & Langrand [1989] for a detailed description of the site). Field techniques largely followed those of Rasamimanana and Rafidinarivo (1993). The individual troops were followed continuously through out the day from 6:00 to 18:00 h, during 110 observation days spread across the four seasons (gestation, birth, lactation, and mating) between 1987–1989. More details on the divisions of these seasons are presented in Rasamimanana and Rafidinarivo (1993).

One of the two study groups (Group F) was located in natural gallery forest and the second group (Group H) occurred in natural forest and areas with introduced plants (see Rasamimanana & Rafidinarivo, 1993). It is important to point out that Group H was occasionally provisioned with various types of food by tourists. Information on the observations made of these two troops are combined in order to generalize the relationship between social organization, food intake, and habitat.

Feeding Protocol

The following protocol was used for feeding observations. The moment the first adult animal was seen feeding, this marked the start of its feeding bout. A feeding session is composed of two main phases: foraging and eating. Searching, sniffing, and tasting (i.e., taking one bite and dropping the item) and vigilance (i.e., holding the food item but not looking directly at it) are included in the foraging category. Having the item in its mouth, chewing, and swallowing it are included in the eating category. A feeding session ended when the animal did not perform any of the above activities for twenty seconds. At this point, I switched observations to the next adult animal of the opposite sex just beginning a feeding bout. I also recorded the duration of each phase of the bout and the number of times leaves and fruits were consumed. The number of individuals feeding in the same patch as the focal animal were counted at the beginning and the end of each bout, in order to define the social organization pattern (defined below). A patch is defined as a group of trees or adjacent food sources of the same species within 5 m diameter of the focal animal. After each session, a sample of the food plant was collected as a voucher specimen. Botanical identifications were conducted on site or when necessary by specialists at the Parc Botanique et Zoologique de Tsimbazaza, Antananarivo.

Statistics

Statistical tests included Student's t-test to compare the averages of food intake parameters between sexes, Spearman's correlation to examine the relationship between plants and food intake, and Chi-square to study the distribution these parameters. The unit of analysis, was the number of feeding bouts in a session instead of the number of individuals. The period of each eating bout (in seconds) was calculated by subtracting the duration of the foraging time from the duration of the feeding bout. The efficiency of feeding was calculated by dividing the duration of eating by the duration of feeding. The rate of feeding was calculated by dividing the number of items consumed by the duration of eating, and then multiplying by 60, giving a per minute unit. The social organization pattern (see below) was defined by both initial and final numbers of individuals around the focal animal per session. A feeding session is composed of a sequence of feeding bouts in the same patch. Each season, I tried to obtain data for each individual in both groups throughout the observations.

Social Organization Patterns

In this study 11 different types of social organization patterns were defined with regard to the focal animal and the food patch:

1) They could feed alone—no other individual was feeding in the patch from the beginning to the end of the feeding session (0-0).

2) At the beginning of the session the animal was alone and then was joined by another individual (0-x).
3) At the beginning the focal animal was in the company of at least one other individual which departed before the end of the session (x-0).
4) At the beginning the focal animal was in the company of at least one other individual and other individuals subsequently joined them (xi-xn).
5) At the beginning the focal animal was in the company of several individuals, several of which departed before the end of the session (xn-xi).
6) At the beginning the focal animal was in the company of several individuals and the number remained constant (x-x).
7) At the beginning the focal animal was in the company of several individuals and then the main group joined them (x-gr).
8) The complete troop was present during the feeding (gr-gr).
9) The complete troop was present at the beginning of the session and left before the end (gr-0).
10) The complete troop was present at the beginning of the session and several individuals departed before the end (gr-x).
11) The focal animal was alone at the start and then was joined by the whole group for the balance of the session (0-gr).

All these 11 cases occurred at different rates during this study. The first three cases (0-0, 0-x, and x-0) have been lumped together as a single pattern referred to as **individual organization pattern**, since the focal animal was alone for at least a portion of the feeding bout. A second combined category (xi-xn, xn-xi, and x-x), referred to as **subgroup organization pattern**, involved only a portion of the troop. The last five cases of the above 11 categories have been combined together in the **whole group organization pattern**, since during at least a portion of the feeding bout the whole group was involved.

RESULTS

On the basis of this study, the individual organization pattern accounted for 35% of the observations, subgroup organization pattern 56%, and whole group organization patterns 9%. There was no significant difference with regard to female and male organization patterns. Males tended to be only slightly and nonsignificantly more inclined towards the individual organization pattern than females (Table 1).

With regards to seasonal variation, most patterns were equally distributed during the four periods (gestation, birth, lactation, and mating) for both sexes. The exception was that the individual organization pattern for females differed between these four periods (Chi-square = 14.27, df = 3, p < 0.01). Specifically, females tended to hold the

Table 1. Rate of occurrence of each social organization pattern within sexes for *Lemur catta*

Organization patterns	Rate of occurrence (%)	
	male	female
Subgroup	56	58
Individual	36	33
Whole	8	9

Table 2. Rate of occurrence of each social organization pattern in *Lemur catta* during different periods of the year

	Individual (%)		Subgroup (%)		Whole (%)	
	male	female	male	female	male	female
Gestation	27	27	64	61	9	12
Birth	35	30	54	61	11	9
Lactation	39	40	49	52	12	8
Mating	36	57	64	43	0	0

individual pattern, as compared to the subgroup pattern, more commonly during the period from lactation to mating. Feeding in the whole group pattern during the mating period did not occur for either sex (Table 2).

Number of Individuals for Each Social Organization Pattern

As mentioned above, both male and female *Lemur catta* feed more commonly in the subgroup organization pattern. The group size was determined by the numbers of individuals feeding in the patch during a portion of the focal animal's feeding bout. The number of individuals making up the subgroups was approximately the same with regard to the sex of the focal animal, the only exception was when the number of animals decreased from the beginning to the end of the feeding bout (i.e., xn-xi). In most cases, for the subgroup organization patterns, the focal animal initiated its feeding when four to eight individuals were already feeding and ended it when four to six individuals had left or joined it. The average group size during the study was 16 individuals (range nine to 24).

In all of these social organization patterns four distinct possibilities exist: 1) the focal animal was either joined and left by others; 2) the focal animal joined and left the group; 3) the focal animal joined the others and subsequently was left until alone; and 4) the focal animal remained alone during the observation session.

The second category, to be joined, occurred rarely for both sexes, whereas the first category, to join, was observed frequently. The focal animal demonstrated the third category more frequently than the fourth category (Table 3). Further, there was a significant difference between males and females when they joined a group (category 2) as compared to when they stayed alone (category 4). Females joined their group mates more frequently (category 2) than males, and females remained alone in the patch more frequently than males (category 4). With regard to the periods of the year, males tended to forage alone more frequently during the gestation period than during the other three periods of the year (Table 2). Females on the other hand, tended to forage as a single individual more frequently during the gestation and mating periods.

Table 3. Rate of occurrence of each social interaction pattern in *Lemur catta*. Values are percentages

To be joined		To join		To stay until alone		Alone	
male	female	male	female	male	female	male	female
3	3	26	39	8	11	4	6

Plants Species Visited during Each Organization Pattern

Lemur catta was observed to consume 38 different plant species during this study. Most of these plants were consumed by both sexes in the context of the individual and subgroup organization patterns. However, when feeding with the whole group pattern, only 38% of these plants were eaten. Focal males were observed to eat in 13% of them and focal females in 16%. Males and females joined the others (category 2) in most of the recorded food sources, but were joined (category 1) in 18% of them. Eighty percent of the plants species consumed in the context of the whole organization system are native to the site, and of these 43% are big trees, 31% bushes, and 26% herbaceous plants or fallen fruits. About 60% of the plants eaten by focal males included native bushes, while 66% of the plants consumed by focal females were introduced bushes.

On the basis of transects carried out in the various habitats within the reserve in 1998 (Dainis & Jolly, pers. com.), some generalities can be provided on the frequencies of the native woody vegetation with Diameter at Breast Height (DBH) greater than 20 cm: *Tamarindus indica* above 30%; *Acacia rovumae* and *Neotina isoneura* 10–20%; *Celtis bifidis* 5–10%; *Rinorea greveana*, *Crateva excelsa*, *Celtis phillippensis*, *Quisivianthe papinae*, and *Cordia varo* below 5%; and *Salvadora angustifolea* and *Azima tetracantha* up to 1%. These values are used later as a measure of tree abundance at the site. *Azima tetracantha* is common in open canopy forest and rarer in my study site; however, most individuals of this plant have a DBH less than 20 cm and thus are misrepresented in the above transects. Across all forested habitats within the reserve, introduced trees, including *Cordia rothii*, *Melia azedarach*, *Azadirachta indica*, *Agave sisalana*, and *Caesalpinia pulcherima*, represented less than 1% of the trees with DBH of greater than 20 cm. In my study area there are considerable numbers of planted ornamentals.

On average the most abundant tree species, based on the above transect, were eaten during displays of the whole organization pattern and the less abundant tree species in the subgroup and individual organization patterns. Another important difference was noted—the largest trees were preferentially chosen by groups using the whole organization pattern, while bushes were more commonly chosen by groups displaying subgroup and individual organization patterns (Table 4).

Types of Food Intake in the Context of the Different Social Organization Patterns

For females feeding on leaves in the individual pattern, there was a negative correlation between abundance of the plant and efficiency ($r_s = -0.76$, n = 14, p = 0.01); in particular they were less efficient in eating the leaves in the most abundant species. Females tended to forage on the more abundant native species than on the less common introduced plants. Further, they fed more in trees than in bushes and on herbaceous plants. For males feeding in the individual pattern, there was no correlation between abundance and efficiency.

In the case of females feeding on fruits, duration and frequency of visits to a given plant species were associated ($r_s = 0.75$, n = 8, p = 0.05), as were the plants abundance and frequency of visits ($r_s = 0.71$, n = 8, p = 0.05), and the plants abundance and duration of visits ($r_s = 0.77$, n = 8, p = 0.05). In simple terms, when feeding on fruits the more often they visited a given plant the longer they stayed, and they tended to visit the most

Table 4. Food sources used by each sex of *Lemur catta* in the various social organization patterns

Plant species	Characteristics[1]		Whole both sexes	Individual male	Individual female	Subgroup both sexes
Tamarindus indica	N	tree	+	+	+	+
Neotina isoneura	N	tree	+	+	+	+
Celtis bifidis	N	tree	+	+	+	+
Rinorea greveana	N	bush	+	+	+	+
Crateva excelsa	N	tree	+	+	+	+
Celtis phillipensis	N	tree	+	+	+	+
Cordia varo	N	bush	+	+	+	+
Azima tetracantha	N	bush	+	+	+	+
Boerhavia diffusa	N	herb	+	+	+	+
Capparis seppiaria	N	bush	+	+	+	+
Commicarpus commersonii	N	herb	+	+	+	+
Caesalpinia pulcherima	IP	bush	+	+	+	+
Cordia rothii	IP	tree	+	+	+	+
Banana	I	fruit	+	+	+	+
Casuarina sp.	IP	tree		+		
Enterospermum pruinosum	N	bush		+		
Hippocratea sp.	N	bush		+		
Maerua filiformis	N	bush		+		
Agave sisalana	IP	bush			+	
Brunfelsia hopeana	IP	bush			+	
Cissampelos sp.	NP	tree			+	
Neodypsis decaryii	NP	tree			+	
Opuntia vulgaris	IP	bush		+		
Orange	I	fruit			+	
Papaya	I	fruit				+
Setcreasea sp.	IP	herb			+	
Citrullus vulgaris	IP	herb				+
Poupartia caffra	NP	tree				+
Flueggea obovata	N	bush				+
Melia azedarach	IP	tree				+
Pennisetum purpureum	IP	herb				+
Quivisianthe papinae	N	tree				+
Salvadora angustifolea	N	bush				+
Eucalyptus sp.	IP	tree				+
Cactus (non-*Opuntia*)	IP	bush				+
Azadirachta indica	IP	tree				+
Acacia rovumae	N	tree				+
Aloe vahombe	NP	bush				+
% of plants used in each organization pattern			37%	13%	16%	34%

[1] N = native, I = introduced, P = planted.

common plants for longer periods. As noted above, within the individual organization pattern, 66% of the plants visited are introduced, and this group of plants is less abundant in the available habitat than native species. However, as far as representation in the diet, native plants were more frequently consumed and visited for longer periods than introduced.

The number of cases of the whole group organization pattern is small, but when both females and males consumed leaves in this context, they were less efficient in feeding in the most abundant trees ($r_s = -1$, n = 4, p = 0.05). Also in this same context, females visited more often the most abundant species for their leaves ($r_s = 1$, n = 4, p = 0.05).

In subgroup organization pattern, the commonest pattern, none of the parameters of abundance of plants, duration, and frequency of visits were associated with one another for either sex. Further, efficiency and frequency of visits to a given plant species were not correlated within this organization pattern. The only exception was that females were more efficient on leaf intake than on fruit.

Fruit intake for females appears to be influenced by the organization pattern, above all for animals feeding as individuals. The duration of feeding bouts for fruits was significantly greater than that for leaves (number feeding on fruits = 41, number feeding on leaves = 47, p = 0.05). Further, females ate fruits for significantly longer periods than males (number of females = 41, number of males = 33, p = 0.05). The period of time females fed on fruits in the individual and subgroup patterns were similar, and in both cases these were shorter than during the whole pattern.

Males and females were more efficient in feeding when displaying the whole pattern than the individual pattern. In all social organization patterns, males were more efficient in feeding on fruits than females. In the case of leaves, males were more efficient than females when feeding in the group patterns and less efficient than females in the individual pattern (Table 5). Duration of feeding on fruit by males during subgroup and whole patterns were similar and were longer as compared to the individual pattern (number of whole pattern = 4, number of individual = 33, p = 0.05) (Fig. 1). In the subgroup pattern, males ate fruit significantly longer than did females. In the whole organization pattern the duration of eating fruit was not statistically different for males and females, despite an apparently large difference between them.

Within the individual pattern males and females fed on leaves for the same duration. Both sexes fed increasingly more on leaves when they shifted from the individual to subgroup to whole social organization patterns. Duration between all these three

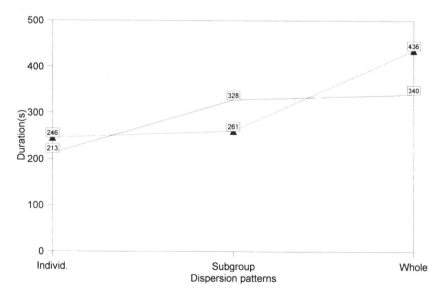

Figure 1. Average duration of fruit eating by both sexes in *Lemur catta*. Males are denoted by the solid line and females by the broken line.

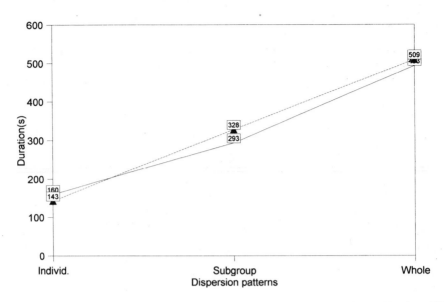

Figure 2. Average duration of leaf eating by both sexes in *Lemur catta*. Males are denoted by the solid line and females by the broken line.

patterns were significantly different (p = 0.05), but males and females within the whole pattern fed on leaves for the same duration (Fig. 2).

Females consumed a greater amount of fruit within the individual pattern than within subgroup or whole group patterns. The amounts of fruit eaten by males within the three patterns are not significantly different. Nevertheless, males consumed more fruit than females within individual and subgroup patterns (Fig. 3).

The amount of leaves consumed showed a different pattern. Females and males consumed an increasing quantity of leaves with the transition from individual to whole organization patterns. The amount of leaves eaten by females and males in the individual pattern is less than that eaten in the subgroup and whole patterns. Males ate significantly more leaves than females in each pattern (Fig. 4).

In summary, males and females behaved quite differently within the three social patterns. In the individual pattern, females fed longer on fruits than males, in the subgroup pattern females fed longer on leaves than males, and in the whole pattern females fed longer on fruits than males. However, within the three organization patterns, males always ate more or at least the same amount of any food item than females. Consequently, males ate fruit faster than females, particularly in the case of the individual pattern (Fig. 5). For leaves, the rate of intake for the two sexes was similar, although the rate was higher in the individual pattern. Male consumption

Table 5. Proportion of leaves in the diet of *Lemur catta* based on the different social organization patterns

	Male	Female
Individual	60	65
Subgroup	70	71
Whole	81	78

Influence of Social Organization Patterns on Food Intake of *Lemur catta*

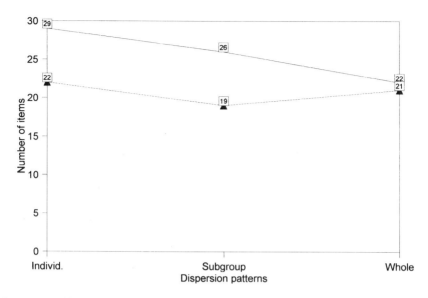

Figure 3. Amount of fruit eaten by both sexes of *Lemur catta*. Males are denoted by the solid line and females by the broken line.

of leaves exceeded that of females, but the difference was not statistically significant (Fig. 6).

Within all three social organization patterns, both sexes ate leaves much faster than fruit. In other words, males compensated for their shorter duration of feeding by eating faster, and achieved as high or higher intake per feeding bout as females.

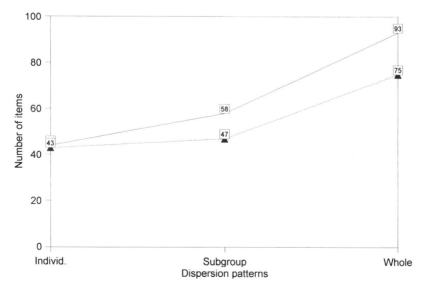

Figure 4. Amount of leaves eaten by both sexes in *Lemur catta*. Males are denoted by the solid line and females by the broken line.

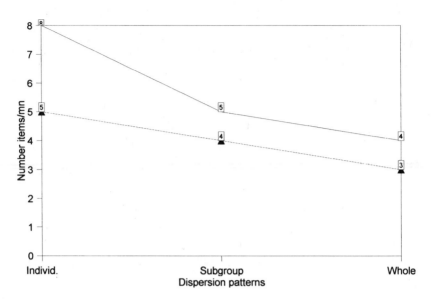

Figure 5. Feeding rate on fruit by both sexes in *Lemur catta*. Males are denoted by the solid line and females by the broken line.

As for the difference between social and individual foraging, the feeding rate on any food item was always faster in the individual pattern than in the subgroups or whole groups (Table 5). Males and females increased their fruit consumption when feeding in the individual pattern or surrounded by few individuals. There appears to be a trend that males in the individual pattern consume proportionately more fruit than females.

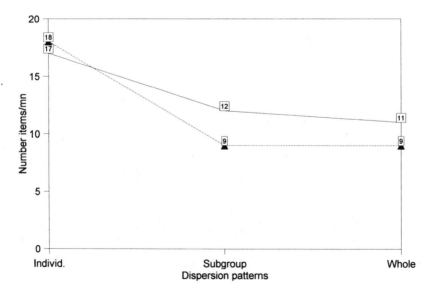

Figure 6. Feeding rate on leaves by both sexes in *Lemur catta*. Males are denoted by the solid line and females by the broken line.

DISCUSSION

Occurrence of Social Organization Pattern and Seasonal Physiological Periods

Subgroup organization was the commonest pattern found in *Lemur catta*, this is particularly notable for males in all seasons. This pattern is similar to many other groups of living primates, which tend to spread out in smaller groups when feeding.

During the mating season, female *Lemur catta* fed more frequently in the individual pattern. The mating season is the "hinge" period of the year when fruits become scarce. The contrasting aspects of low food resources versus the physiological demands of competition between males for access to females and females rebuffing would-be mates poses extreme metabolic demands on the animals. This could explain why the whole group organization pattern was avoided by both sexes during the mating period. Individuals, particularly females, are obliged to seek out the most energetic and nutrition rich foods. Small bushes used as food plants still retain some new leaves and fruit during this period, but on the basis of their size and food resources they cannot accommodate a troop of *L. catta*. If the troop were to feed as a group, only the most aggressive individuals could have access to such bushes. During the mating period, females are very aggressive towards males and other females (Sauther, 1992). They frequently displaced males from any food source, even though they may not take up the place he was foraging (Rasamimanana & Rafidinarivo, 1993). Males always retreat and wait until the female has finished foraging and has left the patch. Males do not exploit introduced species chosen by females, such as *Setcreasea* sp., but rather they depend on native bushes, which are often uncommon, such as *Maerua filiformis*.

During gestation, in contrast, the high rate of both sexes using the subgroup and whole group patterns can be explained by the scarcity of nutritious foods during thedry and cold season (Rasamimanana & Rafidinarivo, 1993). During this season, the only available food for *Lemur catta* is the mature leaves and some fruits of *Tamarindus indica*. *Tamarindus* are generally large and common trees, capable of supporting a largenumber of animals. Thus, feeding in larger social groups during this season is not diadvantageous in the sense of feeding competition and there is also a general decrease in energy expenditure. An alternative strategy for some animals is to feed in the inviual pattern, and exploit the more widely dispersed fruits of plants other than *Tamarindus*.

Interaction Patterns in the Context of Feeding

Females tend to lead the troop when moving between food resources or when searching for a site to sleep or sunbathe. Sometimes males deviate from this trajectory (Sussman, 1974; Jolly, 1985; Sauther, 1992). Within our classification of social interaction patterns it is expected that category 1 (to be joined) has a lower rate occurrence than the other three categories (Table 3). This is because there is only a single dominant female and male in the troop and there are proportionately fewer observations on these animals when they initiate the feeding bout as compared to all other members of the troop. A second explanation could be that I missed some initial interactions of subordinates, which might on occasion initiate feeding and be joined by other troop members. Nevertheless, the percent of category 2 (to join) is higher than any other category. This might be due to social facilitation, dominant or subordinate individuals follow the examples of others (King, 1994).

Female Strategies Cope with High Energy Requirements

Females represent the sex showing a notable change in the three different organization patterns over the course of the seasons. For the females there is an increase in the percentage of introduced plants visited during the period they pursue the individual social organization pattern. It is important to note that in one troop no introduced plants were consumed, however, when data for the two troops are combined, this group of plant species remains important in the visitation rates of females and to a lesser degree for the males. When females showed the whole group social organization pattern, there was a negative correlation between feeding on leaves, abundance of food resources, and efficiency of feeding. Females were less efficient because they spent a considerable amount of time chasing males. Finally, given the negative correlation between abundance and efficiency, there seems to be a paradox in explaining the food choices of these animals. Several explanations are possible. Resources of abundant species may become depleted within each patch, so that the animals spend time choosing the ripest fruit, or fruit not yet attacked by insects. Alternatively, if the patches are large and non-depletable, lemurs may not eat in such a hurry.

Male Feeding Efficiency Counters the Effect of Female Dominance

As shown in this study, males ate more fruits and leaves (Figs. 3 and 4) than females and in general are more efficient at feeding throughout the year regardless of the organization pattern they adopted (Table 6). Male feeding strategies tend to be more homogenous seasonally. The only exception was when the males displayed the whole social organization pattern. In this case, as with females, there was a negative correlation between feeding on leaves, abundance of food resources, and efficiency of feeding. Not only did females lose time chasing males, but males in turn were less efficient in feeding because of their vigilance towards females and searching for other spots to feed. These observations show the feeding strategy adopted by males is simpler than the females, and in the final count may be more efficient resulting in a larger quantity of food obtained.

Our own interpretation of the reason why females are less efficient in feeding than males is that females spend more time choosing between various types of abundant native foods that may have higher nutritional qualities. Further, females have

Table 6. Feeding efficiency for both sexes of *Lemur catta* on different types of food items based on the different organization patterns

	Fruit		Leaves	
	male	female	male	female
Individual	0.69	0.58	0.66	0.70
Subgroup	0.70	0.65	0.77	0.72
Whole	0.84	0.66	0.79	0.75

higher energetic requirements than males, including chasing off non-potential genitors, sustaining pregnancies for 140 days, giving birth, nursing, and carrying the young for at least four months. Thus, females have had to develop different feeding strategies than males, which may include the question of female feeding priority and dominance over males.

Sex Differences in the Use of Introduced Plants

Adult males eat more native plants (in the individual social organization pattern) than females and seem less inclined to feed on novel (introduced) plants. Males ate faster and more than females regardless of the season, so perhaps certain native food resources were partially depleted by the males and potentially as a result the females ate proportionately more exotic plants than males.

Another possibility is that the nutritional content of certain introduced plants is higher than that of native plants and the strategy of a female is to maximize the quality of her diet (see Ganzhorn & Abraham, 1991). Currently, we lack the critical information to test the role of the nutritional content of plants chosen by males and females in explaining their different feeding strategies. Further, differences in secondary compounds between native and introduced plants is unknown, but in general *Lemur catta* seems to choose plants with lower concentrations of phenols and alkaloids (see Simmen et al., this volume). Finally there may be a behavioral difference in the readiness of males and females to adopt new behavioral patterns (Kappeler, 1987).

Tests are needed to assess if the females are actively choosing novel plants in the environment (in this case introduced plants) to start the adoption of new elements in the diet. Potentially once these elements have been incorporated in the diet of females the process of social facilitation will subsequently include these plants into the general palette chosen by this species. Previous research with *Eulemur fulvus* in captivity indicates that these animals chose new food items by sniffing and taking a single bite of the plant, and this may be the same process followed by other prosimians such as *Lemur catta* (Glander, 1983). Given that many of these introduced plants in the Berenty Reserve have only been available to *L. catta* for about a decade (=two *L. catta* generations) the process of incorporating new elements into their diet has not yet reached the troop level. If this was indeed the case, the percentage of introduced plants in the diets of adult males should increase over the next few generations.

In the introduction to this chapter predictions were made that males eat less efficiently than females, that males spend more time alone than females, and that males spend less time feeding per bout when they are with the entire troop than in subgroups or alone. None of these predictions were confirmed by the field observations. Rather, the strategy used by males is simply to maximize food consumption on familiar foods (native plants). For females a counter strategy is taken, which includes feeding priority and dominance over males; and probably the freedom to chose a higher quality diet, with strategies varying by season. These factors apparently provide benefits to females in their form of social relations with males.

ACKNOWLEDGMENTS

Financial support for this study was received from the New York Zoological Society and the T-Shirt Fund of the Wildlife Preservation Trust International. I am grateful to Alison Jolly and Dorothy Fragaszy for their encouragement, help, and advice. For comments on an earlier version of this paper I acknowledge the help of Steve Goodman and Alison Jolly.

REFERENCES

BUDNITZ, N., AND K. DAINIS. 1975. *Lemur catta*: ecology and behavior, pp. 219–235. *In* Tattersall, I., and R. W. Sussman, eds., Lemur Biology. Plenum, New York.
FLEAGLE, J. G. 1988. Primate Adaptation and Evolution. Academic Press, London.
FRAGASZY, D., AND E. VISALBERGHI. 1996. Social learning in monkeys: Primate "Primacy" Reconsidered, pp. 65–85. *In* Heyes, C. M., and B. G. Galef, Jr. eds., Social Learning in Animals: The Roots of Culture. Academic Press, San Diego.
GANZHORN, J. U., AND J.-P. ABRAHAM. 1991. Possible role of plantations for lemur conservation in Madagascar: food for folivorous species. Folia Primatologica, **56**: 171–176.
GLANDER, K. 1983. Food choice from endemic North Carolina tree species by captive Prosimians (*Lemur fulvus*). American Journal of Primatology, **5**: 221–229.
HEYES, C. M. 1994. Social learning in animals: categories and mechanisms. Biological Review, **69**: 207–231.
———, AND B. G. GALEF. 1996. Social Learning in Animals: The Roots of Culture. Academic Press, San Diego.
HRDY, S. B. 1981. The Woman That Never Evolved. Harvard University Press, Cambridge, Massachusetts.
JOLLY, A. 1966. Lemur Behavior. The University of Chicago Press, Chicago.
———. 1985. The Evolution of Primate Behavior. MacMillan, New York.
———. 1993. Territoriality in *Lemur catta* groups during the birth season at Berenty, Madagascar, pp. 85–109. *In* Kappeler, P. M., and J. U. Ganzhorn, eds., Lemur Social Systems and their Ecological Basis. Plenum Press, New York.
KAPPELER, P. M. 1987. The acquisition process of a novel behavior pattern in a group of Ring-tailed Lemurs (*Lemur catta*). Primates, **28**: 225–228.
———. 1990. Female dominance in *Lemur catta*: more than just female feeding priority? Folia Primatologica, **55**: 92–95.
KING, B. J. 1994. The Information Continuum. Evolution of Social Information Transfer in Monkeys, Apes, and Hominids. SAR Press, Santa Fe.
LEFEBVRE, L., AND L.-A. GIRALDEAU. 1996. Is social learning an adaptive specialization?, pp. 107–128. *In* Heyes, C. M., and B. G. Galef, Jr. eds., Social Learning in Animals: The Roots of Culture. Academic Press, San Diego.
PEREIRA, M. E., R. KAUFMAN, P. KAPPELER, AND D. J. OVERDORFF. 1990. Female dominance does not characterize all of the Lemuridae. Folia Primatologica, **55**: 96–103.
RASAMIMANANA, H., AND E. RAFIDINARIVO. 1993. Feeding behavior of *Lemur catta* females in relation to their physiological state, pp. 123–133. *In* Kappeler, P. M., AND J. U. Ganzhorn, eds., Lemur Social Systems and their Ecological Basis. Plenum Press, New York.
SAUTHER, M. L. 1992. The effect of reproductive state, social rank, and group size on resource use among free-ranging ringtailed lemurs (*Lemur catta*) of Madagascar. Ph.D. thesis, Washington University, St. Louis.
———. 1993. A new interpretation of the social organization and mating system of the ringtailed lemur (*Lemur catta*), pp. 111–121. *In* Kappeler, P. M., and J. U. Ganzhorn, eds., Lemur Social Systems and their Ecological Basis. Plenum Press, New York.
SUSSMAN, R. W. 1974. Ecological distinctions in sympatric species of lemurs, pp. 75–108. *In* Martin, R. D., G. A. Doyle, and A. C. Walker, eds., Prosimian Biology. Duckworth, London.
———. 1992. Male life history and intergroup mobility among ringtailed lemurs (*Lemur catta*). International Journal of Primatology, **13**: 395–414.
TAYLOR, L. 1986. Kinship, dominance, and social organization in semi-free ranging group of ringtailed lemurs (*Lemur catta*). Ph.D. thesis, Washington University, St. Louis.
VISALBERGHI, E., AND D. FRAGASZY. 1995. The behavior of Capuchin Monkeys (*Cebus apella*) with novel food: the role of social context. Animal Behavior, **49**: 1089–1095.
WATTS, D. P. 1985. Relations between group size and composition and feeding competition in mountain gorilla groups. Animal Behavior, **33**: 72–85.

11

THE IMPORTANCE OF THE BLACK LEMUR (*EULEMUR MACACO*) FOR SEED DISPERSAL IN LOKOBE FOREST, NOSY BE

Christopher R. Birkinshaw[1]

[1] Missouri Botanical Garden, BP 3391
Antananarivo (101), Madagascar
birkinsh@mbg.mbg.mg

ABSTRACT

In order to classify trees in the Lokobe Forest, Nosy Be, as dispersed, possibly dispersed, or not dispersed by Black Lemurs (*Eulemur macaco*), two Black Lemur groups were habituated and observed during the day and night for all months of the year (total 1219 hours). When fruits were eaten, the species was identified, and the maturity of the fruit and treatment of the seeds noted. Black Lemur droppings were searched for seeds; these were identified and signs of damage noted. Species that had ripe fruit that were not eaten by the Black Lemur were also identified, as were fleshy-fruited species that produced little or no ripe fruits during the study. Other frugivores feeding on the fruits of black lemur-dispersed species were also noted. In order to estimate the proportion of tree species, tree trunks, and trunk basal area in the Lokobe Forest dispersed, possibly dispersed, and not dispersed by Black Lemurs, plots were installed in Lokobe's two forest types and the trees they enclosed were identified and their dbh measured. For the slope forest Black Lemurs dispersed 57% of the represented tree species, 71% of the represented tree trunks, and 73% of the represented trunk basal area. For the ridge forest these values were 49%, 76%, and 88% respectively. Only four of the 38 black lemur-dispersed species had fruits that were also eaten by other frugivore species. These results show that the Black Lemur is very important for seed dispersal in Lokobe Forest.

RÉSUMÉ

Dans le but de classifier les arbres de la forêt de Lokobe à Nosy Be par rapport au Lémur macaco (*Eulemur macaco*) en arbres par lui disséminés, probablement disséminés ou non disséminés par ce lémurien, deux groupes de Lémur macaco ont été habitués et observés de jour et de nuit pendant tous les mois d'une année (1219 heures au total). Pour les fruits consommés, l'identification de l'espèce a été effectuée et ont été notés la maturité des fruits et l'état des graines. Les déjections des lémuriens ont été ramassées pour y rechercher des graines afin de les identifier et pour relever des signes d'altération. Les espèces dont les fruits mûrs n'avaient pas été consommés par Lémur macaco ont également été identifiées ainsi que les espèces à fruits charnus qui produisaient peu voir pas de fruits mûrs pendant l'étude, et enfin les autres frugivores consommateurs des fruits disséminés par Lémur macaco ont également été relevés. Afin de pouvoir estimer la proportion des espèces d'arbres, des troncs et leur aire basale qui sont disséminés, probablement disséminés ou non disséminés par Lémur macaco dans la forêt de Lokobe, des parcelles ont été mises en place dans les deux types de forêts rencontrées à Lokobe, les espèces d'arbres présents ayant été identifiés et leur dhp mesuré. Dans la forêt sur pente, Lémur macaco a dispersé 57% des espèces d'arbres représentés, 71% des troncs représentés et 73% de l'aire basale des troncs représentés. Dans la forêt sur crête, ces proportions ont respectivement été de 49%, 76% et 88%. Sur les 38 espèces disséminées par le Lémur macaco, les fruits de quatre espèces seulement ont également été consommés par d'autres espèces frugivores. Ces résultats montrent que le Lémur macaco est important dans la dissémination de graines dans la forêt de Lokobe.

INTRODUCTION

Seed dispersal is important for plant regeneration (Janzen, 1970; Connell, 1971; Howe & Smallwood, 1982; Augspurger, 1983; Clark & Clark, 1984). In tropical rain forests many plant species have fleshy fruits and are dispersed by frugivorous animals which eat the fruits and deposit the viable seeds away from the parent plant (e.g., Janzen, 1975; Foster, 1982; Gentry, 1982; Howe & Smallwood, 1982; Howe, 1984; Gautier-Hion et al., 1985; Fleming et al., 1987; Willson et al. 1989; Stiles, 1992). Several of Madagascar's lemur species are known, at least sometimes, to disperse seeds (i.e., *Eulemur fulvus*, *E. macaco*, *E. rubriventer*, *Varecia variegata*, *Cheirogaleus medius*, *Microcebus* spp., and *Propithecus verreauxi* (Dew, 1991; Birkinshaw, 1995; Ganzhorn & Kappeler, 1996; Scharfe & Schlund 1996; Ralisoamalala, 1996; Ganzhorn et al., in press) (but see also Overdorff & Strait (1998) for examples of *E. fulvus* and *E. rubriventer* acting as seed predators). However, their importance for seed dispersal at a given site has rarely been quantified (Ganzhorn et al., in press). In this study, the importance of the Black Lemur (*Eulemur m. macaco*) for seed dispersal in the Lokobe Forest is demonstrated by providing estimates of the proportion of the tree species, tree trunks, and trunk basal area dispersed by this species in replicated plots in the forest.

STUDY SITE AND METHODS

The Resérve Naturelle Intégrale de Lokobe is located on a small basalt hill (summit 430 m) on Nosy Be in northwestern Madagascar (13° 23–25′ S, 48° 18–20′ E). The island's climate is characterized by a high equable temperature throughout the year (maximum mean daily temperature of 28°C in January and February, minimum of 23°C in July and August) and a moderately high annual rainfall which is distinctly seasonal (mean of 2,356 mm, most falling between November and May) (White, 1983). The vegetation of Lokobe Forest is classified as primary low altitude evergreen humid forest (Faramalala, 1988), but, on the basis of flora and structure, includes two distinct variants: slope forest and ridge forest. The vertebrate frugivore fauna in the Lokobe Forest (defined as species for which ripe fruit is a major part of their annual diet) includes just five species: the Black Lemur, the Madagascar Bulbul (*Hypsipetes madagascariensis*), the Madagascar Blue Pigeon (*Alectroenas madagascariensis*), the Madagascar Fruit Bat (*Pteropus rufus*), and the Straw-colored Fruit Bat (*Eidolon dupreanum*). In addition, the Madagascar Green Pigeon *(Treron australis*) was seen on Nosy Be, but it was not recorded in the Lokobe Forest.

Study Species

Eulemur macaco is an arboreal, cat-sized, group-living prosimian. Its mean weight is 2.4 kg and its mean head and body length is 41 cm (Tattersall, 1982). Its annual diet at Lokobe consists of 78.0% ripe fruits, 4.7% unripe fruits, 12.7% leaves, and 4.8% flowers, nectar, and invertebrates (Birkinshaw, 1995).

Identifying Plant Species that Are Dispersed, Possibly Dispersed, and Not Dispersed by the Black Lemur

Between November 1991 to March 1993 two Black Lemur groups were habitated and observed for 785 hours during the day and 434 hours at night. When the lemurs ate ripe fruits the plant species was identified and the treatment of the seeds noted (i.e. whether they were chewed, spat out or swallowed whole). Black Lemur droppings were collected whenever possible and examined for seeds. The seeds were identified and their state noted (i.e. whether visibly damaged or not). In addition, other forest tree species with ripe fruits, which were not eaten by the Black Lemur, were identified, as were fleshy-fruited species that produced little or no ripe fruits during the study.

Birkinshaw (1995) tested (by means of germination tests) the viability of samples of visibly undamaged seeds defecated by the Black Lemur for 29 species. Some seeds of all of the species germinated (mean percentage germination = 73.1%). Birkinshaw (1995) also estimated for 16 plant species the percentage of seeds swallowed by Black Lemurs during the day that were deposited away from the parent plant and fruiting conspecifics. This ranges between 0% to 28.5% according to the species. Given these results it is assumed here that species with fruits that were eaten when ripe and with seeds that were swallowed whole and voided visibly undamaged were at least sometimes dispersed by the Black Lemur and are therefore classified as black lemur-dispersed. Whereas, species with fruits which were always eaten when unripe or not exploited at all, or species with seeds which were always chewed or spat out *in situ* are classified as not black lemur-dispersed. This class also includes species without fleshy fruits that were not produced during the study. Species with fleshy fruits that were not produced during the study but that are similar in their characteristics to the fruits of species known to be dispersed by Black Lemurs, are classified as possibly black lemur-dispersed. Finally, species which were not identified and which produced no fruits during the study are placed in a fourth class. Black lemur-dispersed species with fruits that are also eaten by other animal species were identified by means of opportunistic observations of focal fruiting trees made during the course of lemur observations (i.e. the trees observed were fruiting trees in the vicinity of the lemur individual being observed).

Estimating the Proportion of Tree Species, Tree Trunks, and Trunk Basal Area in Lokobe Forest Dispersed, Possibly Dispersed, and Not Dispersed by the Black Lemur

Four 0.125 ha plots (20m × 62.5m) were established at randomly chosen locations in the slope forest and three such plots in the ridge forest. Within each plot, trees with diameter at breast height (dbh) ≥ 10cm were identified and thereby classified as dispersed, possibly dispersed, or not dispersed by Black Lemurs based on the above criteria. The dbh of the trees was measured and used to calculate trunk basal area. The proportion "Σ trunks for a given class/Σ trunks" provides a measure of the relative density of the class, and the proportion "Σ trunk basal area for a given class/Σ trunk basal area" provides a measure of the relative dominance of the class.

RESULTS AND DISCUSSION

Table 1 lists the tree species occurring in the slope forest and ridge forest plots together with their representation in terms of number of trunks and trunk basal area. The species are classified as dispersed by Black Lemurs, possibly dispersed by Black Lemurs, not dispersed by Black Lemurs, or not identified and not producing fruits. These results are summarized in Figures 1 and 2 for the slope forest and ridge forest respectively. For the slope forest the Black Lemur dispersed 57% of the sampled tree species, 71% of the sampled tree trunks and 73% of the total trunk basal area. For the ridge forest these values were 49%, 76%, and 88%, respectively.

These proportions are at the top end of the range of similar estimates for other frugivore and folivore-frugivore animals in other tropical forests, for example: forest Chimpanzees (*Pan troglodytes*) disperse 56% of tree species (> 10cm dbh) identified within belt plots covering 10 ha of the Kibale Forest, Uganda (Wrangham et al., 1994); Elephants (*Loxodonta africana*) are supposed to be the sole dispersers of 30% of 71 tree species that were identified along a 2000 m transect in the Tai Forest, Ivory Coast (Alexandre, 1978); Red Howlers (*Alouatta seniculus*) disperse 26% of species whose fruits were gathered from the forest floor at Nourages, French Guiana (Julliot, 1992); and three species of hornbill (*Ceratogymna* spp.) disperse 22% of the tree species in Dja Reserve, Cameroon (Whitney et al., 1998).

Table 1 also shows species with fruits eaten by other frugivore species. These included only four of the 38 plant species dispersed by Black Lemurs. The Madagascar Bulbul and Madagascar Blue Pigeon are unable to exploit the fruits of most of these species because these fruits are too large for them to swallow whole and these fruits cannot easily be pecked into smaller pieces. The Madagascar Fruit Bat has similar feeding abilities to the Black Lemur and could exploit the fruits of many of the black lemur-dispersed species. However, they were seen exploiting the fruits of just one of these species (*Ficus assimilis*, Moraceae). Indeed with the exception of times when they exploited the fruit of *F. assimilis* and nectar from the flowers of *Parkia madagascariensis* (Fabaceae), the Madagascar Fruit Bat was not seen feeding within the

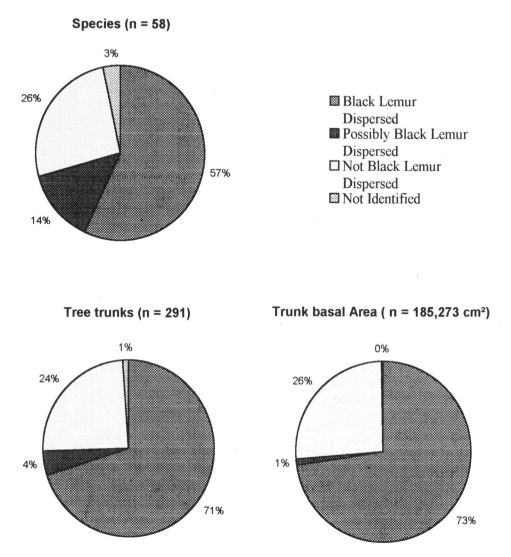

Figure 1. Proportion of tree species, tree trunks, and trunk basal area dispersed by the Black Lemur in the slope forest of the Lokobe Forest.

Lokobe Forest. Rather, they were quite frequently seen feeding around the maritime fringe of the forest (on fruits from naturalized mangoes and *Terminalia catappa* (Combretaceae), and elsewhere on the island (on the nectar from *Ceiba pentandra* (Bombaceae) flowers and fruits from figs and cultivated fruit trees). The Straw-colored Fruit Bat, although considerably smaller than the Black Lemur, could also theoretically exploit the fruits of some of the black lemur-dispersed species, but like the Madagascar Fruit Bat it was rarely seen inside the Lokobe Forest (i.e., only when exploiting nectar from the flowers of *Parkia madagascariensis*). Outside the forest it was also seen exploiting nectar from *Ceiba pentandra* flowers.

A large proportion of Lokobe's tree species, tree trunks, and trunk basal area are dispersed by the Black Lemur and a large proportion of black lemur-dispersed species

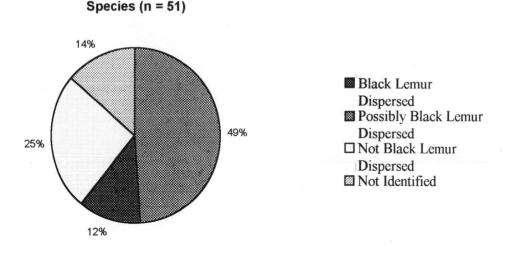
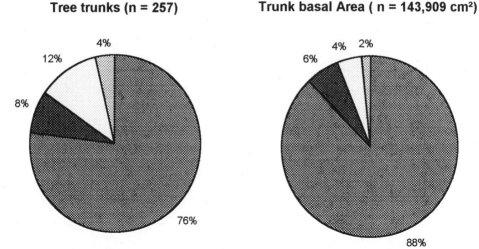

Figure 2. Proportion of tree species, tree trunks, and trunk basal area dispersed by the Black Lemur in the ridge forest of the Lokobe Forest.

have fruits which seem to be exploited only by this animal. Several factors explain these results. First, ripe fruit is a very important part of the Black Lemur diet. Indeed, with an estimated 78.0% of its annual diet (in terms of time spent feeding) consisting of ripe fruit, the Black Lemur at Lokobe is one of the most frugivorous primates known (Birkinshaw, 1995). Secondly, Black Lemurs have a wide repertoire of foraging and feeding methods which allows them to exploit fruits with diverse morphologies (e.g., small, medium, and large fruits and fruits with and without a thick husk) and diverse spatial and temporal distribution patterns (e.g., fruits which occur at high densities for a short period of time and fruits which occur at low densities for long periods of time) (Birkinshaw, 1995). Thirdly, Black Lemurs in Lokobe Forest disperse the seeds of most of the fruit species that they exploit. During this study Black Lemurs were seen to

Table 1. Classification of tree species in the slope forest and ridge forest in the Lokobe Forest based on plot data. Plant lists are presented according to whether they are dispersed, possibly dispersed, or not dispersed by Black Lemurs or not identified

Family	Species	Slope forest # of trunks	Slope forest Basal area [cm^2]	Ridge forest # of trunks	Ridge forest Basal area [cm^2]
A. Dispersed by Black Lemurs					
Flacourtiaceae	*Aphloia theiformis*	1	79	4	1,242
Rubiaceae	*Breonia* sp. 1	2	1,276	1	154
Menispermiaceae	*Burasaia* sp. 1	11	1,940	2	129
Burseraceae	*Canarium madagascariense,*	29	67,445	2	1,973
Sapotaceae	*Chrysophyllum bovinianum*	9	2,042	0	0
Sapotaceae	*Chrysophyllum sambiranensis*	5	1,240	1	177
Fabaceae	*Corydala madagascariensis*	1	1,886	0	0
Lauraceae	*Cryptocarya* sp. 1	2	2,165	0	0
Ebenaceae	*Diospyros clusiifolia*	32	13,041	2	845
Liliaceae	*Dracaena* sp. 1	3	306	0	0
Arecaceae	*Dypsis ampasindavae*	3	674	1	346
Arecaceae	*Dypsis madagascariensis*	1	133	1	380
Moraceae	*Ficus assimilis*[1,2,3]	1	1,075	0	0
Clusiaceae	*Garcinia* sp. 1	8	1,172	11	2,523
Clusiaceae	*Garcinia* sp. 2	3	1,142	11	2,331
Tiliaceae	*Grewia* sp. 1	1	380	2	664
Sapindaceae	*Macphersonia madagascariensis*	2	668	0	0
Oleaceae	*Noronhia* sp. 1	1	227	0	0
Oleaceae	sp. 1	2	190	3	269
Pandanaceae	*Pandanus androcephalanthus*	1	79	1	95
Fabaceae	*Parkia madagascariensis*	10	13,970	9	8,375
Rubiaceae	*Peponidium horridum*	0	0	2	233
Sapindaceae	*Plagioscyphus* sp. 1	4	1,896	0	0
Annonaceae	*Polyalthia richardiana*	6	2,713	2	13,497
Burseraceae	*Protium madagascariense*[1]	13	4,889	12	4,293
Rubiaceae	sp. 2	0	0	2	211
Flacourtiaceae	*Scolopia hazomby*	7	817	5	740
Moraceae	*Streblus dimepate*[1]	12	3,363	10	1,751
Moraceae	*Streblus mauritianus*	2	402	0	0
Menispermiaceae	*Strychnopsis thouarsii*	7	1,214	0	0
Loganiaceae	*Strychnos* sp. 1	5	1,957	0	0
Monimiaceae	*Tambourissa* sp. 1	5	856	0	0
Combretaceae	*Terminalia calophylla*	0	0	1	79
Moraceae	*Treculia africana*	11	2,115	0	0
Moraceae	*Trophis montana*[1]	2	1,741	2	368
Euphorbiaceae	*Uapaca louveli*	0	0	109	97,356
Annonaceae	*Xylopia* sp. 1	2	1,222	1	113
Annonaceae	*Xylopia* sp. 2	0	0	1	177
TOTALS		204	134,314	198	126,170
B. Possibly Dispersed by Black Lemurs					
Clusiaceae	*Calophyllum* sp. 1	0	0	2	228
Liliaceae	*Dracaena reflexa*[1]	1	113	0	0
Arecaceae	*Dypsis nossibensis*	1	177	1	133
Annonaceae	*Isolona* sp. 1	2	249	11	1,979
Lauraceae	sp. 1	4	569	0	0
Flacourtiaceae	*Ludia scolopioides*	1	133	0	0
Sapindaceae	*Macphersonia radlkoferi*	0	0	1	133
Myrtaceae	sp. 1	1	133	1	3,526
Lauraceae	*Ocotea* sp. 1	1	531	0	0
Clusiaceae	*Symphonia* sp. 1	2	157	4	2,994
TOTALS		13	2,061	20	8,992

(*continued*)

Table 1. (Continued)

Family	Species	Slope forest # of trunks	Slope forest Basal area [cm^2]	Ridge forest # of trunks	Ridge forest Basal area [cm^2]
C. Not dispersed by Black Lemurs					
Fabaceae	*Albizzia* sp. 1	1	95	1	177
Rhizophoraceae	*Cassipourea gummiflua*	0	0	2	208
Fabaceae	*Dalbergia* sp. 1	0	0	2	228
Melastomataceae	*Dichaetanthera heteromorpha*	0	0	2	192
Ebenaceae	*Diospyros gracipes*	2	314	2	157
Anacardiaceae	*Gluta tourtour*	1	4,186	0	0
Chrysobalanaceae	*Grangeria porosa*	30	16,814	4	750
Flacourtiaceae	*Homalium micranthum*	4	9,240	0	0
Euphorbiaceae	*Lautembergia coriacea*	3	488	0	0
Euphorbiaceae	*Lautembergia* sp. 1	0	0	1	95
Euphorbiaceae	*Macaranga* sp. 1	0	0	3	236
Rhizophoraceae	*Macarisia lanceolata*	12	6,906	1	804
Rubiaceae	*Mussaenda* sp. 1	1	227	0	0
Passifloraceae	*Paropsis obscura*	1	79	4	820
Araliaceae	*Polyscias nossibensis*[1]	1	177	1	113
Strelitziaceae	*Ravenala madagascariensis*	1	616	4	1758
Fabaceae	*Senna lactia*	1	452	0	0
Apocynaceae	*Stephanostegia megalocarpa*[1]	5	3,123	3	907
Sterculiaceae	*Sterculia perrieri*[1]	6	4,968	0	0
Rutaceae	*Zanthoxylum* sp. 1	2	538	0	0
TOTALS		71	48,222	30	6,444
D. Not identified					
Euphorbiaceae	sp. 1	0	0	1	79
Fabaceae	sp. 1	0	0	1	133
Rubiaceae	sp. 1	2	500	0	0
Rubiaceae	sp. 3	0	0	1	201
Rubiaceae	sp. 4	0	0	2	211
Sapindaceae	sp. 1	0	0	1	452
Unidentified	sp. 1	0	0	1	661
Unidentified	sp. 2	1	177	2	566
TOTALS		3	677	9	2,302

Fruits also eaten by:
[1]Madagascar Bulbul (*Hypsipetes madagascariensis*),
[2]Madagascar Blue Pigeon (*Alectroenas madagascariensis*),
[3]Madagascar Fruit Bat (*Pteropus rufus*).

consume fruits of 70 plant species (including trees, liana and shrubs). Of these 70 species, the seeds of 57 species were generally swallowed and voided visibly undamaged, the seeds of five species were sometimes swallowed and voided visibly undamaged and sometimes predated or wasted, the seeds of five species were generally predated or wasted, and the fate of the seeds of three species was uncertain. Fourthly, there are very few frugivore species in the Lokobe Forest. This is the result of the species-poor frugivore fauna of Madagascar in general (Dewar, 1984; Fleming et al., 1987; Langrand, 1990; Goodman & Ganzhorn, 1997) being diminished further in the Lokobe Forest by its small size and island location. Fifthly, it is possible that the Black Lemurs themselves, being the dominant element in a minuscule frugivore community, have largely controlled which species in the Lokobe Forest regenerate.

The Black Lemur at Lokobe has a stable population and is not threatened by hunting. However, there are many examples of forests elsewhere in Madagascar where this species or other seed-dispersing lemur species are unnaturally rare or already extinct because of hunting. For example in the Tampolo Forest, Fenerive-Est, as a result of hunting, the formerly plentiful Brown Lemur (*Eulemur fulvus*) is now extremely rare (Goodman & Rakotondravony, 1998). The results from the Lokobe Forest suggest that in such forests a large proportion of the tree species would suffer from reduced seed dispersal. This could lead to their poor regeneration and, with time, a change in the plant species composition of the forest. In fragments of dry deciduous forest in western Madagascar, Ganzhorn et al. (in press) have demonstrated that lemur-dispersed trees in fragments lacking *E. fulvus* have significantly poorer regeneration compared to fragments where this species is present. A similar result of the increasing rarity or extinction of primates has been found in Kibale Forest, Uganda (Chapman & Onderdonk, 1998). Compared to the disappearance of primates from tropical forests elsewhere in the world, the extinction of seed-dispersing lemurs from Madagascar's forests is likely to have a greater impact on forest regeneration because of this country's relatively species-poor frugivore fauna.

For some lemur-dispersed species it is likely that some seed dispersal will continue even after the extinction of their lemur dispersers because of seeds from fallen decomposed fruits being transported by rodents or by ground wash following heavy rain. However, for other species, in particular those with seeds trapped within thick-husked fruits, there may be little of no seed dispersal following lemur extinction. For example, Baum (1995) suggests that the present day poor regeneration of *Adansonia* spp. (Bombaceae) may be related to the extinction of the large, apparently frugivorous, and potentially seed-dispersing *Archeolemur* and the inability of Madagascar's extant frugivore fauna to process the large, thick-husked fruits of these species.

ACKNOWLEDGMENTS

I would like to thank Patrice Antilahimena and Josephine Andrews for their assistance in the field; and Christian Camara, for the translation of the Abstract. I am also grateful to the Association Nationale pour la Gestion des Aires Protégées, the Ministère de l'Enseignement Supérieur, and the Ministère des Eaux et Forêts for authorization to conduct this study. This research was begun as a doctoral study at University College London supervised by Dr. F. B. Goldsmith.

REFERENCES

ALEXANDRE, D. Y. 1978. Le rôle disséminateur des elephants en Forêt de Tai, Cote d'Ivoire. La Terre et la Vie, **32**: 47–72.

AUGSPURGER, C. K. 1983. Seed dispersal of the tropical tree, *Platypodium elegans*, and the escape of its seedlings from fungal pathogens. Journal of Ecology, **71**: 759–771.

BAUM, D. 1995. A systematic review of *Adansonia* (Bombaceae). Annals of the Missouri Botanical Garden, **82**: 440–470.

BIRKINSHAW, C. R. 1995. The importance of the black lemur, *Eulemur macaco* (Lemuridae, Primates), for seed dispersal in Lokobe Forest, Madagascar. Doctoral Thesis, University College London, London.

CHAPMAN, C. A., AND D. A. ONDERDONK. 1998. Forests without primates: primate/plant codependency. American Journal of Primatology, **45**: 127–141.

CLARK, D. B., AND D. A. CLARK. 1984. Spacing dynamics of a tropical rain forest tree, evaluation of the Janzen-Connell model. American Naturalist, **124**: 769–788.
CONNELL, J. H. 1971. On the role of natural enemies in preventing competitive exclusion in some marine animals and in rain forest trees, pp. 2948–3100. *In* den Boer, P. J., and G. R. Gadwell, eds., Dynamics of Populations. Center for Agricultural Publishing and Documentation, Wageningen.
COUSENS, J. 1974. An Introduction to Woodland Ecology. Oliver and Boyd, Edinburgh.
DEW, J. L. 1991. Frugivory and seed dispersal in Madagascar's eastern rainforest. Masters thesis, Duke University, Durham, NC.
DEWAR, R. E. 1984. Extinctions in Madagascar: the loss of the subfossil fauna, pp. 574–593. *In* Martin, P. S., and R. S. Klein, eds., Quarternary Extinctions: A Prehistoric Revolution. University of Arizona Press, Tucson.
FARAMALALA, M. H. 1988. Etude de la végétation de Madagascar à l'aide des données spatiales. Thèse de Doctorat, l'Université Paul Sabatier de Toulouse.
FLEMING, T. H., R. BREITWISCH, AND G. H. WHITESIDES. 1987. Patterns of tropical vertebrate diversity. Annual Review of Ecology and Systematics, **18**: 91–109.
FOSTER, R. B. 1982. The seasonal rhythm of fruitfall on Barro Colorado Island, pp. 151–172. *In* Leigh, E. G., A. S. Rand, and D. M. Windsor, eds., The ecology of a tropical forest—seasonal rhythms and long-term changes. Smithsonian Institution Press, Washington, D. C.
GANZHORN, J.U., J. FIETZ, E. RAKOTOVAO, D. SCHWAB, AND D. ZINNER. In press. Lemurs and the regeneration of dry deciduous forest in Madagascar. Conservation Biology.
GANZHORN, J. U., AND P. M. KAPPELER. 1996. Lemurs of the Kirindy Forest, pp. 257–274. *In* Ganzhorn, J. U., and J.-P Sorg, eds., Ecology and Economy of a Tropical Dry Forest in Madagascar. Primate Report special issue, **46–1**: 1–382.
GAUTIER-HION, A., J.-M. DUPLANTIER, R. QURIS, F. FEER, C. SOURD, J.-P. DECOUX, G. DUBOST, L. EMMONS, C. ERARD, P. A. HECKETSWEILER, A. MOUNGAZI, C. ROUSSILHON, AND J.-M THIOLLAY. 1985. Fruit characters as a basis of fruit choice and seed dispersal in a tropical forest vertebrate community. Oecologia, **65**: 324–337.
GENTRY, A. H. 1982. Patterns of neotropical species diversity. Evolutionary Biology, **15**: 1–84.
GOODMAN, S. M., AND J. U. GANZHORN. 1997. Rarity of figs (*Ficus*) on Madagascar and its relationship to a depauperate frugivore community. Revue d'Ecologie (Terre Vie), **52**: 321–329.
GOODMAN, S. M., AND D. RAKOTONDRAVONY. 1998. Les lémuriens, pp. 213–221. *In* Ratsirarson J., and S. M. Goodman, eds., Inventaire biologique de la forêt littorale de Tampolo (Fenoarivo Atsinanana). Recherches pour le développement, Série Sciences Biologiques, **14**: 1–261.
HOWE, H. F. 1984. Implications of seed dispersal by animals for tropical reserve management. Biological Conservation, **30**: 261–281.
——, AND J. SMALLWOOD. 1982. Ecology of seed dispersal. Annual Review of Ecology and Systematics, **13**: 201–228.
JANZEN, D. H. 1970. Herbivores and the number of tree species in tropical forest. American Naturalist, **104**: 501–528.
——. 1975. The Ecology of Plants in the Tropics. E. Arnold, London.
JULLIOT, C. 1992. Utilisation des ressources alimentaires par le singe Hurleur Roux, *Alouatta seniculus* (Atelidae, Primates), en Guyane: impact de la dessemination des graines sur la regéneration forestière. Thèse de Doctorat, l'Université de Tours.
LANGRAND, O. 1990. Guide to the Birds of Madagascar. Yale University Press, New Haven.
OVERDORFF, D. J., AND S. G. STRAIT. 1998. Seed handling by three prosimian primates in southeastern Madagascar: implications for seed dispersal. American Journal of Primatology, **45**: 69–82.
RALISOAMALALA, R. C. 1996. Role de *Eulemur fulvus rufus* (Audeberg 1799) et de *Propithecus verreauxi verreauxi* (A Grandier 1867) dans les dessemination des graines, pp. 285–293. *In* Ganzhorn, J. U., and J.-P. Sorg, eds., Ecology and Economy of a Tropical Dry Forest in Madagascar. Primate Report, **46–1**: 1–382.
SCHARFE, F., AND W. SCHLUND. 1996. Seed removal by lemurs in dry deciduous forest in western Madagascar, pp. 295–304. *In* Ganzhorn, J. U., and J.-P. Sorg, eds., Ecology and Economy of a Tropical Dry Forest in Madagascar. Primate Report, **46–1**: 1–382.
STILES, E. W. 1992. Animals as seed dispersers, pp. 87–104. *In* Fenner, M., ed., Seeds: The Ecology of Regeneration in Plant Communities. CAB International, Wallingford.
TATTERSALL, I. 1982. The Primates of Madagascar. Columbia University Press, New York.
WHITE, F. 1983. The Vegetation of Africa, A Descriptive Memoir to Accompany the UNESCO/AETFAT/UNSO Vegetation Map of Africa. UNESCO, Paris.

WHITNEY, K. D., M. K. FOGIEL, A. M. LAMPERTI, K. M. HOLBROOK, D. J. STAUFRFER, B. D. HARDESTY, V. T. PARKER, AND T. B. SMITH. 1998. Seed dispersal by *Ceratogymna* hornbills in Dja Reserve, Cameroon. Journal of Tropical Ecology, **14**: 351–371.

WILLSON, M. F., A. K. IRVINE, AND N. G. WALSH. 1989. Vertebrate dispersal syndromes in some Australian and New Zealand plant communities, with geographical comparisons. Biotropica, **21**: 133–147.

WRANGHAM, R. W., C. A. CHAPMAN AND L. C. CHAPMAN. 1994. Seed dispersal by forest chimpanzees in Uganda. Journal of Tropical Ecology, **10**: 355–368.

12

TASTE DISCRIMINATION IN LEMURS AND OTHER PRIMATES, AND THE RELATIONSHIPS TO DISTRIBUTION OF PLANT ALLELOCHEMICALS IN DIFFERENT HABITATS OF MADAGASCAR

Bruno Simmen,[1] Annette Hladik,[1] Pierrette L. Ramasiarisoa,[2] Sandra Iaconelli,[1] and Claude M. Hladik[1]

[1] CNRS/UMR 9935, Laboratoire d'Ecologie Générale
Muséum National d'Histoire Naturelle
4 av. du Petit Château, 91800 Brunoy, France
hladik@ccr.jussieu.fr (corresponding author)
[2] CNRE, BP 1739, Antananarivo (101), Madagascar

ABSTRACT

This chapter deals with the adaptation of taste responses of lemurs and other primates to different environments, in relation to primary and secondary compounds in potential foodstuffs. Emphasis is placed on the relationship between taste sensibility to sugars and energy expenditure across species. In the most specialized species, the adaptive trends are inferred according to the importance of the deviation from such allometric relationship. The signification of sugar mimics present in some fruits is discussed in terms of coevolution of plants and tasting ability of primates, that, for lemurs, parallels that of platyrrhine monkeys.

Taste responses towards other tastants such as sodium chloride are examined in relation to potential risks of deficiency and/or toxicity. Sensitivity to tannins has been investigated in different species, with a two-bottle preference test. We observed large variations that are likely to be adaptive to the concentrations in plant species in various environments. For instance, the rejection threshold for a mixture of tannin and fructose is much higher in *Propithecus verreauxi* (above 170 g/l) than in *Microcebus murinus* (0.54 g/l). Recognition thresholds can also vary slightly between human populations, in relation to ancient or recent food practices. There is also a wide range of taste sensitivity towards quinine, without any correlation, in this case, with body mass or other factors related to energy expenditure.

Different habitats of Madagascar are compared according to the results of screening tests on tannins and alkaloids. The eastern rain forest (at Andasibe) present slightly lower proportion of plants with alkaloid-like reaction, and a significantly higher proportion of tannin-rich plants than both the gallery forest and the Didiereaceae bush in the south (at Berenty). The results have been related to the gustatory ability of lemur species having to cope with these secondary compounds, and the food niche of the different species.

RÉSUMÉ

Nous présentons, dans ce chapitre, les adaptations de la sensibilité gustative des lémuriens et des autres primates à différents environnements, en fonction de la teneur des aliments potentiels en composés primaires et secondaires. Nous montrons d'abord l'importance d'une relation d'allométrie entre la sensibilité aux sucres et les besoins énergétiques des différentes espèces. Les tendances vers un régime alimentaire spécialisé se traduisent par une déviation par rapport à la tendance moyenne rapportée au poids corporel de l'espèce. Nous montrons également des exemples de coévolution entre les possibilités de perception des produits sucrés par les primates et l'apparition de substances dont le goût mime celui des sucres dans différents environnements. Dans ce cas, il existe un parallélisme entre les possibilités de perception des primates platyrrhiniens et celle des lémuriens.

Les réponses vis-à-vis d'autres substances auxquelles réagissent les organes de la gustation, comme par exemple le chlorure de sodium, sont discutées en fonction des risques de carence ou des possibles effets toxiques. La sensibilité aux tannins a été étudiée chez différentes espèces, en fonction d'un test comportemental de préférence-évitement. De ce point de vue, il existe d'importantes différences entre primates, susceptibles de correspondre aux possibilités d'adaptation aux concentrations des tannins dans les espèces végétales des différents milieux. Par exemple, le seuil d'évitement d'un mélange de tannin et de fructose est beaucoup plus élevé chez le propithèque, *Propithecus verreauxi* (plus de 170 g/l) que chez le microcèbe, *Microcebus murinus* (0.54 g/l). Les seuils de reconnaissances des tannins peuvent également varier, mais dans une moindre mesure, chez les populations humaines, en relation avec une adaptation ancienne ou relativement récente des pratiques alimentaires. De la même façon, on observe des différences de sensibilité à la quinine; mais dans ce cas il n'existe aucune relation avec la masse corporelle ou tout autre paramètre relatif à la dépense énergétique.

Nous avons comparé, dans différents habitats de Madagascar, les fréquences des tannins et des alcaloïdes, en fonction des résultats de tests préliminaires (*screening*). Dans la forêt dense humide de l'est (à Andasibé), nous avons trouvé une proportion sensiblement inférieure à celle des forêts du sud (forêt galerie et bush à Didiereaceae de Berenty) de plantes susceptibles de contenir des alcaloïdes. Au contraire, en ce qui concerne les teneurs en tannins, la proportion est nettement plus élevée dans la forêt humide de l'est que dans les deux autres environnements étudiés. Ces résultats ont été rapportés à ce que nous savons des adaptations gustatives et des comportements alimentaires des différentes espèces de lémuriens confrontées aux produits secondaires de ces habitats.

INTRODUCTION

Recent advances in the field of taste physiology have revealed that most of the taste bud sensory cells of the primate tongue respond to several substances, having higher affinities for some of them (Faurion, 1987). The shape of the signal elicited on gustatory nerves is the result of the combined firing of all these cells. It is a kind of "signature" differing more or less in the various tasting substances, an evidence that left to rest the old idea of the "four basic tastes" (Faurion, 1993; Hladik & Simmen, 1996).

There are, nevertheless, categories of substances (sugars, acids, etc.) that elicit taste signals with such resembling shapes that, for non human primates and humans, it is not easy to discriminate among different products within each category, even at high concentration. The occurrence of such classes of tasty substances among natural products is related to the evolutionary trends in food nutrient content and toxicity, and linked to sensory perception of potential consumers.

As diets have evolved in past and present environments, tastes have responded adaptively, especially in order to maximize energy intake. In turn, food plants have evolved nutrients and toxins in relation to the tasting abilities of consumers. These compounds can be beneficent or harmful in various environments and at different concentrations, as shown by the examples discussed in this chapter.

Taste abilities of lemurs and other primates are presented in terms of thresholds and above-threshold responses to potential foods. The method of investigating taste ability in non-human primates is based on a standard behavioral testing procedure: the "two-bottle test" (Glaser, 1979; Simmen & Hladik, 1988). Consumptions of a tastant solution and tap water presented simultaneously are measured using various concentrations of the tastant, to determine the lowest concentration that is "discriminated". Although the behavioral thresholds of some primate species were quite similar to thresholds obtained by directly recording signals on a peripheral taste nerve (Glaser & Hellekant, 1977), the results must be carefully interpreted because the test provides information on both taste discrimination and preference.

SUGAR DISCRIMINATION

Lemurs, as most non-human primates, include fruits in their diet. The form and function of the digestive system (including taste perception) have been shaped in parallel to the evolution of fruit-bearing plants following the Mesozoic (Hladik & Chivers, 1994; Simmen, 1994). Although variable across plant species and in relation to ripening (Bollard, 1970), fruit composition generally includes soluble forms of sugars—mainly fructose, glucose, and sucrose.

The thresholds for sucrose, which are known for 33 non-human primate species (Simmen & Hladik, 1998), vary between 6 and 330 millimoles/l (that is 2 to 113 g/l). Since these "behavioral thresholds" are the minimum concentrations that remain attractive, most ripe fruits have a sugar content that can actually be tasted and produce a sensory reward in most primate species.

For lemurs, as for other primates, including humans (Fig. 1), the threshold for sucrose is correlated with species body mass, the larger the species, the better taste acuity (i.e., low threshold). There is a similar correlation between taste ability and body mass for fructose, although less data are available regarding thresholds (Simmen &

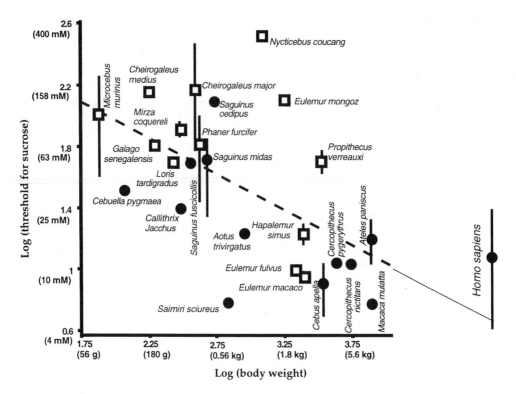

Figure 1. Allometric relationship between taste sensitivity to sucrose and body mass in lemurs (squares) and other primates (circles); data from Simmen and Hladik (1998); Hladik and Pasquet (in press).

Hladik, 1998). These relationships may reflect the importance of taste acuity to improve foraging efficiency, since large body-sized primates perceive a wide range of sugar concentrations as palatable and can use a wide array of foodstuffs.

The adaptive trends are revealed by shifts from the regression line, although the wide scatter of data is partly due to inaccuracy or differences between the methods used to measure thresholds. Among primate species differing noticeably from the common pattern (i.e. located outside of the regression line), the Slow Loris, *Nycticebus coucang*, exhibits a high taste threshold, presumably corresponding to a generalized decrease in taste sensitivity. This allows the use of pungent insects and other prey unpalatable to most primates (Hladik, 1979). In contrast, the eclectic frugivorous diet of the Squirrel Monkey, *Saimiri sciureus*, would necessitate a taste acuity better than predicted by the allometric function to cope with the high energy expenditure of foraging in extremely large home ranges (Terborgh, 1983).

Furthermore, there is a dichotomy in taste ability for peculiar sweeteners (protein sugar mimics such as monellin and thaumatin) between New World and Old World primates, including humans, and lemurs appear, in this respect, as close to platyrrhines (Glaser et al., 1978). For instance, the fruit of a rain forest species of west Africa, *Thaumatococcus daniellii*, has a very sweet pulp around the seeds, but almost no sugar. The corresponding sweetener, thaumatin, is tasted by Old World primates, but not by lemurs and New World primates.

Species differences in the ability to discriminate the very strong sweet taste (as perceived by humans) of such natural sweeteners are most likely explained by different binding mechanisms on chemoreceptors. Protein evolution in taste receptors would have followed species diversification, after catarrhine and platyrrhine primates evolved separately on the Old World and American continental plates respectively. In their corresponding rain forest environments, flowering plants competing for seed dispersal evolved fruits containing large amounts of sugars; the more sugar, the more efficient their dispersal by consumers. As a result, genes coding for the fortuitous emergence of proteins with tastes mimicking those of sugars would have been selected for. Primates feeding on these fruits of the African rain forest are "tricked" by the plant species for which they work as seed dispersers without receiving any energy in return (although they obtain a sensory reward).

In Madagascar, from this viewpoint, prosimian taste perception remain close to that of the present platyrrhine primates of the New World, and the plants bearing fruits with sugar mimics have not been observed in the various Malagasy habitats. Nevertheless, one can wonder whether homologous forms have evolved, that have not yet been detected, since sugar mimics tasted by prosimians would be tasteless for humans whose taste buds have typical characteristics of catarrhine primates.

Taste sampling of soluble sugars allows high energy intake through immediate preference; but this example cannot be generalized to other high-calory foodstuffs. Indeed, several nutritious foods have little taste, including most plant parts containing starch or fat (such as tubers, nuts, and grains), the staple foods of human populations. The apparently imperfect taste response to these highly nutritious compounds (as compared to clear-cut responses to soluble sugars) could be related to the relatively recent radiation of flowering plants. Whereas sugars—always present in plant metabolic pathways—may have been concentrated in fruits of the early angiosperms, fatty fruits seem to be the result of a more recent and sophisticated evolutionary process (McKey et al., 1993). In this case, the trend towards reduction in the size of the fruits is compensated by a high caloric density, and the reward (in terms of energy intake), although determined by a delayed response of the organism, can be associated to other cues of taste perception for an immediate sensory reward.

DISCRIMINATION OF OTHER TASTANTS

The positive responses to sodium chloride of most mammal species have been considered as adaptive. However, mineral deficiencies are unlikely to occur among wild primates, especially in forest environments where available foods provide higher dietary supplies than estimated requirements (Hladik & Gueguen, 1974). Sodium chloride (which is harmful only if ingested in too large amounts) is present at low concentration in most plant parts (less than 0.5% of the dry weight, that is below 20mM concentration). The resulting salty taste is unperceivable for most primates, which have thresholds ranging between 5mM and 500mM. In this context of low risk of mineral deficiency, one may question whether geophagy plays a role in mineral nutrition. Indeed, clay and other phyllitous soil materials eaten by primates can also work as adsorbent of tannins of the stomach content. This beneficent effect is the most likely explanation for geophagy during the periods of intense feeding on mature leaves that contain digestibility reducers such as tannins.

Tannins are widespread in plants (Bate-Smith, 1974), known for their role as a chemical defence preventing destruction by predators (Swain, 1979). The biological effect derives from: (1) a repellant taste, that renders the plant tissues unpalatable, (2) affinity to bind with proteins and to form insoluble complexes, reducing the digestibility of protein (see review in Haslam, 1989). Several primate species select plant parts with low levels of tannins (Ganzhorn, 1988), whereas other species appear to tolerate large amounts (Struhsaker et al., 1997).

Recently, a gallotannin has been shown to elicit a signal on a branch of the chorda tympani—the proper nerve which conveys only gustatory signals—of *Microcebus murinus* (Hellekant et al., 1993). The same tannin produces responses in the neurones of the orbitofrontal cortex (secondary taste area) of *Macaca fascicularis* (Critchley & Rolls, 1996). The results suggested that astringency corresponds to one or several taste qualities.

In terms of plant adaptive strategies, tannins are efficient only when large amounts are present to deter herbivores. Condensed tannins in fruits—and their distasteful taste—tend to decline during maturation, simultaneously with the increase of sugars; the taste response is necessarily directed towards the resulting mixtures. For instance, Simmen (1994) showed that *Callithrix jacchus* and *Callimico goeldii*, which have similar perception of fructose, tolerate tannin/fructose mixtures, but reject them when the tannic acid reaches 4% of the fructose content (that is 0.4g/l for a moderately sweet solution). Nevertheless, the more sugar in the mixture, the more tannin tolerated.

In *Microcebus murinus*, the behavioral method "two-bottle test" was used to measure differences of consumption between a solution of fructose at 100mM versus the same solution added with tannins (tannic acid, oak tannin). The inhibition threshold was defined as the lowest tannin concentration for which the mean difference of consumption between mixture and sweet solution is significant (paired-sample t-test). The inhibition threshold corresponds to 0.54g/l for tannic acid and to 2.0g/l for oak tannin, that is between 3 and 11% of the weight of fructose added to the solution (Fig. 2). This is a level corresponding to tannin concentration in many unripe fruits, a concentration that varies throughout the ripening process (van Buren, 1970). For example, immature fruits eaten by chimpanzees, may contain 5% (12g/l assuming 80% moisture in fruits) of condensed tannins (Wrangham & Waterman, 1983). As demonstrated by Simmen et al. (in press), the tolerance of tannins is dependant upon the concentration of sugar, which corroborates the idea of a trade-off between acceptable levels of tannin and nutrient content, mediated by oropharyngeal sensations. The electrophysiological recordings obtained from the chorda tympani of *Microcebus murinus* proper nerve showed that tannic acid elicits a reponse at 0.34g/l and no response at 0.21g/l (Hellekant et al., 1993). Since astringency of tannic acid may partly be masked by sweetness (Lyman & Green, 1990), the results, using either electrophysiological or behavioral methods, are concordant.

The recognition thresholds were also determined for humans, as part of a European Union program, with a blind test during which tannic acid, oak tannin, and various non-tannin substances, were presented at random, starting from the weakest concentrations (Iaconelli et al., 1998). The individual recognition threshold is the lowest concentration for which the taste can be described according to standard quality labels (sweet, sour, salty, bitter or astringent). The recognition thresholds for tannins varies among European populations between 0.32 and 0.79g/l for oak tannin, and between 0.22 and 1.15g/l for tannic acid. Significant differences ($p < 0.05$) for both tannins have

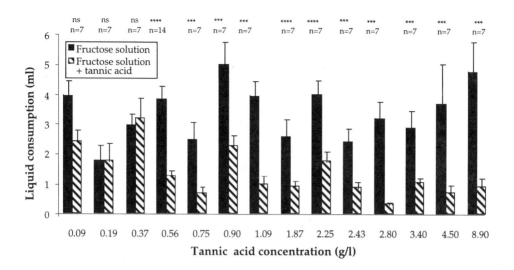
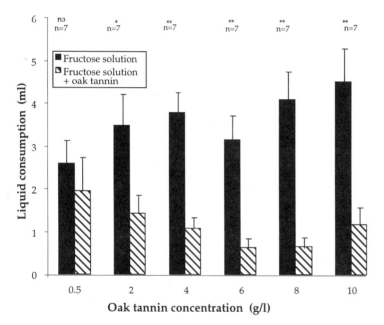

Figure 2. Ingestive responses of male *Microcebus murinus* towards binary solutions of fructose and tannin (upper: tannic acid; lower: oak tannin) versus pure fructose solution in a two-bottle test. The fructose concentration is held constant at 100 mM whereas tannin concentration is varied in each test. The inhibition threshold is defined as the lowest tannin concentration for which the mean difference of consumption between mixture and sweet solution is significant (paired-sample t-test).

been found between the north samples (France) and the south samples (Italy and Spain), the latter having higher thresholds. Alimentary inquiries indicate that the proportion of astringent products in the diet was much higher in mediterranean populations (astringent vegetables, olive oil, oak acorn, chestnut, red wine, lemon, and grape). Either dietary or genetic factors may influence the recognition threshold for astringent taste.

TASTE AND FEEDING SELECTIVITY IN VARIOUS ENVIRONMENTS

Taste thresholds for quinine vary widely—from 0.8 to 800 micromoles per liter (μM)—among non-human primates; but in contrast to what was observed for sugars, no relationship could be found between the taste sensitivity to quinine and the body mass of different species (Fig. 3). A wide range in sensitivity may reflect the adaptations of different primate species to different nutritional environments, as exemplified by the two marmosets, *Callithrix argentata*, living on white-sand riverine forests, and *Cebuella pygmaea*, inhabiting the interior of the rain forest. Both species feed mainly on the gum exuded by a tree bark after they have gouged it with their incisors. These primates are in contact with bark substances evolved by tree species as chemical defences (for instance quinine is a chemical substance in cinchona bark). However, due to the peculiar environment where these marmosets live in, the alkaloid content and toxicity of the bark is likely to vary. Contrary to rain forests, where there is little risk of eating bitter plants—because most alkaloids, such as caffein, are not likely to be highly toxic—forests with less diversified flora (such as that inhabited by *Callithrix argentata*) present a higher risk, that can be avoided by an extreme sensitivity to quinine.

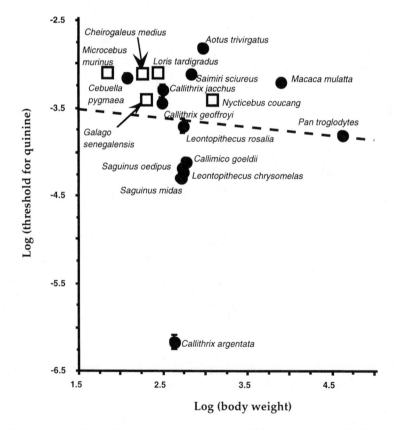

Figure 3. Relationship (without allometry) between taste sensitivity to quinine and body mass in lemurs (squares) and other primates (circles); data from Simmen and Hladik (1998).

In fact, plant parts are frequently selected for on the basis of low content of alkaloid and/or tannin. But this is not a general rule. In the Gabon rain forest, for example, where 14% of the plant specimens (among 382 species tested) react positively to the alkaloid test, the Chimpanzee includes in its diet a similar proportion (15%) of plants likely to have a high alkaloid content. Accordingly, since chimps have no particular detoxification system, most alkaloids in this environment can be compared to caffein in their weak toxic effect (Hladik & Hladik, 1977). Furthermore, the observations of Huffman and Seifu (1989) of Chimpanzees, in a montane forest, "curing themselves" with a bitter plant species, *Vernonia amygdalina*, usually discarded by healthy individuals, provide evidence that the aversive response to alkaloids can be reversed.

Our recent observations in various environments of Madagascar, also provide evidence of the variation of tannin and alkaloid occurrence, with which lemurs have to cope, but the question of whether Malagasy prosimian species differ in their tolerance of plant secondary compounds has been little investigated so far. In the eastern rain forest of Andasibe, niche partitioning has been inferred on the basis of species ability to feed on plants containing alkaloids and tannins (Ganzhorn, 1988). In captivity, primate species exhibit distinct taste discriminative thresholds for quinine hydrochloride and tannic acid (see above).

Plants were collected in the eastern rain forest at Andasibe and in the gallery forest as well as in the spiny Didiereaceae bush in southern Madagascar (Berenty). The screening of alkaloids was performed using Mayer's and Dragendorff's reagents. For phenolic compounds, we used ferric chloride and salted gelatine. The occurrence of these secondary compounds is determined according to the precipitate obtained when adding the reagents to solubilized leaf samples. The amount of the precipitate is expressed on a scale ranging from 0 to +++ (Table 1; see detailed results in Appendix).

The distribution of phenolics in plants at Andasibe differs significantly from that found in the two forest types at Berenty. Plants responding ++ and +++ account for much of the difference. Conversely, the plants at Andasibe are more frequently poor in alkaloids compared with the other two forests. It is noticeable however that, in all three sites, proportions of plants that give strong positive responses to alkaloid reagents are higher than those found in many other primate habitats, including lemur habitats (e.g., dry deciduous forest of Morondava; A. Hladik, 1980). It must be stressed that the reagents used to detect these substances are not totally specific (preliminary results are presented here). In addition, the high alkaloid content found in both the rainforest (Andasibe) and gallery forest (Berenty) can be explained by the large number, in our

Table 1. Comparison of a mid-montane rain forest of the east of Madagascar (Andasibe) with two southern habitats, the gallery forest and the Didieraceae Bush (at Berenty), according to the screening of phenolics and alkaloids in mature leaves. Significance levels are derived from Chi-square tests (**$p \leq 0.01$; ***$p \leq 0.001$)

Site	Phenolics (% of plant species)				Alkaloids (% of plant species)			
	0	±/+	++/+++	n	0	±/+	++/+++	n
Mid-montane rain forest (Andasibe)	10.8	13.2	76.0	129	36.7	25.0	38.3	128
Gallery forest (Berenty)	27.0	27.8	45.2	115***	15.4	27.9	56.7	104***
Spiny bush (Berenty)	28.6	35.7	35.7	42***	11.8	27.5	60.8	51**

Table 2. Comparison of plants eaten by three lemur species at Berenty with a random sample of plants from the gallery forest at Berenty, according to tests for the presence or absence of phenolics and alkaloids. Significance levels are based on Chi-square tests (*p ≤ 0.05). The proportion of plants with alkaloids in the diet of *Lemur catta* differ almost significantly from the occurrence of plants with alkaloids in the gallery forest (p = 0.07)

	Phenolics (% of plant species)				Alkaloids (% of plant species)			
Lemur species	0	±/+	++/+++	n	0	±/+	++/+++	n
Propithecus verreauxi	14.7	38.2	47.1	34	15.6	31.2	53.1	32
Lemur catta	55.6	27.8	16.7	18	27.8	44.4	27.8	18
Eulemur fulvus[1]	75.0		25.0	8	62.5		37.5	8
Gallery forest (Berenty)	27.0	27.8	45.2	115	15.4	27.9	56.7	104

[1] *E. fulvus* has a low dietary diversity in the gallery forest. Therefore results of "0" and "±/+" were pooled to allow application of the Chi-square test, corrected for continuity.

samples, of species living in open habitats (pioneer species, or introduced plants). The open parts of a rainforest, with a relatively low plant diversity, present a high frequency of alkaloid-rich plants (Hladik & Hladik, 1977), as in the case of white-sand riverine forests (see above).

In May/June 1998, in the gallery forest of Berenty, the diet of sympatric groups of three species, *Lemur catta*, *Eulemur fulvus* (introduced population), and *Propithecus verreauxi* was studied in terms of the relative proportions of ingested matter. Forty-six leaf species was observed to be eaten by *Propithecus*, versus 22 by *Lemur catta*, and 9 by *Eulemur fulvus*.

Table 2 shows the alkaloid and phenolic contents (according to screening tests) of a subset of the leaves selected by each of the three species, as compared with a random sample of plants available in the habitat. Plants tested accounted for more than 70% of the diets (by weight). It may be seen that only *Lemur catta* choose plants with low phenolics and tends to avoid plants containing alkaloids (the difference, however, is not significant, with p = 0.07). The distribution of these secondary compounds in the diet of the other two species do not differ significantly from that of the random sample. For instance, *Propithecus verreauxi* can feed on the tannin-rich leaves of *Vernonia pectoralis* (Asteraceae) in the gallery forest of Berenty (Fig. 4).

Phenolic compounds include tannins as well as non-tannins molecules, and, as discussed above, the occurrence of alkaloids and tannins does not necessarily imply toxic or digestibility-reducing effects. For instance, one of the most common fruit at Berenty, in December, *Rinorea greveana*, that responded positively to the alkaloid reagents, was eaten by all lemur species including *Lemur catta* (Fig. 5). However, the fact that *L. catta* tends to avoid many plants that respond positively suggests that leaves actually have a deterrent effect, probably mediated by alkaloids and/or tannins. We also observed a few cases of geophagy in *L. catta*, a behavior that, besides other beneficent effects, may efficiently reduce tannin activity through adsorption by earth (Johns & Duquette, 1991; Setz et al., in press).

When testing solutions of pure tannic acid with the tannin reagents, large precipitates were obtained for concentrations higher than 0.4 g/l. Accordingly, this might correspond to the tolerance threshold of *Lemur catta*. In the case of *Propithecus verreauxi*, much larger amounts are required (above 170 g/l) to depress the ingestion of mixtures of sucrose with tannic acid (Dennys, 1991). In the eastern rain forest

Figure 4. *Propithecus verreauxi* feeding on the tannin-rich leaves of *Vernonia pectoralis* (Asteraceae) in the gallery forest of Berenty (photo B. Simmen, June 1998).

Figure 5. *Lemur catta* feeding on fruits of *Rinorea greveana* in the gallery forest of Berenty (photo C. M. Hladik, December 1997).

of Andasibe, the tolerance of condensed tannins and alkaloids in *Eulemur fulvus* (Ganzhorn, 1988), a phenomenon that is apparently similar in Berenty, might also be related to a low taste sensitivity.

Such data provide evidence that different abilities to taste bitter or astringent compounds may explain food choices of different species living in the same habitat. It is likely that lemur species having evolved distinct sensibilities also have to adjust food choices in relation to the relative abundance of plant secondary compounds in different habitats.

ACKNOWLEDGMENTS

We are grateful to the Association Nationale pour la Gestion des Aires Protégées (ANGAP, Madagascar) and the Direction de la Gestion Durable des Ressources Forestières (DGDRS, Ministère des Eaux et Forêts, Madagascar), which allowed us to carry out our research program in protected areas of Madagascar. We are also indebted to M. Jean De Heaulme for giving us permission to study the lemurs of the Réserve de Berenty, and for his very kind support in this research.

REFERENCES

BATE-SMITH, E. C. 1974. Phytochemistry of proanthocyanidins. Phytochemistry, **14**: 1107–1113.
BOLLARD, E. G. 1970. The physiology and nutrition of developping fruits, pp. 387–425. *In* Hulme, A. C., ed., The Biochemistry of Fruits and their Products, vol. 1. Academic Press, London.
CRITCHLEY, H., AND E. ROLLS. 1996. Responses of primate taste cortex neurons to the astringent tastant tannic acid. Chemical Senses, **21**: 135–145.
DENNYS, V. 1991. Approche du rôle de la perception gustative dans la différenciation et la régulation du comportement alimentaire des lémuriens. Ph.D., Université Paris XIII, Villetaneuse.
FAURION, A. 1987. Physiology of the sweet taste, pp. 130–201. *In* Otosson, D., ed., Progress in Sensory Physiology. Springer Verlag, Heidelberg.
———. 1993. Why four semantic taste descriptors and why only four? 11th International Conference on the Physiology of Food and Fluid Intake. Oxford, July 1993: 58.
GANZHORN, J. U. 1988. Food partitioning among Malagasy primates. Oecologia, **75**: 436–450.
GLASER, D. 1979. Gustatory preference behavior in primates, pp. 51–61. *In* Kroeze, J. H. A., ed., Preference Behavior and Chemoreception. I. R. L., London.
———, AND G. HELLEKANT. 1977. Verhaltens und electrophysiologische Experimente über den Geschmackssinn bei *Saguinus midas* (Callitrichidae). Folia Primatologica, **28**: 43–51.
———, G. HELLEKANT, J. N. BROUWER, AND H. VAN DER WEL. 1978. The taste responses in primates to the proteins thaumatin and monellin and their phylogenetic implications. Folia Primatologica, **29**: 56–63.
HASLAM, E. 1989. Plant Polyphenols. Vegetable Tannins Revisited. Cambridge University Press, Cambridge.
HELLEKANT, G., C. M. HLADIK, V. DENNYS, B. SIMMEN, T. W. ROBERTS, D. GLASER, G. DUBOIS, AND D. E. WALTERS. 1993. On the sense of taste in two Malagasy primates (*Microcebus murinus* and *Eulemur mongoz*). Chemical Senses, **18**: 307–320.
HLADIK, A. 1980. The dry forest of the west coast of Madagascar: climate, phenology, and food available for prosimians, pp. 3–40. *In* Charles-Dominique, P., H. M. Cooper, A. Hladik, C. M. Hladik, E. Pagès, G. F. Pariente, A. Petter-Rousseaux, J. J. Petter, and A. Schilling, eds., Nocturnal Malagasy Primates: Ecology, Physiology, and Behavior. Academic Press, New York.
———, AND C. M. HLADIK. 1977. Signification écologique des teneurs en alcaloïdes des végétaux de la forêt dense: résultats des tests préliminaires effectués au Gabon. Revue d'Ecologie (Terre et Vie), **31**: 515–555.
HLADIK, C. M. 1979. Diet and ecology of Prosimians, pp. 307–357. *In* Doyle, G. A., and R. D. Martin, eds., The Study of Prosimian Behavior. Academic Press, New York.
———, AND D. J. CHIVERS. 1994. Foods and the digestive system, pp. 65–73. *In* Chivers, D. J., and P. Langer, eds., The Digestive System in Mammals: Food, Form and Function. Cambridge University Press, Cambridge.
HLADIK, C. M., AND L. GUEGUEN. 1974. Géophagie et nutrition minérale chez les primates sauvages. Comptes Rendus de l'Académie des Sciences de Paris, III, **279**: 1393–1396.
HLADIK, C. M., AND P. PASQUET. In press. Evolution des comportements alimentaires: adaptations morphologiques et sensorielles. Bulletins et Mémoires de la Société d'Anthropologie de Paris.
HLADIK, C. M., AND B. SIMMEN. 1996. Taste perception and feeding behavior in nonhuman primates and human populations. Evolutionary Anthropology, **5**: 161–174.
HUFFMAN, M. A., AND M. SEIFU. 1989. Observations on the illness and consumption of a possibly medicinal plant, *Vernonia amygdalina* (Del.), by a wild chimpanzee in the Mahale Mountains National Park, Tanzania. Primates, **30**: 51–63.
IACONELLI, S., C. M. HLADIK, P. PASQUET, AND B. SIMMEN. 1998. Tannin perception: comparative studies on taste thresholds in a non-human primate and among samples of human populations. Poster presented at the 17th Congress of the International Primatological Society, Antananarivo.
JOHNS, T., AND M. DUQUETTE. 1991. Detoxification and mineral supplementation as functions of geophagy. American Journal of Clinical Nutrition, **53**: 448–456.
LYMAN, B., AND B. GREEN. 1990. Oral astringency: effects of repeated exposure and interactions with sweeteners. Chemical Senses, **15**: 151–164.
MCKEY, D., O. F. LINARES, C. R. CLEMENT, AND C. M. HLADIK. 1993. Evolution and history of tropical forests in relation to food availability — Background, pp. 17–24. *In* Hladik, C. M., A. Hladik, O. F. Linares, H. Pagezy, A. Semple, and M. Hadley, eds., Tropical Forests, People, and Food. Biocultural Interactions and Applications to Development. UNESCO/Parthenon, Paris.
SETZ, E. Z. F., J. ENZWEILER, V. N. SOLFERINI, M. P. AMÊNDOLA, AND R. S. BERTON. In press. Geophagy in the golden-faced saki monkey, *Pithecia pithecia chrysocephala*, in the Central Amazon. Journal of Zoology.
SIMMEN, B. 1994. Taste discrimination and diet differentiation among New World primates, pp. 150–165. *In*

Chivers, D. J., and P. Langer, eds., The Digestive System in Mammals: Food, Form and Function. Cambridge University Press, Cambridge.
——, AND C. M. HLADIK. 1988. Seasonal variation of taste threshold for sucrose in a prosimian species, *Microcebus murinus*. Folia Primatologica, **51**: 152–157.
——. 1998. Sweet and bitter taste discrimination in primates: scaling effects across species. Folia Primatologica, **69**: 129–138.
SIMMEN, B., B. JOSSEAUME, AND M. ATRAMENTOWICZ. In press. Frugivory and taste responses to fructose and tannic acid in a prosimian primate and a didelphid marsupial. Journal of Chemical Ecology.
STRUHSAKER, T. T., D. O. COONEY, AND K. S. SIEX. 1997. Charcoals consumption by Zanzibar red colobus monkeys: its function and ecological and demographic consequences. International Journal of Primatology, **18**: 61–72.
SWAIN, T. 1979. Tannins and lignins, pp. 657–682. *In* Rosenthal, G. A., and D. H. Janzen, eds., Herbivores, their Interaction with Secondary Plant Metabolites. Academic Press, New York.
TERBORGH, J. 1983. Five New World Primates. A Study in Comparative Ecology. Princeton University Press, Princeton.
VAN BUREN, J. 1970. Fruit phenolics, pp. 269–304. *In* Hulme, A. C., ed., The Biochemistry of Fruits and their Products, vol. 1. Academic Press, London.
WRANGHAM, R. W., AND P. G. WATERMAN. 1983. Condensed tannins in fruits eaten by chimpanzees. Biotropica, **15**: 217–222.

Appendix

List of the specimens tested for phenolics and alkaloids, in the mid-montane forest (at Andasibe) and, (at Berenty) in the thorny bush (B) and the gallery forest (F), eventually planted (Pl). Plant specimens have been collected by A. Hladik (Ref. AH), and by B. Simmen and P. Ramasiarisoa (Ref. M), with tentative identification for sterile specimens.

Andasibe

FAMILY	Species	Ref.	Phenolics Salt. gel.	Phenolics FeCl$_3$	Alkaloids Mayer	Alkaloids Drag
ACANTHACEAE	*Strobilanthes sp.*	M 74	0	0	0	+
" (?)	unidentified	AH 6229	0	++	0	0
ANACARDIACEAE	*Protorhus cf. ditimena* Perr.	M 29	±	++	0	0
"	*Protorhus thouvenotii* Lecomte	M 25	++	++	0	+
ANNONACEAE	*Artabotrys sp.*	M 60	++	++	++	++
"	*Xylopia sp.*	M 35	+++	+	++	++
"	unidentified	M 65	+	++	++	±
APOCYNACEAE	*Carissa edulis* Vahl	M 26	++	+	0	±
"	unidentified	AH 6213	0	+	0	0
AQUIFOLIACEAE	*Ilex mitis* (L.) Radlk.	M 1	+	++	++	++
ARALIACEAE	*Polyscias sp.*	M 41	+++	+++	0	0
"	*Schefflera sp.*	AH 6167	++	0	0	0
"	cf. *Cussonia*	AH 6179	±	±	0	0
ASTERACEAE	*Ageratum conyzioides* L.	AH 6190	±	++	+	+
"	*Emilia humifusa* D.C.	AH 6029	±	+++	+	+
"	*Emilia sp.*	AH 6192	0	++	0	0
"	*Psiadia altissima* Benth. & Hook.	AH 6027	+	+++	±	++
"	*Vernonia sp.*	AH 6161	+	+++	0	0
BURSERACEAE (?)	unidentified	AH 6209	+	++	0	0
CHLAENACEAE	*Rhodolaena bakeriana* Baill.	M 31	+++	+++	±	++
CONVOLVULACEAE	*Merremia tridentata* (L.) Hallier	AH 6195	+++	+++	0	0
CUNNIONACEAE	*Weinmannia bojeriana* Tul.	M 52	+++	+++	0	0
"	*Weinmannia rutenbergii* Engl.	M 16	++	++	0	0
CYPERACEAE	*Carex sp.*	AH 6228	0	0	0	0
DIOSCOREACEAE	*Dioscorea sp.*	AH 6039	++	++	±	±
EBENACEAE	*Diospyros sp.*	M 54	+++	+++	0	+
"	*Diospyros sp.*	M 63	+++	+++	±	++
ERICACEAE	*Philippia sp.*	AH 6263	+++	++	0	0
ERYTHROXYLACEAE	*Erythroxylum nitidulum* Bak.	AH 6217	+++	+++	++	++

FAMILY	Species	Ref.	Phenolics		Alkaloids	
			Salt. gel.	FeCl$_3$	Mayer	Drag
EUPHORBIACEAE	*Suregada cf. laurina* Baill.	M 33	0	+	±	+
"	*Blotia sp.*	M 19	0	0	0	0
"	*Bridelia tulasneana* Baill.	AH 6186	+	+++	0	++
"	*Croton mongue* Baill.	AH 6211	+	+	0	0
"	*Lautembergia sp.*	M 59bis	0	0	+++	++
"	*Macaranga alnifolia* Bak.	M 5	+++	+++		
"	*Macaranga alnifolia* Bak.	AH 6005			±	++
"	*Macaranga obovata* Bak.	AH 6004	++	+++	0	++
"	*Macaranga cf. ankafinensis* Baill.	AH 6180	++	++	±	+
"	*Macaranga sp.*	AH 6202	++	+++	0	0
"	*Uapaca densifolia* Bak.	AH 6182	++	++	0	0
"	*Uapaca sp.*	AH 6183	+	++	0	0
FLACOURTIACEAE	*Aphloia thaeiformis* Benn.	M 3	±	+++	0	±
GUTTIFERAE	*Garcinia chapeleri* (Planch. & T.) Perr.	M 28	++	++	0	++
"	*Mammea sp.*	M 36	++	++	0	+
"	*Symphonia louvelii* Jum. & Perr.	M 27	+++	+++	±	±
"	*Symphonia tanalensis* Jum. et Perr.	M 11	++	++	0	0
HYPERICACEAE	*Harungana madagascariensis* Choisy	M 7	++	+	0	++
"	*Psorospermum androsaemifolium* Bak.	M 57	++	±	++	+++
LAURACEAE	*Ocotea similis* Kosterm.	M 18	0	+	+	±
"	*Ocotea sp.*	AH 5820			+	+
"	*Ocotea sp.*	AH 6212	+++	+++	+++	±
"	*Ravensara crassifolia* (Bak.) Danguy	M 37			+	+
"	*Ravensara ovalifolia* Danguy	M 46	++	+		
"	*Ravensara sp.*	AH 6003	0	0	0	0
LILIACEAE	*Dianella ensifolia* (L.) Redouté	M 45	±	±	0	0
"	*Dianella sp.*	M 67	++	++	0	0
LOGANIACEAE	*Anthocleista madagascariensis* Bak.	M 10	0	0		
"	*Anthocleista rhizophoroides* Bak.	M 55	0	0	0	±
"	*Anthocleista sp.*	AH 6045	±	±	++	++
"	*Buddleia sp.*	AH 6246	+	+	+++	++
MALVACEAE	*Sida rhombifolia* L.	AH 6036	0	+	0	0
"	*Urena lobata* L.	AH 6037	++	+++	0	0
MELASTOMATACEAE	*Clidemia hirta* G. Don	M 8	+++	+++	+	+++
"	*Dichaetanthera oblongifolia* Bak.	AH 6016	+	++	0	0
"	*Dichaetanthera sp.*	M 73	+	+++	0	+
"	*Medinilla cf. occidentalis* Naud.	M 49	++	+++	0	±
"	*Medinilla sp.*	M 22			0	0
"	*Medinilla sp.*	M 34	++	+	0	±
"	*Tristemma mauritianum* Gmel.	M 72	++	+++	±	++
MIMOSACEAE	*Acacia delbeata* Link.	AH 6063	+++	+++	+++	+++
"	*Albizia gummifera* (Gmel.) G.A. Smith	AH 6002	0	0	++	++
"	*Albizia chinensis* (Osb.) Merr.	AH 6042	0	++	++	++
"	*Dichrostachys cf. tenuifolia* Benth.	AH 6006	±	+	++	+
MONIMIACEAE	*Tambourissa trichophylla* Bak.	M 38	++	0		
"	*Tambourissa purpurea* (Tul.) A. DC.	AH 6196	+++	+++	+	±
MORACEAE	*Bosqueia sp.*	AH 6197	+++	+++	++	++
"	*Ficus sp.*	AH 6039bis	++	++	0	0
"	*Ficus sp.*	AH 6201	±	±	±	±
"	*Pachytrophe dimepate* Bur.	M17	+++	+++	0	0
"	unidentified	AH 6214	+	+++	0	0
MYRICACEAE	*Myrica spathulata* Mirbel	AH 6218	+	++	0	0
MYRSINACEAE	*Oncostemum sp.*	M 15	++	+		
"	*Oncostemum sp.*	M 32			++	++
"	*Oncostemum sp.*	M 42	++	++	0	0
"	*Oncostemum sp.*	M 47	++	++	+	++
MYRTACEAE	*Eucalyptus sp.*	AH 6021	++	+++	+++	+++
"	*Eugenia goviala* H. Perr.	M 20	++	+++	±	++
"	*Eugenia sp.*	M 58	++	+++		
"	*Eugenia sp.*	AH 6200	++	+++	0	0
"	*Psidium cattleyanum* Sabine	M 71	+	+++	±	++
"	*Psidium guayava* Berg	M 70	++	+++	0	++
OCHNACEAE	*Campylospermum lanceolatum* (Bak.)Perr.	M 39	0	0	0	0
"	*Campylospermum anceps* (Bak.)Perr.	M 13	++	++	+	+
OENOTHERACEAE	*Jussiaea sp.*	AH 6024	++	+++	0	0

(continued)

Appendix (*Continued*)

FAMILY	Species	Ref.	Phenolics		Alkaloids	
			Salt. gel.	FeCl$_3$	Mayer	Drag
OLEACEAE	*Noronhia sp.*	M 12	++	++	0	++
PANDANACEAE	*Pandanus sp.*	AH 6166	0	++	±	+
PAPILIONACEAE	*Dalbergia monticola* Bosser & Rabe.	M 48	++	+++	+	++
PASSIFLORACEAE	*Passiflora foetida* L.	AH 6035	0	+++	++	++
"	*Passiflora incarnata* L.	AH 6026	0	+	+++	+++
RHIZOPHORACEAE	*Cassipourea sp.*	AH 6199	0	++	0	0
ROSACEAE	*Rubus roridus* Lindl.	AH 6019	+++	+++	+	++
"	*Rubus rosaefolius* Smith	AH 6007	++	+++	+	+++
RUBIACEAE	*Canthium sp.*	M 33bis	++	++	++	++
"	*Coffea sp.*	AH 6243	++	+++	++	+
"	*Danais sp.*	AH 6203	±	±	0	0
"	*Enterospermum sp.*	M 9	0	0	++	+
"	*Gaertnera macrostipula* Lam.	M 4	++	++	0	+
"	*Rothmannia sp.*	M 24	0	±	+	+
"	*Sabicea diversifolia* Pers.	AH 6040	+++	+++	+	++
"	unidentified	AH 6216	+	++	0	0
"	*Pyrostria sp.*	AH 6230	0	0	++	+
" (?)	unidentified	AH 6215	+	+++	0	0
SAPINDACEAE	*Allophyllus cobbe* (L.) Raeusch.	M 53	+	0	0	0
"	*Filicium decipiens* (W.&A.) Thw.	AH 6163	++	++	0	±
SAPOTACEAE	*Gambeya boiviniana* (Pierre) Aubrév.	M 171	++	++	+	+
SMILACACEAE	*Smilax kraussiana* Meissn.	AH 6032	+	++	0	0
SOLANACEAE	*Solanum auriculatum* Ait	M 68	0	0	++	++
"	*Solanum sp.*	M 64	+	+++	+	+
STERCULIACEAE	*Dombeya sp.*	AH 6012	++	++	0	0
"	*Dombeya sp.*	AH 6014	0	0	±	+
STRELIZIACEAE	*Ravenala madagascariensis* Gmel.	AH 6018	++	0	+	+
THEACEAE	*Camellia thaeiformis* Hance	AH 6015	+++	+++	+	++
ULMACEAE	*Trema orientalis* Bl.	M 69	++	+++	++	++
VACCINACEAE	*Vaccinium sp.*	M 21	+++	+++	0	++
VERBENACEAE	*Clerodendron sp.*	AH 6177	+	+	++	0
"	*Lantana camara* L.	AH 6040bis	++	+++	0	0
"	*Stachytarpheta jamaicensis* Vahl	AH 6043	0	0	0	0
ZINGIBERACEAE	*Aframomum angustifolium* K. Schum.	AH 6033	++	0	0	++
"	*Hedychium coronarium* Kœniz	AH 6033bis	++	+	0	+
"	unidentified	M 51	++	++	++	++
Pteridophytes:						
DENNSTAEDTIACEAE	*Pteridium aquilinum* (L.) Kühn.	AH 6030	++	+++	0	±
GLEICHENIACEAE	*Dicranopteris linearis* (Burm.) Under.	AH 6176	+++	+++	0	0
"	*Sticherus flagellaris* (Bory) St John	AH 6164	±	+	0	0
SCHIZAEACEAE	*Lygodium lanceolatum* Desv.	AH 6189	0	++	0	0

Berenty

FAMILY	Species	Ref.	B/F	Phenolics		Alkaloids	
				Salt. gel.	FeCl$_3$	Mayer	Drag.
ACANTHACEAE	*Crossandra poissonii* R. Ben.	M175	F	0	++	0	+
"	*Hypoestes sp.*	M261	F	±	++	0	+
" (?)	unidentified	M131	B	0	++	+	++
AGAVACEAE	*Sanseveria sp.*	M269	F/Pl	0	0	0	+
AMARANTHACEAE	*Aerva madagassica* Suess.	M308	B	±	0	++	+
ANACARDIACEAE	*Operculicarya cf. decaryi* H. Perr.	M338	F/Pl	++	+	0	++
"	*Poupartia caffra* (Sond.) H. Perr.	M306	B/Pl	+++	+++	+++	+++
"	*Poupartia minor* (Boj.) Marchand	M208	F/Pl	++	0	0	++
ANNONACEAE	*Annona sp.*	M244	F/Pl	0	0	++	++
APOCYNACEAE	*Hazunta modesta* (Bak.) Pichon	M255	F	0	+++	+++	+++
"	*Catharanthus roseus* (L.) G. Don	M333	F	0	±	++	++
"	*Plectaneia sp.*	M316	B	0	0	++	+++
ARISTOLOCHIACEAE	*Aristolochia sp.*	M235	F	0	0		
"	*Aristolochia sp.*	M229	F			+++	+++
ASCLEPIADACEAE	*Leptadenia madagascariensis* Decne.	M134	B	±	+	++	++
"	*Leptadenia sp.*	M225	F	0	0	+	0
"	*Leptadenia sp.*	M285	F	±	+++	++	++

FAMILY	Species	Ref.	B/F	Phenolics		Alkaloids	
				Salt. gel.	FeCl$_3$	Mayer	Drag.
"	Pervillea decaryi (Choux) Klack.	M144	B	±	±	++	+++
"	Ceropegia sp.	M336	F	0	+		
"	Cynanchum sp.	AH 5905	B			+	±
"	cf. Marsdenia sp.	M319	B	0	0	0	+
"	cf. Secamone sp.	M288	F	±	+++		
"	unidentified	M195	F			+++	++
"	unidentified	M195bis	F	0	++		
"	unidentified	M238	F	0	0	0	±
ASTERACEAE	Vernonia pectoralis Bak.	M205	F	+	+++	0	0
"	cf. Vernonia sp.	M286	F	±	+++		
"	unidentified	M188	F	+	+++	+	++
"	unidentified	M280	F	±	±	±	±
BIGNONIACEAE	Fernandoa madagascariensis (Bak.) Gentry	M332	F/Pl	++	++	+	+
BORAGINACEAE	Cordia rothii Roem. & Schult.	M107	F/Pl	±	+++	0	0
"	Cordia sp.	M304	F	±	0		
"	Cordia sp.	M165	F			++	++
BURSERACEAE	Commiphora sp.	M122	B	+	++	±	+++
"	Commiphora sp.	M130	B	+	+	0	+
"	cf. Commiphora sp.	M196	F	+	++	0	0
CACTACEAE	Opuntia vulgaris Miller		F/Pl	0	0	+	+
CAESALPINIACEAE	Tetrapterocarpon geayi H. Humb.	M125	B	0	+	++	+++
"	Bauhinia grandidieri Baill.	M132	B	+	+	++	+++
"	Bauhinia sp.	M158	F	+	++		
"	Caesalpinia bonduc Roxb.	M259	F	0	0	++	++
"	Cassia siamea Lam.	M236	F/Pl	0	+++	±	+++
"	Cassia sp.	M232	F/Pl	0	0	++	+++
"	Delonix regia (Hook.) Raf.	AH 5986	F/Pl	++	+++	+	++
"	Tamarindus indica L.	M234	F	++	+	0	+
CAPPARIDACEAE	Boscia longifolia Hadj Moust.	M133	B			0	+
"	Boscia longifolia Hadj Moust.	M143	B	0	0	+	+++
"	Cadaba virgata Boj.	M267	F	±	±		
"	Cadaba virgata Boj.	AH 5976	F			++	+++
"	Capparis sepiaria L.	M102	F	0	0		
"	Capparis sp.	M270	F			+	+
"	Crataeva greveana Baill.	M262	F	0	++		
"	Crataeva sp.	M105	F			++	++
"	Crataeva sp.	M263	F	0	0		
"	Maerua filiformis Drake	M295	F/B	0	+++		
"	Maerua filiformis Drake	AH 5968	F/B			0	++
CELASTRACEAE	Evonymiopsis longipes Perr.	M317	B	±	0	+	+
COMBRETACEAE	Combretum sp.	M192	F	+++	+++	++	++
"	cf. Terminalia sp.	M313	B	++	+++	0	0
COMMELINACEAE	Commelina sp.	M257	F	±	++	+	+
CONVOLVULACEAE	Hildebrandtia promontorii Deroin	AH 5969	B			±	+
"	Hildebrandtia valo Deroin	M310	B	0	++	+	+
"	Ipomoea cairica (L.) Sweet	M201	F	0	±	0	0
CRASSULACEAE	Kalanchoe beauverdii Hamet	M283	F	++	++	0	0
"	Kalanchoe beharensis Drake		F/Pl	0	0	0	0
CUCURBITACEAE	Xerosicyos decaryi Guill.	M282	F	+	+	+++	+++
"	cf. Zehneria sp.	M264	F	0	±	+	++
DIDIEREACEAE	Alluaudia ascendens Drake	M149	B	0	0	0	0
"	Alluaudia procera Drake	M146	B	0	0	0	0
DIOSCOREACEAE	Dioscorea fandra Perr.	AH 5980	B			0	0
"	Dioscorea nako Perr.	M345	F	0	0		
"	Dioscorea nako Perr.	AH 5978	B			±	++
EUPHORBIACEAE	Acalypha sp.	M109bis	F	++	+++	+++	+++
"	Acalypha sp.	M109	F	+	+++	++	+++
"	Croton sp.	M178	F	++	+	++	++
"	Euphorbia sp.	M142	B	0	0	0	+
"	Euphorbia sp.	M213	F/Pl	±	0	+	+
"	Phyllanthus casticum Willem.	M186	F	+++	+++		
"	Phyllanthus casticum Willem.	M110	F			++	+++
"	Securinega cf. capuronii Léandri	AH 5971	B			++	+++
"	Croton sp.	M138	B			0	+
"	Croton sp.	M324	B			++	+
"	cf. Sclerocroton melanostictus	M116	B			±	++

(*continued*)

Appendix (*Continued*)

FAMILY	Species	Ref.	B/F	Phenolics		Alkaloids	
				Salt. gel.	FeCl$_3$	Mayer	Drag.
FLACOURTIACEAE	*Flacourtia indica* (Burm.) Merril.	M217	F/Pl	++	+++	±	±
"	*Physena sessiliflora* Tul.	M206	F	+	+	+	+
HIPPOCRATEACEAE	cf. *Hippocratea sp.*	AH 5903	B	++	±	++	++
"	cf. *Loesneriellia sp.*	M243	F	+++	+++	+++	+++
"	cf. *Loesneriellia sp.*	M273	F	±	+	++	++
"	unidentified	AH 5907	F			++	++
LILIACEAE	*Aloe vahombe*		F/Pl	0	0	0	0
"	*Aloe capitata*		F/Pl	0	0	±	±
LOGANIACEAE	*Strychnos sp.*	M252	F	±		+++	+++
"	*Strychnos sp.*	M294	B	0	0	+++	+++
MALPIGHIACEAE	*Tristellateia sp.*	M337	F	0	0		
MALVACEAE	*Abutilon pseudocleistoganum* Hochr.	M239	F	0	0	0	±
"	*Hibiscus sp.*	M180	F	+++	+++	±	+
"	*Hibiscus sp.*	M318	B	±	+	0	++
"	*Hibiscus sp.*	M117	B	0	+++	++	+
"	*Hibiscus sp.*	M127	B	+	++	++	++
"	*Megistostegium nodulosum* (Drake) Hochr.	AH 5988	B	+	++	0	0
MELIACEAE	*Azadirachta indica* Jussieu	M166	F/Pl	0	0	++	++
"	*Melia azedarach* L.	M330	F/Pl	0	0	++	+++
"	*Quivisianthe papinae* Baill.	M223	B/F	+	0		
"	*Quivisianthe papinae* Baill.	M137	B/F			++	++
"	cf. *Cedrelopsis grevei*	M124	B	0	±		
"	cf. *Turraea sp.*	M245	F	0	0	±	+
MENISPERMACEAE	unidentified	M341	F	+	++		
MIMOSACEAE	*Acacia rovumae* Oliv.	M214	F	+	±	++	++
"	*Acacia sp.*	M342	F	0	0		
"	*Albizzia polyphylla* Fourn.	M293	F	++	++	++	++
"	*Dicrostachys sp.*	M118	B	+	++	0	+
"	*Leucaena glauca* (L.) Benth.	M211	F/Pl	±	+	++	+
"	*Pithecellobium dulce* Benth.	M170	F/Pl	++	+		
"	*Pithecellobium dulce* Benth.	M212	F/Pl			+++	+++
"	unidentified	M275	F	++	0	+	+
MORACEAE	*Ficus cf. trichopoda* Bak.	M329	F	0	0	0	0
"	*Ficus cf. cocculifolia* Bak.	M305	F	+++	+++		
"	*Ficus cf. megapoda* Bak.	M179	F	0	0	0	++
"	*Ficus cf. grevei* Baill.	M299	F			0	+
"	*Ficus cf. grevei* Baill.	M300	F	+	+++		
"	*Ficus sp.*	M301	F	+++	±	±	+
"	*Ficus sp.*	M198	F	0	+	0	+
MYRTACEAE	*Eucalyptus sp.*	M216	F/Pl	++	+++	++	++
NYCTAGINACEAE	*Bougainvillaea spectabilis* Willd.	M231	F/Pl	+	+	+	+
"	*Commicarpus commersonii* Cav.	M265	F	0	+	++	++
PAPILIONACEAE	*Clitoria heterophylla* Lam.	M128	B			++	++
"	*Mundulea scoparia* R. Viguier	M115	B	0	±	++	++
"	unidentified	M284	F	0	+++	++	++
POACEAE	*Phragmites mauritianus* Kunth.	M276	F	0	++	++	+++
RHAMNACEAE	*Colubrina sp.*	M291	F	±	±	+	±
"	*Zizyphus sp.*	M254	F	0	+++	++	++
RUBIACEAE	*Enterospermum sp.*	M221	F	0	±		
"	*Enterospermum sp.*	M181	F			+++	+++
"	*Enterospermum sp.*	M260	F	0	++		
"	unidentified	M260bis	F	±	±		
"	unidentified	M292	F	+	+++	+++	+++
RUTACEAE (?)	unidentified	M164	F	+	±	0	++
"	unidentified	M302	F	0	0	+	+
SALVADORACEAE	*Azima tetracantha* Lam.	M162	F	0	0	+	++
"	*Salvadora angustifolia* Turrill.	M272	F/B	0	0	+	++
SAPINDACEAE	*Neotina isoneura* (Radlk.) Capuron	M187	F	+	0	0	0
SOLANACEAE	*Solanum batoides* d'Arcy & Rak.	M120	B	0	0		
"	*Solanum croatii* d'Arcy & Keat.	AH 5972	F/B	+	++	+++	++
STERCULIACEAE	cf. *Byttneria sp.*	M161	F	+	+	+	++
TILIACEAE	*Grewia grevei* Baill.	M126	B	++	+++	±	++
"	*Grewia sp.*	M321	F/B	0	0	0	±
"	*Grewia sp.*	M242	F	±	0	++	++
"	*Grewia sp.*	M343	F	0	++		

Taste Discrimination in Lemurs and Other Primates

FAMILY	Species	Ref.	B/F	Phenolics		Alkaloids	
				Salt. gel.	FeCl$_3$	Mayer	Drag.
ULMACEAE	*Celtis bifida* Leroy	M230	F	+	+	0	0
"	*Celtis philippensis* Blanco	M114	F	+	+	++	++
"	*Trema orientalis* Blume	M340	F	0	0		
VERBENACEAE	*Clerodendron sp.*	M322	B	+	±		
" (?)	*cf. Clerodendron sp.*	M312	B			++	++
VIOLACEAE	*Rinorea greveana* H. Bn.	M154	F	+	++	++	+++
VITACEAE	*Cissus quadrangularis* L.		F	±	±	+	+
	unidentified	M251	F	0	+	++	+
	unidentified	M246	F	0	+	++	+
	unidentified	M290	F	+	+	0	++
	unidentified	M227	F	0	0	0	0
	unidentified	M326	F	0	0		
	unidentified	M193	F			0	0
	unidentified	M147	B	0	0	+++	+++
	unidentified	M328	F	0	±	++	++
	unidentified	AH 5901	B			++	±
	unidentified	M129	B	+	+++		
	unidentified	M135	B	+	+		
	unidentified	M287	F	++	++	++	±
	unidentified	M121	B/F	+	++	0	0
	unidentified	M258	F	±	+++		
	unidentified	M256	F	+	+++	±	±
	unidentified	M250	F	0	0	+++	+++
	unidentified	M279	F			0	0
	unidentified	M281	F			0	0
	unidentified	M311	B	0	0	±	+
	unidentified	M331	F	+++	+++	0	++
	unidentified	M320	B	0	+	+++	+++

13

FOLIVORY IN A SMALL-BODIED LEMUR

The Nutrition of the Alaotran Gentle Lemur (*Hapalemur griseus alaotrensis*)

Thomas Mutschler[1]

[1] Anthropological Institute and Museum
University of Zurich
Winterthurerstrasse 190, 8057 Zurich, Switzerland
tom@aim.unizh.ch

ABSTRACT

Hapalemur griseus alaotrensis, a relatively small-bodied (1240 ± 140 g, n = 58), folivorous lemur, was studied over a period of 15 months at Lac Alaotra. The study showed that *H. g. alaotrensis* had an exclusively folivorous diet and mainly fed on leaves and stems of grasses and sedges. However, since folivorous diets are known to be poor sources of readily available energy and small-bodied animals generally have high metabolic requirements, *H. g. alaotrensis* is expected to have adaptations at several levels (morphological, physiological, behavioral) to resolve this conflict. The present study on *H. g. alaotrensis* showed that dietary diversity was extremely low and food choice highly selective, but chemical composition of the food items yielded no evidence that *H. g. alaotrensis* selected higher quality foods than larger folivores, nor was there evidence that *H. g. alaotrensis* minimized energy expenditure at the behavioral or physiological level. This lack of behavioral or physiological adaptations to folivory imply that the digestive ability of *H. g. alaotrensis* may be higher than predicted for an animal of its size.

RÉSUMÉ

Hapalemur griseus alaotrensis, un lémurien d'assez petite taille (1240 ± 140 g, n = 58) a été étudié au cours d'une période de 15 mois au Lac Alaotra à Madagascar. L'étude a montré que le régime alimentaire d'*H. g. alaotrensis* est exclusivement folivore, se composant surtout des feuilles et des tiges d'espèces d'herbe et de laîche. Dans la mesure oú les régimes alimentaires folivores sont connus pour fournir des ressources pauvres du point de vue de l'énergie disponible d'une part et que des animaux de petite taille ont généralement des exigences métaboliques importantes d'autre part, il est attendu qu'*H. g. alaotrensis* montre des adaptations à divers niveaux (morphologique, physiologique, de comportement) pour accommoder cette contradiction. Cette étude a cependant montré que la diversité des aliments consommés par *H. g. alaotrensis* aire était extrêmement faible, le choix des aliments très sélectif en même temps que la composition chimique des éléments consommés ne permettait pas de conclure qu'*H. g. alaotrensis* favorise des aliments de meilleure qualité que ceux consommés par les plus grands folivores, et aucune preuve n'a été trouvé pour montrer qu'*H. g. alaotrensis* réduirait ses dépenses énergétiques aux niveaux du comportemental ou physiologique. En l'absence d'adaptations à ces deux niveaux pour accomoder le régime folivore, la capacité digestive d'*H. g. alaotrensis* doit être plus importante encore que celle avancée pour un animal de cette taille.

INTRODUCTION

The Alaotran Gentle Lemur (*Hapalemur griseus alaotrensis*) is a relatively small-bodied lemur (1240 ± 140g, n = 58), living in the marshes of Lac Alaotra, Madagascar (Mutschler & Feistner, 1995). Together with *Avahi* (body mass of about 1000g), *Lepilemur* (body mass of 500–1000g) and other *Hapalemur* spp. (body mass of 700–2400g), they belong to a group of small-bodied prosimians relying mainly on folivorous diets (Petter & Peyrieras, 1970; Hladik, 1978; Ganzhorn et al., 1985; Pollock, 1986; Wright, 1986; Ganzhorn, 1988; Glander et al., 1989; Wright & Randriamanantena, 1989; Harcourt, 1991; Overdorff et al., 1997; Nash, 1998; Mutschler et al., 1998). It is hypothesized that small-bodied folivores may face the problem of obtaining energy and nutrients out of diets dominated by plant structural tissues, which usually contain low levels of readily available energy (McNab, 1978; Milton, 1984). Moreover, such foods need extensive digestion with the assistance of symbiotic microbes for fermentation, and the efficiency of this process is related to gut capacity and gut passage rate (Parra, 1978). Due to their small body size, these lemurs are predicted to have relatively high energetic requirements per unit of body mass (Kleiber, 1961; Martin, 1990) and relatively small digestive tract capacities (Parra, 1978; Demment & van Soest, 1985). Nevertheless, these lemurs successfully pursue a folivorous foraging strategy. Different physiological, morphological, and behavioral adaptations potentially enable them to do so:

1) Reduction of the basal metabolic rate (BMR). A reduction of BMR has the same effect as increasing body mass; both result in a reduction of energetic needs per unit body mass, and therefore modify the tolerance for low-energy food (Martin, 1990).
2) Specialization of the digestive tract. In folivores, an increased complexity of the alimentary tract is generally found, involving enlargement of either the foregut (stomach) or the hindgut (caecum, colon). Folivorous lemurs are assumed to have enlarged hindguts (fermentation site) which, in combination with long gut passage times, increase an animals digestive capability for fibrous foods (Bauchop, 1978; Demment & van Soest, 1985).
3) Selectivity for high-quality foods. Small-bodied folivores should select for relatively low-fiber foods to compensate for their high metabolic requirement/gut

capacity (MR/GC) ratio. High-quality foods would allow them to combine energy transfer from enzymatic digestion in the foregut with fermentation in the hindgut (Demment & van Soest, 1985; Justice & Smith, 1992).

4) Coprophagy. Nutrient losses (mainly protein and vitamins) due to the location of the fermentation site (hindgut) could potentially be compensated through ingestion of faeces (Bauchop, 1978).

5) Reduction of energy expenditure at the behavioral level (energy minimizing). Voluntary energy expenditure can be reduced by increasing the proportion of resting in the activity budget (Milton, 1978; Nash, 1998).

6) Cathemeral activity. Small-bodied folivores have difficulties processing foods in bulk, due to their limited gut capacity. By adopting a cathemeral activity pattern, a small-bodied folivore could spaced foraging bouts to minimize time in which there is no food being processed (Engqvist & Richard, 1991).

In recent years, data on nutrition and feeding behavior of *Hapalemur griseus* have accumulated from both field and captive studies (Pollock, 1986; Wright, 1986; Glander et al., 1989; Santini-Palka, 1994; Cabre-Vert & Feistner, 1995; Fidgett et al., 1996; Overdorff et al., 1997; Mutschler et al., 1998). The genus *Hapalemur* differs from other small-bodied folivorous lemurs in that its members feed preferentially on different parts of bamboos. However, the Alaotran Gentle Lemur is an exception. It does not feed on bamboo because no bamboo grows in the marshes of Lac Alaotra. These lemurs have replaced bamboo with tall grasses and sedges like reeds (*Phragmites*) and papyrus (*Cyperus*) (Petter & Peyrieras, 1970; Pollock, 1986; Mutschler et al., 1998). Nevertheless, information on the diet of members of the genus *Hapalemur* (Wright, 1986; Glander et al., 1989; Wright & Randriamanantena, 1989; Overdorff et al., 1997; Mutschler et al., 1998) is still sparse and there is a need for more basic nutritional data in order to understand the foraging strategies of members of the genus *Hapalemur*. We need to know how *Hapalemur* spp. and other relatively small-bodied folivores cope with the constraints imposed by their folivorous diets. This paper presents data on diet, feeding, and the chemical composition of the diet of wild Alaotran Gentle Lemurs. Results are compared to other folivorous primates and adaptations to folivory are discussed.

METHODS

The study was carried out in the marshes of Lac Alaotra around Andreba (17°38′ S, 48°31′ E). Lac Alaotra is very shallow, fluctuating between 1 m and 4 m in depth, and is surrounded by up to 80,000 ha of inundated marsh and irrigated rice fields (Moreau, 1987; Pidgeon, 1996). The climate is characterized by pronounced seasonality. A dry season with moderate temperatures occurs in May—November, while a wet season with elevated temperatures occurs in December—April (Mutschler et al., 1998).

Behavioral data base on observations of two different groups (A and K). Group A (9 members; 3 adult males, 2 adult females, 4 non-adults) was observed from February to July 1996, September to October 1996, and January to February 1997 (study sites were not accessible by canoe during November and December due to low water level). Group K (3 members; 1 adult male, 1 adult female, 1 non-adult) was observed from January to July 1996, and January to February 1997. All animals were individually marked (radio and color collars). In the home ranges of both groups, channels through the vegetation were cut and maintained, to allow the observer to follow the animals closely in a canoe.

Groups were habituated through the daily presence of the author in December 1995 and January 1996. Habituation was considered successful when animals did not flee on the arrival of the observer and observation distances fell below 15 m. After habituation, a total of 1334 h of observation was carried out. Each group was observed on five consecutive days (05:00–18:00 h) per month. In addition, group A was also observed on one night (18:00–05:00 h; split into two sessions) per month. Two nights (March, October) are excluded from the results, because the author was unable to approach the animals closely enough to identify activity.

During habituation it became obvious that visibility was problematic in the dense marsh vegetation. Therefore, behavior was identified through a combination of direct and indirect clues, both visual (e.g., movement of vegetation caused by the animal's activities) and auditory (e.g. chewing noises). This method was largely feasible for both day and night sessions. However, to surely identify the location of the focal animal radio-tracking information was needed and thus radio-collared animals (adult male in group A; adult female in group K) were used as focal animals throughout the study. Their behavior was sampled instantaneously (Altmann, 1974). Since groups were highly cohesive and behaved mostly synchronously (Mutschler, unpubl. data), the behavior of the sampled individuals represent their groups as a whole.

During days, the focal animal's activity was recorded instantaneously at sample intervals of 2 min, and during nights at sample intervals of 5 min. The sampled behaviors were: rest, feed, travel, other, and not determined. Definitions and descriptions of the sampled behaviors are given in Table 1.

Because data sets for diurnal and nocturnal activity are based on different sample sizes (diurnal: 2-min intervals over five days per month; nocturnal: 5-min intervals for one night per month), activity records for both light cycles were normalized (numbers of records for the individual behavioral categories divided by the total number of records × 100). Activity records are hence given as percentages, and statistical comparisons are based on percentages.

Although unmistakable foraging and/or feeding noises allowed identification of the consumed food plant even without visual contact and during nights, the list of consumed food items was based on direct, visual observation. Food items eaten by any member of the observed groups were noted and then samples were collected for analysis during the same month in which their consumption was observed. Care was taken to ensure collection of the precise part eaten. In addition, samples of soil and a 'non-food' item were collected and analyzed (papyrus pith fraction directly above the eaten portion).

Table 1. Definition and description of sampled behaviors

Behavior	Definition and description
Rest	Animal is inactive, or no noise and no movement of the vegetation is detectable at the location of the focal animal.
Feed	Animal is eating or processing food, or biting or chewing noises are heard at the location of the focal animal.
Travel	Animal is moving directional, or movements of vegetation indicates that focal animal advances.
Not detectable	Location of focal animal not known, or distance to focal animal greater than 15 m.
Other	Grooming, vocalizing, scent-marking, and other activities, not fitting into one of the former categories.

For each item (including soil) 20 to 150 g of wet material were collected. Samples were weighed wet using electronic scales (Neolab), dried in the sun and stored in opaque plastic containers with a small envelope of silica gel. Chemical analysis was carried out by the Swiss Federal Research Station for Animal Production (Eidgenössische Forschungsanstalt für Nutztiere, Posieux, Switzerland). The samples were analyzed for dry matter (gravimetric determination of water loss during 3 h at 105°C), gross energy (bomb calorimetry), crude protein (Kjeldahl method; conversion factor 6.25), crude lipids (Twisselman extraction, distillation, gravimetry), ash (4 h at 550°C, gravimetry), calcium (ash boiled in 25% HCl, photometry), crude fibre (Weende method; boiled in acid and alkaline solution, undissolved residue filtered, washed with water, and acetone, dried, weighed, and combusted), acid detergent fibre (van Soest method; 1 h boiled in acid detergent solution using Fibretec equipment, undissolved residue filtered, washed, dried, weighed, and combusted), neutral detergent fibre (van Soest-Wine method; 1 h boiled in neutral detergent solution using Fibretec equipment, undissolved residue filtered, washed, dried, weighed, and combusted), and amino acids (HCl-hydrolysis of proteins, photometry).

Statistics were computed using the StatView version 4.02 software (Abacus Concepts Inc.). Nonparametric tests (specified when used) were used with $p < 0.05$ regarded as significant.

RESULTS

The Alaotran Gentle Lemurs were active both during daylight hours (05:00–18:00 h) and during nights (18:00–05:00 h). Table 2 shows the diurnal and nocturnal activity records of focal animal A (observed during days and nights) for all observation months. Although night observation was carried out on one night per month only, and thus did not account for nightly variation, data clearly show that *Hapalemur g. alaotrensis* exhibits substantial amounts of night activity including all major behavioral categories (feed, forage, travel, groom, and other).

Activity budgets for diurnal (for groups A and K) and nocturnal (for group A) activity are shown in Figure 1. The total number of diurnal feeding (including foraging) records did not differ significantly between the focal animals ($U = 1280$, $Z = -1.33$,

Table 2. Total counts of individual activity records of focal animal A during days and nights

Month	Diurnal activity records [total counts for 5 days/month]		Nocturnal activity records [total counts for 1 night/month]	
	active*	inactive (= rest)	active*	inactive (= rest)
Feb. 1996	841	1109	—	—
Mar. 1996	572	1378	—	—
Apr. 1996	697	1253	33	99
May 1996	689	1261	37	95
Jun. 1996	848	1102	39	93
Jul. 1996	807	1143	31	101
Sep. 1996	838	1112	3	129
Oct. 1996	851	1099	—	—
Jan. 1997	785	1165	41	91
Feb. 1997	704	1246	49	83

*sum of records for feed, travel, other, and not determined.

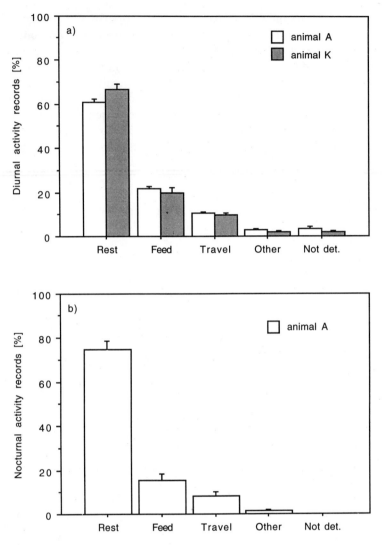

Figure 1. Diurnal and nocturnal activity budget (mean percentages ± SEM) of *Hapalemur griseus alaotrensis*. a) day-time budget (n = 10 months); b) night-time budget (n = 7 months).

p = 0.184; Mann-Whitney U-test [MWU]). On average, animals A and K spent 21.8% and 20.1% of the day, respectively, on feeding and foraging. In addition, animal A spent 15.4% of the night feeding and foraging. Thus, it can be estimated (Tyler, 1979) that animal A (observed during both light cycles) spent on average 170 min of daylight hours (5:00–18:00 h) and 102 min of night hours (18:00–5:00 h) feeding and foraging.

In order to investigate seasonal variation, data were split into wet season (December–April) and dry season (May–November). Both focal animals had more diurnal feeding records in the dry season than in the wet season, although seasonal variation did not reach statistical significance (A: U = 5.50, Z = 1.46, p = 0.144; K: U = 2.00, Z = 1.81, p = 0.071; MWU). Nocturnal feeding did not vary significantly with season either (U = 1.00, Z = –1.77, p = 0.077; MWU), but this may well be because the night data were based on only one night per month. However, if feeding records of animal A (observed during both light cycles) for the wet and dry seasons are compared, a trend towards increased diurnal feeding time and decreased nocturnal feeding time in the dry and cold months was found (Fig. 2).

Mean feeding records throughout the 24-hour period are shown in Figure 3. No significant variation was found in the nocturnal pattern, but diurnal variation was significant for both focal animals (DF = 12, H = 85.28, p < 0.001; Kruskal-Wallis), with peaks of feeding activity at 6:00–7:00 h and at 17:00–18:00 h and minimal feeding activity between 12:00 and 14:00 h. Feeding pattern did not differ significantly between the seasons, but a trend towards reduced feeding activity between 8:00 and 12:00 h and increased nocturnal feeding activity after midnight was found during the wet season (compared to the dry season).

The Alaotran Gentle Lemur fed on 11 different plant species (Table 3). Food diversity was extremely low. Focal animals fed on a maximum of six different plant species per month. Moreover, the focal animals invested more than 95% (4042 feeding records) of diurnal and 99% (140 feeding records) of nocturnal feeding time on plants of the families Cyperaceae (*Cyperus madagascariensis*) and Poaceae (*Phragmites communis*, *Leersia hexandra*, and *Echinochloa crusgalli*) (Table 4). These four species were therefore considered as dietary staples.

Both animals spent most diurnal feeding time on *Cyperus* (A: 65%; B: 45%). During nights, group A preferentially fed on *Phragmites* (63%) and on *Leersia* (24%). By comparing the focal animals, it was found that animal A spent significantly more diurnal feeding time on *Cyperus* than animal K (U = 15, Z = –2.45, p = 0.014; MWU), which fed more on *Phragmites* and *Echinochloa* instead (ns, p = 0.060 and p = 0.683, respectively; MWU).

Sixteen different parts of the 11 consumed plant species were eaten (Table 5). Food choice was highly selective; only small fractions of food plants were eaten. The plants of the family Poaceae were all consumed the same way: the top part of the stem (shoot) was bitten off or pulled out, and turned upside-down. The tender base of this

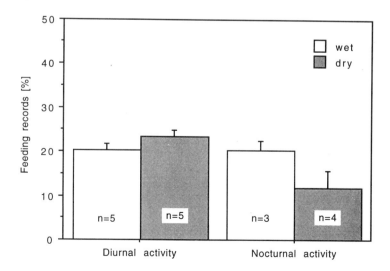

Figure 2. Seasonal variation of diurnal and nocturnal feeding records (mean percentage per month ± SEM) of focal animal A.

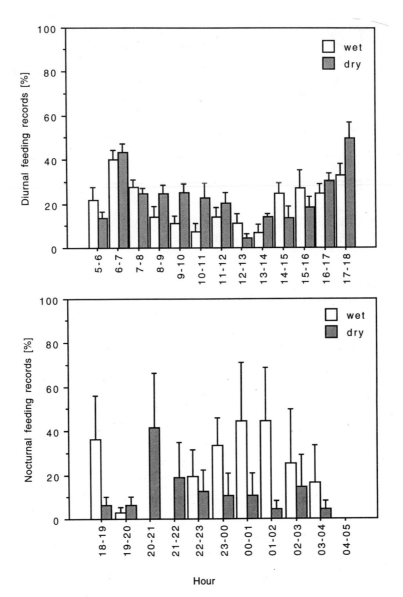

Figure 3. Variation of diurnal and nocturnal feeding records (mean percentage ± SEM) of focal animal A in relation to hour and season.

section was then eaten (combined with stripping away old leaves and sheaths). Infrequently, the leaves of these grasses were also consumed. With some plants, several parts were eaten (e.g. papyrus, *Cyperus*: white pith at the very base of the stem, base of the leaves, buds, seeds/ears), whereas with other plants only one part was eaten (e.g. *Polygonum*: leaf-tips). A complete list of selected plant parts is given in Tables 3 and 5. Apart from the plant foods, the animals of group A were also occasionally observed to feed on soil.

Results from biochemical analyses of food items are given in Tables 5 and 6. Mean energy and nutrient values of food items (n = 16) on a dry matter basis are: 17.3 ± 1.7

Table 3. Plant species and parts eaten by *Hapalemur griseus alaotrensis*

Species	Family	Local name	Consumed parts
Cyperus madagascariensis	Cyperaceae	*zozoro*	-pith of stem base -base of leaves -bud -ear/seed
Phragmites communis	Poaceae	*bararata*	-stem -leaves
Leersia hexandra	Poaceae	*karangy*	-stem -leaves
Echinochloa crusgalli	Poaceae	*vilona*	-stem with leaves
Cyclosorus gongylodes	Polypodiaceae	*tritra*	-tip of leaves
Polygonum glabrum	Polygonaceae	*tamboloana*	-tip of leaves
Argyreia vahibora	Convolvulaceae	*vahankelana*	-shoot -tip of leaves
Cuscuta sinensis	Convolvulaceae	*tsy hita fototra*	-stem
Ethulia conyzoides	Astraceae	*revaka*	-stem
Nymphaea lotus	Nymphaceae	*tatamo*	-stem of leaves
Eichhornia crassipes	Pontederiaceae	*tsikafona*	-stem -flower

MJ/kg gross energy; 131.6 ± 56.7 g/kg crude protein; 14.2 ± 6.4 g/kg crude lipid; 222.1 ± 72.5 g/kg crude fibre; 361.3 ± 67.2 g/kg ADF (acid detergent fibre); and 102.9 ± 35.6 g/kg ash.

With respect to the biochemical composition of the food items, general trends were observed: protein content was positively related to the amino-acid contents ($0.703 < r < 0.937$, $p < 0.01$; Fisher's R to Z-test) and negatively related to crude fibre content ($r = -0.642$, $p = 0.031$; Fisher's R to Z-test) and ADF content ($r = -0.601$, $p = 0.050$; Fisher's R to Z-test); moreover, gross energy content was negatively related to ash content ($r = -0.758$, $p = 0.005$; Fisher's R to Z-test). However, no correlation between energy or nutrient contents and the time spent (or the rank order of feeding time) on the different food items was found. Only if food items were split into three categories by the rank order of mean feeding time (MFT) per 24 h (major: MFT > 20 min; minor: MFT = 1–20 min; marginal: MFT < 1 min), was a significant difference in the gross energy content found between the categories (DF = 2, H = 8.73, $p = 0.01$; Kruskal-Wallis). However, a nonparametric Tukey-like multiple comparison (post-hoc test; Zar, 1984) showed that 'minor food items' differed significantly from 'marginal food items' but not from 'major food items', and that there was no significant difference in energy content of the 'major' and 'minor' food items. Moreover, no other components (nutrients) differed significantly between categories.

Table 4. Percentage of diurnal and nocturnal feeding time spent on different food plants

Period	Group	TFR [counts]	Feeding records					
			C. m.	*P. c.*	*E. c.*	*L. h.*	*A. v.*	Other
day	A	4255	65.4%	12.7%	13.4%	4.0%	2.4%	2.1%
day	K	3518	45.1%	30.7%	20.5%	3.3%	0.1%	0.2%
night	A	142	10.6%	63.4%	1.4%	23.9%	0.7%	0%

TFR = Total feeding records; *C. m.* = *Cyperus madagascariensis*; *P. c.* = *Phragmites communis*; *E. c.* = *Echinochloa crusgalli*; *L. h.* = *Leersia hexandra*; *A. v.* = *Argyreia vahibora*.

Table 5. Feeding and food composition (focal animal A)

Species	Part	Feeding				Food analysis (dry matter)							
		D [% of TFR]	N [% of TFR]	MFT [min]	Category	GE [MJ/kg]	CP [g/kg]	CL [g/kg]	CF [g/kg]	ADF [g/kg]	NDF [g/kg]	Ash [g/kg]	Ca [g/kg]
Cyperus madagascariensis	pith	48.3	10.6	93	major	16.4	123	14	178	332	n.d.	121	1.4
Phragmites communis	stem	12.7	63.4	86	major	16.6	247	12	203	256	n.d.	156	3.4
Leersia hexandra	stem	4.0	23.9	31	major	15.7	143	12	265	298	572	103	4.0
Echinochloa crusgalli	stem	13.4	1.4	24	major	16.9	115	20	366	403	711	96	1.9
Cyperus madagascariensis	leaves	12.6	—	21	major	18.2	119	10	298	403	659	63	2.3
Cyperus madagascariensis	bud	2.3	—	4	minor	18.2	81	13	331	498	n.d.	59	1.0
Cyperus madagascariensis	seeds	2.1	—	4	minor	19.1	84	30	241	384	n.d.	71	1.4
Argyreia vahibora	shoot	1.6	0.7	3	minor	19.1	184	11	172	398	n.d.	89	3.3
Argyreia vahibora	leaves	0.8	—	1	minor	19.5	184	13	148	282	297	84	8.3
Polygonum galbrum	leaves	1.6	—	3	minor	19.3	208	7	108	325	230	46	3.4
Ethulia conyzoides	stem	0.1	—	0	marginal	15.9	39	8	271	351	366	107	15.9
Eichhornia crassipes	stem	0.1	—	0	marginal	16.0	104	n.d.	163	n.d.	n.d.	141	18.1
Eichhornia crassipes	flower	0.1	—	0	marginal	n.d.	161	n.d.	162	406	n.d.	n.d.	n.d.
Cuscuta sinensis	stem	eaten	—	0	marginal	17.9	118	20	209	n.d.	n.d.	88	1.2
Nymphea lotus	stem	eaten	—	0	marginal	13.9	64	n.d.	216	n.d.	n.d.	151	6.5
Cyclosorus gongylodes	leaves	eaten	—	0	marginal	n.d.	n.d.	n.d.	n.d.	n.d.	n.d.	n.d.	n.d.
Soil		0.3	—	1		n.d.	n.d.	n.d.	20	574	288	319	8.1
Cyperus madagascariensis	stem*	—	—	—	non-food	16.5	54	7	405	461	701	85	1.3

D = Day; N = Night; TFR = Total feeding records; MFT = Mean feeding time/24 hr; GE = Gross Energy; CP = Crude Protein; CL = Crude Lipid; CF = Crude Fiber; ADF = Acid Detergent Fiber; NDF = Neutral Detergent Fiber; Ca = Calcium; *fraction not eaten; n.d. = not determined.

The protein-to-fibre ratio (crude protein/acid detergent fibre), which is occasionally used as a measure of dietary quality (e.g., Milton, 1979; Dasilva, 1994; Waterman & Kool, 1994), did not differ significantly between the categories, but the means of this ratio did gradually decrease from 'major' to 'marginal food items'. Thus, although no significant relationship between biochemical composition of food items and food choice was evident, there was a trend towards a preference for food items with higher protein-to-fibre ratios.

Comparing individual food items, reed shoots (*Phragmites*) had the highest crude protein (247 g/kg DM), highest ash (156 g/kg DM) and lowest ADF (256 g/kg DM) contents of all food plants. Extremely high calcium levels (15.9 and 18.1 g/kg DM) were found in two 'marginal food items' (*Ethulia* and *Eichhornia*; stems), which even greatly exceeded the calcium level in ingested soil (8.1 g/kg DM). The only 'non-food' item collected and analyzed (papyrus pith fraction directly above the eaten part), showed lower crude protein, crude fat and ash contents, and higher crude fibre and ADF contents than the eaten papyrus fractions.

Amino acid composition is given in Table 6. Chemical scores (CS; measure of protein quality) of the essential amino acids, where amino acid composition of the food items is referred to the amino acid balance in an 'ideal protein' (lean muscle protein; Agricultural Research Council, 1981), were calculated for the five most important food items for *Hapalemur g. alaotrensis*. In all five lysine was the limiting essential amino acid, with an average chemical score of 55 ± 9 (% of content in ideal protein). Protein quality was highest in *Echinochloa* (CS = 66) and lowest in *Leersia* (CS = 43).

DISCUSSION

The Alaotran Gentle Lemur was found completely to rely on a diet composed of leaves and grasses throughout the year. Food diversity was low (11 plant species; 16 plant parts) and more than 95% of feeding time was spent on only four different plant species. Thus, the Alaotran Gentle Lemur should be considered as one of the most specialized (herbivorous) feeders, comparable to its congeners *Hapalemur g. griseus* (Pollock, 1986; Overdorff et al., 1997), *H. aureus*, and *H. simus* (Wright & Randriamanantena, 1989), as well as to the Giant Panda *Ailuropoda melanoleuca*, (Schaller et al., 1984). Food diversity of other primate folivores, such as *Lepilemur leucopus* (Hladik & Charles-Dominique, 1974; Nash, 1998), *Avahi laniger* (Ganzhorn et al., 1985; Harcourt, 1991), *Colobus guereza* (Oates, 1977), *Colobus satanas* (McKey et al., 1981), *Presbytis senex* (Hladik, 1978), *Gorilla gorilla beringei* (Watts, 1988) is clearly higher than in *Hapalemur*.

Gross energy content (mean = 17.3 MJ/kg DM) of *H. g. alaotrensis* foods was at the lower end of the range of good-quality browses for farm animals (17–20 MJ/kg DM) (McDonald et al., 1988; Robbins, 1993) and below the mean value (21.5 kJ/g DM) found in the photosynthetic parts of foods eaten by *Colobus polykomos* (Dasilva, 1994). The only available data on energy content of the food of a lemur folivore (*Avahi laniger*, Ganzhorn et al., 1985) are not comparable to the results of this study, because of the different methods applied (Ganzhorn calculated energy contents indirectly on the basis of protein and sugar contents). Crude protein content (131.6 ± 56.7 g/kg DM) in food items of the Alaotran Gentle Lemur was at the lower range of other folivore primate diets and closest to foods of *Indri indri* (Table 7). Fibre content (ADF) of *H. g. alaotrensis* foods was higher than reported for *H. g. griseus* foods (Ganzhorn et al., 1985), but

Table 6. Amino acid composition of food items

ID	Part	Amino acids [g/kg dry matter]																	
		LYS	MET	CYS	ALA	ARG	ASP	GLU	GLY	HIS	ILE	LEU	PHE	PRO	SER	THR	TYR	VAL	TRP
C. m.	pith	4.4	1.3	1.5	4.6	3.7	16.2	8.8	3.9	1.8	3.4	6.3	4.7	3.7	3.9	3.5	2.8	4.5	n.d.
P. c.	stem	9.4	3.1	2.9	8.9	8.8	18.9	20.0	8.8	3.3	6.5	12.5	7.0	7.6	7.5	7.1	5.4	9.1	n.d.
L. h.	stem	4.3	1.5	1.9	5.7	4.8	21.0	11.5	4.3	1.9	3.0	6.0	4.3	7.2	2.5	3.4	3.2	4.3	n.d.
E. c.	stem	5.3	1.8	1.5	5.4	4.9	12.9	12.5	4.8	2.1	3.6	7.6	5.4	6.3	4.2	4.4	3.6	5.1	1.8
C. m.	leaves	5.0	1.1	1.6	5.4	3.8	18.7	11.2	4.2	2.1	3.6	6.6	4.6	4.0	4.2	4.5	2.8	4.7	n.d.
C. m.	bud	3.6	b.t.	1.1	3.3	3.6	8.2	6.3	3.2	1.3	2.5	4.8	3.4	3.4	2.7	3.0	2.3	3.4	n.d.
C. m.	seeds	4.0	1.2	1.5	4.1	4.1	8.0	7.7	4.0	1.4	3.3	5.9	3.9	4.2	3.4	3.5	2.8	4.0	n.d.
A. v.	shoot	7.3	1.9	2.5	6.1	7.5	14.6	11.3	6.4	2.8	5.7	9.7	5.6	5.6	5.0	5.6	4.7	6.7	n.d.
P. g.	leaves	12.2	2.6	2.7	9.4	10.8	16.4	20.1	9.5	4.4	8.0	15.4	8.9	7.7	7.0	7.8	6.5	10.1	3.2
A. v.	leaves	8.3	1.3	2.4	7.8	7.9	18.5	15.4	7.5	2.8	6.4	12.3	7.6	7.0	5.4	6.7	5.5	8.4	2.9
E. con.	stem	2.1	b.t.	b.t.	1.8	2.1	4.0	3.7	2.0	b.t.	1.4	2.7	2.1	5.8	1.7	1.6	1.4	1.9	n.d.
E. cra.	stem	4.0	1.3	1.7	4.7	4.9	14.8	8.4	4.3	1.9	3.1	5.7	4.2	3.5	3.3	3.6	2.3	4.1	n.d.
E. cra.	flower	6.3	2.7	2.4	8.5	6.6	13.4	15.3	7.9	2.6	5.9	10.9	6.4	5.4	5.3	6.1	4.2	7.6	n.d.
C. s.	stem	5.5	1.5	2.0	4.4	4.9	9.0	9.3	5.1	2.3	4.1	7.0	4.6	4.1	4.7	4.1	4.1	5.2	n.d.
N. l.	stem	3.3	b.t.	1.2	3.2	3.3	5.6	5.7	3.0	1.4	2.4	4.7	3.7	2.9	2.3	2.5	2.4	3.0	n.d.
C. m.	stem*	2.5	b.t.	b.t.	2.2	2.4	7.7	4.5	2.1	1.1	2.0	3.0	2.8	2.5	2.2	1.9	1.4	1.9	n.d.

C. m. = Cyperus madagascariensis; P. c. = Phragmites communis; E. c. = Echinochloa crusgalli; L. h. = Leersia hexandra; A. v. = Argyreia vahibora; P. g. = Polygonum galbrum; E. con. = Ethulia conyzoides; E. cra. = Eichhornia crassipes; C. s. = Cuscuta sinensis; N. l. = Nymphea lotus; *fraction not eaten; n.d. = not determined; b.t. = below threshold.

lower than in the diets of all other lemur folivores (Table 7). Thus the nutrient analyses revealed that, compared to other folivorous primates, *H. g. alaotrensis* subsists on a low-energy, low-protein diet with moderate fibre content.

The protein-to-fibre ratio is occasionally used as an estimate of food quality (e.g., Milton, 1979; Dasilva, 1994; Waterman & Kool, 1994), although this ratio does not take secondary plant compounds into account. In terms of this ratio, *Hapalemur g. alaotrensis* was found to have a medium 'food quality' compared to other folivorous lemurs (Table 7). Although no significant relation between nutrient or energy content and the importance of the food items (in terms of feeding time) was found in the present study, the protein-to-fibre ratio decreased gradually from 'major' to 'marginal' (and 'non-food') items. Thus, there was a trend towards a preference for food items with high protein-to-fibre ratios. This trend, however, cannot explain the distinct selectivity in food choice observed at the behavioral level and it can be speculated that secondary plant compounds may be responsible for the selectivity in food choice.

In addition, Table 7 shows that the Alaotran Gentle Lemur and other small-bodied folivorous lemurs do not select for distinctly higher-quality foods than larger-bodied folivorous primates (e.g., *Colobus, Presbytis, Gorilla*). Thus, the question arises as to how else these small-bodied folivores meet their theoretically predicted higher energy demands and compensate for the effects of lower gut capacity.

During their resting phase (i.e., during daytime), *Lepilemur ruficaudatus* has been found to have a markedly reduced resting metabolic rate (only 40% of Kleiber value) during the day (Schmid & Ganzhorn, 1996), implying a kind of daily torpor. No BMR measures are available for *Avahi* and *Hapalemur*, but *H. g. alaotrensis* slept very superficially, opening their eyes whenever there were noises near the sleeping site, suggesting a lack of deep rest or torpor. However, although the existence of a torpor-like phase during the resting period seems unlikely, this does not exclude the possibility of a reduced BMR in *H. g. alaotrensis*.

Energetic expenditure can also be reduced at the behavioral level. By increasing resting time and decreasing other activities (travelling, feeding, foraging), an animal can save considerable amounts of energy. Such behavioral "energy minimizing" strategy has been proposed to occur in *Lepilemur* (Hladik & Charles-Dominique, 1974; Nash, 1998; Warren & Crompton, 1998) and *Alouatta* spp. (Milton, 1978). Indeed, most folivorous primates have a high percentage of resting in their activity budget (reviewed in Nash, 1998), with most extreme values in *Alouatta* spp. (60–80%) (Milton, 1978; Mendes, 1985), some *Colobus* spp. and *Presbytis* spp. (around 60%) (Oates, 1977; McKey & Waterman, 1982; Ruhiyat, 1983; Stanford, 1991), *Avahi laniger* (about 62%) (Harcourt, 1991), and *Propithecus verreauxi* during the dry season (60–70%) (Richard, 1978). At first glance, the time budget of the Alaotran Gentle Lemur (with more than 60% of resting time during days and 75% during nights) fits fairly well into the picture, since it suggests a very low activity level. However, because *H. g. alaotrensis* is cathemeral, such that its activity budget is distributed throughout the 24h cycle, the high percentage of inactivity is misleading. The activity budgets of other folivore primates cited above are all related to one light period only (either day or night). When activity duration is estimated per 24h (Table 8), the picture is quite different and the Alaotran Gentle Lemur turns out to be one of the most active folivorous primates! It also seems important to mention the report of a cathemeral activity pattern in *Alouatta pigra* (Dahl & Hemingway, 1988). Closer examination of night activity might well explain the low diurnal activity level found in *Alouatta*. However, it might be that behavioral

Table 7. Gross energy, crude protein, fibre (ADF) contents, and protein-to-fibre ratio of photosynthetic plant parts eaten by different primate species

Species	GE ± SD [MJ/kg DM]	CP ± SD [g/kg DM]	ADF ± SD [g/kg DM]	CF ± SD [g/kg DM]	CP/ADF ± SD	Source
Hapalemur g. alaotrensis	17.3 ± 1.7 (n = 14)	131.6 ± 56.7 (n = 15)	361.3 ± 67.2 (n = 12)	222.1 ± 72.5 (n = 15)	0.42 ± 0.2 (n = 12)	this study
Hapalemur g. alaotrensis	—	118.5 ± 36.1 (n = 2)	268 ± 11* (n = 2)	—	0.44 (n = 2)	Pollock (1986)
Hapalemur g. griseus	—	300.1 ± 75.1 (n = 8)	302.5 ± 7.7 (n = 8)	—	1.05 ± 0.3 (n = 8)	Ganzhorn (1988)
Avahi l. occidentalis	—	192.2 ± 110.4 (n = 9)	450.7 ± 152.7 (n = 9)	—	0.57 ± 0.6 (n = 9)	Ganzhorn (1988)
Avahi l. laniger	—	165.2 ± 69.6 (n = 18)	462.5 ± 82.8 (n = 18)	—	0.39 ± 0.2 (n = 18)	Ganzhorn (1988)
Lepilemur mustelinus	—	142.1 ± 56.8 (n = 12)	454.2 ± 99.7 (n = 13)	—	0.33 ± 0.1 (n = 13)	Ganzhorn (1988)
Lepilemur edwardsi	—	167.4 ± 68.2 (n = 12)	411.7 ± 122.6 (n = 13)	—	0.42 ± 0.2 (n = 12)	Ganzhorn (1988)
Indri indri	—	139.5 ± 53.8 (n = 11)	469.8 ± 126.3 (n = 11)	—	0.34 ± 0.2 (n = 11)	Ganzhorn (1988)
Colobus polykomos	21.5 ± 1.8 (n = 13)	183.5 ± 73.8 (n = 13)	346.8 ± 106.9 (n = 14)	—	0.67 ± 0.5 (n = 13)	Dasilva (1994)
Colobus satanas	—	146 (n = 10)	546.8 (n = 28)	—	0.27**	McKey et al. (1981)
Papio anubis	—	277 (n = 16)	171 (n = 16)	—	1.62 ± 0.2 (n = 16)	Barton et al. (1993)
Presbytis spp.	—	182 (n = 28)	386 (n = 28)	—	0.47 (n = 28)	in Barton et al. (1993)
Gorilla g. gorilla	—	183.7 (n = 28)	289.2 (n = 16)	—	0.64 (n = 18)	Rogers et al. (1990)

GE = Gross Energy; CP = Crude Protein; ADF = Acid Detergent Fibre; CF = Crude Fibre; *ADF content estimated as sum of cellulose and lignin; **ratio of means.

Table 8. Calculated duration of activity during a 24 h-period of different folivorous primates

Species	Activity pattern	Activity/24 hrs [Min]*	Reference
Hapalemur g. alaotrensis	cathemeral	472	this study
Hapalemur g. griseus	diurnal	350 (w)	Overdorff et al. (1997)
		456 (d)	
Lepilemur leucopus	nocturnal	403 (w)	Nash (1998)
		331 (d)	
Avahi laniger	nocturnal	292	Harcourt (1991)
Propithecus verreauxi (northern population)	diurnal	360 (w)	Richard (1978)
		277 (d)	
Propithecus verreauxi (southern population)	diurnal	311 (w)	Richard (1978)
		220 (d)	
Colobus satanas	diurnal	266	McKey and Waterman (1982)
Colobus guereza	diurnal	266	Oates (1977)
Presbytis pileata	diurnal	432	Stanford (1991)
Presbytis aygula	diurnal	289	Ruhiyat (1983)
Alouatta palliata	diurnal	245	Milton (1978)
Alouatta fusca	diurnal	202	Mendes (1985)
Brachyteles arachnoides	diurnal	281	Milton (1984)
Brachyteles arachnoides	diurnal	365	Strier (1987)
Gorilla gorilla	diurnal	472	Watts (1988)

*mean percentage of activity records multiplied by daily observation time; w = wet season; d = dry season.

"energy minimizing" is a strategy pursued by some folivorous primates (e.g., *Alouatta*, *Colobus*, Indriidae), although it is not shown by *H. g. alaotrensis*.

Another possible behavioral adaptation to deal with the constraints imposed by a high MR/GC-ratio could be cathemerality. Engqvist and Richard (1991) found that in species showing cathemeral activity only seasonally or at certain study sites (*Aotus trivirgatus*, Wright 1985; *Eulemur rubriventer*, Overdorff 1988; *E. fulvus mayottensis*, Tattersall 1982), cathemerality was associated with a shift away from frugivory towards higher consumption of leaves and/or flowers. Engqvist and Richard (1991) suggested that cathemerality could be an adaptation for folivory in animals without specialized digestive tracts, in the sense that food harvesting is spaced to minimize time in which no food is being processed. *H. g. alaotrensis* is not a good model to test this hypothesis, as it is year-round folivorous and cathemeral, but other studies (Overdorff & Rasmussen, 1995; Curtis, 1997), comparing diet and activity pattern in cathemeral species have questioned the hypothesis. Both Overdorff and Rasmussen (1995) and Curtis (1997) concluded that there is some influence of diet on cathemerality, but they did not find evidence that the amount of difficult-to-digest foods (e.g., unripe fruit or fibre content) consumed correlated with nocturnal activity. However, it might be that the crucial variable is not fibre content but energetic returns. If dietary energy is low, and animals (restricted by gut capacity) could increase the total amount of captured energy by (an) additional feeding bout(s) during the resting period, a cathemeral activity pattern could in fact be advantageous for them. However, other folivorous lemurs (*Avahi*, *Lepilemur*) in the body-size range of *Hapalemur g. alaotrensis* are not cathemeral (but nocturnal), whereas other cathemeral lemur species (*Eulemur* spp.) are not (specialized) folivores.

Thus, *Hapalemur g. alaotrensis* does select relatively poor-quality diets, but does not reduce energy expenditure at the behavioral level. Moreover, there is no evidence so far that *H. g. alaotrensis* has a distinctly reduced metabolism nor that a cathemeral activity pattern is an adaptation to folivory. Does this mean that the digestive abilities of the Alaotran Gentle Lemur are higher than assumed? Although *Hapalemur* was originally reported to show no morphological specialization in the digestive tract (Hill, 1953), Robert (in Fidgett et al., 1996), and Fleagle and Wunderlich (in Overdorff et al., 1997) reported enlarged hindguts (mainly caecum) in *H. griseus*. In addition, long gut passage times between 18 hr and 36 hr (Cabre-Vert & Feistner, 1995; Overdorff & Rasmussen, 1995; Tan, 1998) and a high ability to digest dietary fibre (Klein, 1991) have been found in *H. griseus*. Hence, it seems that *H. griseus* is an efficient intestinal fermenter. Moreover, Justice and Smith (1992) showed that digestive capability of small-bodied folivores (wood rats) with hindgut fermentation can be much higher than theoretically predicted. Thus, perhaps the restrictions imposed by the fibrous, low-energy diet have been overestimated and the digestive capacity of *H. g. alaotrensis* is higher than assumed.

In conclusion, this study found that the Alaotran Gentle Lemurs are highly specialized feeders, relying for more than 95% of feeding time on only four different plant species. On the behavioral level food choice was highly selective, although this selectivity did not result in the ingestion of high-quality foods with regards to primary plant compounds. The nutrient analyses revealed that, compared to other folivorous primates, *Hapalemur g. alaotrensis* subsists on a low-energy, low-protein diet with moderate fibre content. Moreover, there is no evidence that voluntary energy expenditure was minimized nor that *H. g. alaotrensis* has a distinctly reduced basal metabolic rate. Although cathemerality could theoretically resolve gut-capacity restrictions in small-bodied folivores, there is so far no evidence that this activity pattern reflects indeed an adaptation to folivory. Perhaps the digestive capability of *H. g. alaotrensis* are higher than theoretically predicted, and thus the restrictions imposed by the fibrous, low-energy diet have been overestimated.

ACKNOWLEDGMENTS

Field work was carried out under the Accord de Collaboration between the Jersey Wildlife Preservation Trust and the Government of Madagascar through the Commission Tripartite. I thank the governmental institutions of Madagascar for permission to conduct the study. I am grateful to the supervisors of the study, A. T. C. Feistner (Jersey Wildlife Preservation Trust) and R. D. Martin (Anthropological Institute, University of Zurich), for advice and helpful comments on this manuscript. I thank my field assistants Richard Rasolonjatovo and Meline Raharinosy of Andreba. Many thanks go to D. Glaser (Anthropological Institute, University of Zurich), D. Guidon (Swiss Federal Research Station for Animal Production, Posieux), and A. Fidgett (Division of Environmental & Evolutionary Biology, University of Glasgow) for help with the interpretation of food analysis. The study was funded by Jersey Wildlife Preservation Trust, G. & A. Claraz-Donation, Goethe-Foundation for Art and Science, A. H. Schultz-Foundation, Janggen-Poehn-Foundation, and an anonymous contribution. Food analyses were financed by the EU-Project AIR3-CT94-2107 (D. Glaser).

REFERENCES

AGRICULTURAL RESEARCH COUNCIL. 1981. The Nutrient Requirements of Pigs. Commonwealth Agricultural Bureaux, Farnham Royal.

ALTMANN, J. 1974. Observational study of behaviour: sampling methods. Behaviour, **49**: 227–267.
BARTON, R. A., A. WHITEN, R. W. BRYNE, AND M. ENGLISH. 1993. Chemical composition of baboon plant foods: Implications for the interpretation of intra- and interspecific differences in diet. Folia Primatologica, **61**: 1–20.
BAUCHOP, T. 1978. Digestion of leaves in vertebrate arboreal folivores, pp. 193–204. *In* Montgomery, G. G., ed., The Ecology of Arboreal Folivores. Smithsonian Institution Press, Washington, D.C.
CABRE-VERT, N., AND A. T. C. FEISTNER. 1995. Comparative gut passage time in captive lemurs. The Dodo, Journal of the Wildlife Preservation Trusts, **31**: 76–81.
CURTIS, D. J. 1997. The Mongoose Lemur (*Eulemur mongoz*): A study in behaviour and ecology. Ph.D. Thesis. University of Zurich, Zurich.
DAHL, J. F., AND C. A. HEMINGWAY. 1988. An unusual activity pattern for the Mantled Howler Monkey of Belize. American Journal of Physical Anthropology, **75**: 201.
DASILVA, G. L. 1994. Diet of *Colobus polykomos* on Tiwai Island: Selection of food in relation to its seasonal abundance and nutritional quality. International Journal of Primatology, **15**: 655–680.
DEMMENT, M. W., AND P. J. VAN SOEST. 1985. A nutritional explanation for body-size patterns of ruminant and nonruminant herbivores. The American Naturalist, **125**: 641–672.
ENGQVIST, A., AND A. RICHARD. 1991. Diet as a possible determinant of cathemeral activity patterns in primates. Folia Primatologica, **57**: 169–172.
FIDGETT, A. L., A. T. C. FEISTNER, AND H. GALBRAITH. 1996. Dietary intake, food composition and nutrient intake in captive Alaotran Gentle Lemurs *Hapalemur griseus alaotrensis*. The Dodo, Journal of the Wildlife Preservation Trusts, **32**: 44–62.
GANZHORN, J. U. 1988. Food partitioning among Malagasy primates. Oecologia, **75**: 436–450.
———, J. P. ABRAHAM, AND M. RAKOTOMALALA-RAZANAHOERA. 1985. Some aspects of the natural history and food selection of *Avahi laniger*. Primates, **26**: 452–463.
GANZHORN, J. U., AND P. C. WRIGHT. 1994. Temporal patterns in primate leaf eating: The possible role of leaf chemistry. Folia Primatologica, **63**: 203–208.
GLANDER, K. E. 1982. The impact of plant secondary compounds on primate feeding behavior. Yearbook of Physical Anthropology, **25**: 1–18.
———, P. C. WRIGHT, D. S. SEIGLER, V. RANDRIANASOLO, AND B. RANDRIANASOLO. 1989. Consumption of cyanogenic bamboo by a newly discovered species of bamboo lemur. American Journal of Primatology, **19**: 119–124.
HARCOURT, C. S. 1991. Diet and behaviour of a nocturnal lemur, *Avahi laniger*, in the wild. Journal of Zoology London, **223**: 667–674.
HILL, W. C. O. 1953. Primates: Comparative Anatomy and Taxonomy. Vol. 1. Strepsirhini. Edinburgh University Press, Edinburgh.
HLADIK, C. M. 1978. Adaptive strategies of primates in relation to leaf eating, pp. 373–395. *In* Montgomery, G. G., ed., The Ecology of Arboreal Folivores. Smithsonian Institution Press, Washington, D.C.
———, AND P. CHARLES-DOMINIQUE. 1974. The behaviour and ecology of the Sportive Lemur (*Lepilemur mustelinus*) in relation to its dietary peculiarities, pp. 23–37. *In* Martin, R. D., G. A. Doyle, and A. C. Walker, eds., Prosimian Biology. Duckworth, London.
JUSTICE, K. E., AND F. A. SMITH. 1992. A model of dietary fiber utilization by small mammalian herbivores, with empirical results for *Neotoma*. The American Naturalist, **139**: 398–416.
KLEIBER, M. 1961. The Fire of Life: An Introduction to Animal Energetics. John Wiley, New York.
KLEIN, M. 1991. Digestibility of dietary fibre and passage rates in Brown Lemurs and Gentle Lemurs. M. S. Thesis. Duke University, Durham.
MARTIN, R. D. 1990. Primate Origins and Evolution: A Phylogenetic Reconstruction. Princeton University Press, Princeton.
MCDONALD, P., R. A. EDWARDS, AND J. F. D. GREENHALGH. 1988. Animal Nutrition. Longman Scientific and Technical, Essex.
MCKEY, D. B., J. S. GARTLAN, P. G. WATERMAN, AND G. M. CHOO. 1981. Food selection by Black Colobus Monkeys (*Colobus satanas*) in relation to plant chemistry. Biological Journal of the Linnean Society, **16**: 115–146.
MCKEY, D. B., AND P. G. WATERMAN. 1982. Ranging behaviour of a group of Black Colobus (*Colobus satanas*) in the Douoala-Edea Reserve, Cameroon. Folia Primatologica, **39**: 264–304.
MCNAB, B. K. 1978. Energetics of arboreal folivores: Physiological problems and ecological consequences of feeding on an ubiquitous food supply, pp. 153–162. *In* Montgomery, G. G., ed., The Ecology of Arboreal Folivores. Smithsonian Institution Press, Washington, D.C.
MENDES, S. L. 1985. Uso do espaco, padroes de atividades diaries e organizacao social de *Alouatta fusca* em Caratinga. Masters Thesis. Universidade de Brasilia, Brasilia.

MILTON, K. 1978. Behavioral adaptations to leaf-eating by the Mantled Howler Monkey (*Alouatta palliata*), pp. 535–549. *In* Montgomery, G. G., ed., The Ecology of Arboreal Folivores. Smithsonian Institution Press, Washington, D.C.

———. 1979. Factors influencing leaf choice by Howler Monkeys: a test of some hypotheses of food selection by generalist herbivores. The American Naturalist, **114**: 362–378.

———. 1984. Habitat, diet, and activity patterns of free-ranging Woolly Spider Monkeys (*Brachyteles arachnoides* E. Geoffroy 1806). International Journal of Primatology, **5**: 491–514.

MOREAU, J. 1987. Madagascar, pp. 595–606. *In* Burgis, M. J., and J. J. Symoens, eds., African Wetlands and Shallow Water Bodies. ORSTOM, Paris.

MUTSCHLER, T., AND A. T. C. FEISTNER. 1995. Conservation status and distribution of the Alaotran Gentle Lemur *Hapalemur griseus alaotrensis*. Oryx, **29**: 267–274.

———, AND C. M. NIEVERGELT. 1998. Preliminary field data on group size, diet, and activity in the Alaotran Gentle Lemur *Hapalemur griseus alaotrensis*. Folia Primatologica, **69**: 325–330.

NASH, L. T. 1998. Vertical clingers and sleepers: Seasonal influences on the activities and substrate use of *Lepilemur leucopus* at Beza Mahafaly Special Reserve, Madagascar. Folia Primatologica, 69, Supplement **1**: 204–217.

OATES, J. F. 1977. The Guereza and its food, pp. 276–321. *In* Clutton-Brock, T. H., ed., Primate Ecology: Studies of Feeding and Ranging Behaviour in Lemurs, Monkeys and Apes. Academic Press, London.

OVERDORFF, D. 1988. Preliminary report on the activity cycle and diet of the Red-bellied Lemur (*Lemur rubriventer*) in Madagascar. American Journal of Primatology, **16**: 143–153.

———, AND M. A. RASMUSSEN. 1995. Determinants of nighttime activity in "diurnal" lemurid primates, pp. 61–74. *In* Alterman, L., G. A. Doyle, and M. K. Izard, eds., Creatures of the Dark: The Nocturnal Prosimians. Plenum Press, New York.

OVERDORFF, D. J., S. G. STRAIT, AND A. TELO. 1997. Seasonal variation in activity and diet in a small-bodied folivorous primate, *Hapalemur griseus*, in southeastern Madagascar. American Journal of Primatology, **43**: 211–223.

PARRA, R. 1978. Comparison of foregut and hindgut fermentation in herbivores, pp. 205–230. *In* Montgomery, G. G., ed., The Ecology of Arboreal Folivores. Smithsonian Institution Press, Washington, D.C.

PETTER, J.-J., AND A. PEYRIERAS. 1970. Observations éco-éthologiques sur les lémuriens malgaches du genre *Hapalemur*. La Terre et la Vie, **24**: 365–382.

PIDGEON, M. 1996. An ecological survey of Lake Alaotra and selected wetlands of central and eastern Madagascar in analysing the demise of Madagascar Pochard *Aythya innotata*. Unpublished Report, Missouri Botanical Gardens, Antananarivo.

POLLOCK, J. I. 1986. A note on the ecology and behaviour of *Hapalemur griseus*. Primate Conservation, **7**: 97–100.

RICHARD, A. F. 1978. Behavioral Variation: Case Study of a Malagasy Lemur. Bucknell University Press, Lewisburg.

ROBBINS, C. T. 1993. Wildlife Feeding and Nutrition. Academic Press, San Diego.

ROGERS, M. E., F. MAISELS, E. A. WILLIAMSON, M. FERNANDEZ, AND C. E. G. TUTIN. 1990. Gorilla diet in the Lopé Reserve, Gabon: A nutritional analysis. Oecologia, **84**: 326–339.

RUHIYAT, Y. 1983. Socio-ecological study of *Presbytis aygula* in West Java. Primates, **24**: 344–359.

SANTINI-PALKA, M.-E. 1994. Feeding behaviour and activity patterns of two Malagasy Bamboo Lemurs, *Hapalemur simus* and *Hapalemur griseus*, in captivity. Folia Primatologica, **63**: 44–49.

SCHALLER, G. B., H. JINCHU, P. WENSHI, AND Z. JING. 1984. The Giant Pandas of Wolong. The University of Chicago Press, Chicago.

SCHMID, J., AND J. U. GANZHORN. 1996. Resting metabolic rates of *Lepilemur ruficaudatus*. American Journal of Primatology, **38**: 169–174.

STANFORD, C. B. 1991. The Capped Langur in Bangladesh: Behavioural Ecology and Reproductive Tactics. Karger, Basel.

STRIER, K. B. 1987. Activity budgets of Woolly Spider Monkeys, or Muriquis (*Brachyteles arachnoides*). American Journal of Primatology, **13**: 385–395.

TAN, C. L. 1998. Comparison of food passage time in three species of *Hapalemur*. American Journal of Physical Anthropology, Supplement **26**: 215.

TATTERSALL, I. 1982. The Primates of Madagascar. Columbia University Press, New York.

TYLER, S. 1979. Time-sampling: a matter of convention. Animal Behaviour, **27**: 801–810.

WARREN, R. D., AND R. H. CROMPTON. 1998. Diet, body size, and energy costs of locomotion in saltatory prosimians. Folia Primatologia, 69, Supplement **1**: 86–100.

WATERMAN, P. G., AND K. M. KOOL. 1994. Colobine food selection and plant chemistry, pp. 251–284. *In* Davies,

A. G., and J. F. Oates, eds., Colobine Monkeys; Their Ecology, Behavior and Evolution. Cambridge University Press, Cambridge.

WATTS, D. 1988. Environmental influences on Mountain Gorilla time budgets. American Journal of Primatology, **15**: 195–211.

WRIGHT, P. C. 1985. Costs and benefits of nocturnality for *Aotus* (the Night Monkey). Ph.D. Thesis. City University of New York, New York.

———. 1986. Diet, ranging behavior, and activity pattern of the Gentle Lemur (*Hapalemur griseus*) in Madagascar. American Journal of Physical Anthropology, **69**: 283.

———, AND M. RANDRIAMANANTENA. 1989. Comparative ecology of three sympatric Bamboo Lemurs in Madagascar. American Journal of Physical Anthropology, **78**: 327.

ZAR, J. H. 1984. Biostatistical Analysis. Prentice-Hall, Inc., Englewood Cliffs.

14

CONSERVATION OF THE ALAOTRAN GENTLE LEMUR

A Multidisciplinary Approach

Anna T. C. Feistner[1]

[1] Jersey Wildlife Preservation Trust
Les Augrès Manor, Trinity, Jersey JE3 5BP
Channel Islands, British Isles
afeistner@jwpt.org

ABSTRACT

The Alaotran Gentle Lemur *Hapalemur griseus alaotrensis* is restricted to the marshes around Lac Alaotra. It is threatened by habitat destruction and fragmentation and by hunting for food and pets. Since 1990 a conservation program for this lemur—now a flagship for the Alaotran wetlands—has been developed. It has a multidisciplinary approach involving captive breeding (*in* and *ex situ*), research on captive animals (infant development, nutrition), field research (distribution, census, conservation status, behavioral ecology, ranging, diet, social organization), socioeconomic studies, community education, and genetics. The development of this program is described and illustrates how a small-scale yet holistic approach to species conservation can be effective.

RÉSUMÉ

L'Hapalémur du Lac Alaotra (*Hapalemur griseus alaotrensis*) a une distribution limitée aux marais périphériques du Lac Alaotra. Il est menacé par la destruction et la fragmentation de son habitat et par les chasseurs qui recherchent du gibier ou des animaux de compagnie. Depuis 1990 un programme de conservation est mené pour ce lémurien, qui est à présent devenu un porte étendard pour les zones humides du Lac Alaotra. Ce programme a une méthode pluri-disciplinaire incluant la reproduction en captivité (*in* et *ex situ*), la recherche sur des animaux en captivité (croissance des jeunes, nutrition), des recherches sur le terrain (distribution, recensements, statuts de conservation, écologie comportementale, domaine vital, régime alimentaire, organisation sociale), des études socio-économiques, l'éducation des communautés humaines et les études génétiques. Ce programme est décrit ici et illustre comment une étude à petite échelle encore holistique dans la conservation des espèces peut être efficace.

INTRODUCTION

Conservation biology is a relatively new multidisciplinary science that has developed to deal with the crisis confronting biological diversity (Primack, 1993). It has two main aims: to investigate human impact on biological diversity and to develop practical approaches to prevent the extinction of species (Soulé, 1985, 1986; Wilson, 1992). The biodiversity crisis has emphasized not only the need for conservation action but also the requirement for information on which to base rational remedial actions. Research is thus a crucial aspect of any effort to save species, communities or ecosystems, and should form the basis on which conservation decisions are made (Ryder & Feistner, 1995). Both the basic and applied aspects of conservation science are integral to successful conservation programs and the definition, summarized as "find out what's wrong and how to fix it" puts people as well as animals firmly on the agenda.

In this paper the development of a conservation program for a single primate taxon—the Alaotran Gentle Lemur *Hapalemur griseus alaotrensis*—is described. This lemur lives in an isolated marsh habitat around Lac Alaotra, one of Madagascar's main rice-growing regions, an area with a large and expanding human population and one in which the rice and fishing industries have large impacts on the environment (Pidgeon, 1996). In being one of Madagascar's only non-forest lemur, and living so close to local populations that one can hear the progress of a local football match while collecting behavioral data, conserving this lemur presents a special challenge and an interesting case study in primate conservation.

The Alaotran Gentle Lemur

Hapalemur griseus alaotrensis is restricted to the marsh vegetation around Lac Alaotra, Madagascar's largest lake. Within the last decade the international profile of this lemur has increased as its conservation status has given cause for concern. In 1988 a captive breeding program was considered of the highest priority (Wharton et al., 1988) as part of the conservation strategy for the lemur. In 1990 the IUCN Lemur Red Data Book classed it as endangered (Harcourt & Thornback, 1990) and in 1992 the IUCN/SSC Primate Specialist Group gave it highest conservation priority in the Lemur Action Plan (Mittermeier et al., 1992). This classification designated priority status using the following criteria: degree of threat—total population probably between 100 and 1000; its taxonomic status—one of a small number of closely related forms that together are clearly distinct from other species; and its level of protection—*H. g. alaotrensis* does not occur in any protected area. More recently, Mittermeier et al. (1994) have designated it as critically endangered, using the new IUCN (1996) criteria.

Initial Conservation Activities

The Jersey Wildlife Preservation Trust (JWPT) became involved with this taxon in 1990, when, following the St. Catherine's Workshop recommendations (Wharton et al., 1988), an expedition to capture Alaotran Gentle Lemurs to found an *ex-situ* captive breeding program was undertaken (Durrell, 1991). This endeavor was part of a long-term collaboration with the Government of Madagascar, begun with the signing of a joint protocol of collaboration in 1983 (Bloxam & Durrell, 1985). Ten animals were exported, and all remain the property of the Malagasy Government. Given the limited knowledge and assumed very small population size at the time, the development of an *ex-situ* captive breeding program was an initial step to establish a "safety-net" for the lemur. In 1992 one of the wild-caught males was sent on breeding loan to Duke University Primate Center (DUPC, USA) to join their single female (Feistner & Beattie, 1997). These two institutions were the only ones at that time to maintain this lemur in captivity.

The captive lemurs at JWPT first bred in August 1993, about three years after their arrival. This was mainly a consequence of many of the initial imports being subadult individuals and the time required for animals from Madagascar to adjust to the northern hemisphere. The single offspring's development was studied to augment the sparse knowledge on the taxon (Steyn & Feistner, 1994). At the time the captive breeding program was initiated, very little was known about the Alaotran Gentle Lemur, either in terms of its status in the wild or its management in captivity. The genus *Hapalemur*, was little studied in the wild and uncommon in zoos so few comparative data were available.

Successful captive management relies on knowledge about a species' biology and how it lives in its natural habitat. Work on the taxon was very limited: in July 1969 Petter & Peyrieras (1970) had accumulated 20 hours of observation in the area of the lake and in 1984 Pollock (1986) spent three days studying the lemur. It therefore became a priority to undertake fieldwork to establish the conservation status of the lemur and find out more about its biology.

In 1993 and 1994 three studies were undertaken investigating the distribution, population status, threats to, and basic biology of the species (Feistner, 1993, 1994; Feistner & Rakotoarinosy, 1993; Mutschler et al., 1994). All were carried out under the protocol of collaboration and with the cooperation of the Ministère des Eaux et Forêts. These studies established the following: that the total population size was about 7500 individuals, divided into two subpopulations (Mutschler & Feistner, 1995); that the main threats were habitat destruction, by conversion of marshland to rice paddies and burning in the dry season, and hunting (Feistner, 1993, 1994; Feistner & Rakotoarinosy, 1993; Mutschler et al., 1995); that the lemurs lived in small groups and were folivorous (Mutschler et al., 1998); and that the lemurs appeared to defend small territories (Nievergelt et al., 1998a, 1998b).

The lemur field studies and work on the endangered (possibly extinct) Madagascar Pochard *Aythya innotata* (Young & Smith, 1989; Pidgeon, 1993) and their habitat indicated that heightening awareness about the unique and threatened nature of Alaotra's fauna would be beneficial to biodiversity conservation in the region. Although local fishermen knew the wildlife well, people often had no idea that the Alaotran Gentle Lemur was endemic to their area, or even that lemurs were only found in Madagascar. Therefore a poster featuring the lemur and pochard was produced, in collaboration with the World Wide Fund for Nature (WWF) and the Malagasy authorities, explaining briefly the special nature of these animals, and some 2000 examples were distributed around the lake. An informal follow-up two years later indicated that awareness had increased. Most people questioned had seen the poster (a consequence of its having been displayed in shops, bars, schools, and government buildings).

Thus by the end of 1994 *ex-situ* captive breeding, research on infant development, field research, and some basic conservation education were the main activities being undertaken, all with the ultimate goal of securing the continued survival of the lemur and its habitat. However, all these aspects were in the early stages and needed development.

Program Development—Biological Aspects

Captive Breeding. The captive population of *Hapalemur griseus alaotrensis* at JWPT bred annually from 1993 and a pair was sent on breeding loan to Bristol Zoological Gardens (UK) in 1995 (Feistner & Beattie, 1997). Due to the limited number of founders, a further import of wild-caught animals was authorized and undertaken in March 1997, and at the same time pairs of lemurs were also established at two Malagasy institutions— the Parc Botanique et Zoologique de Tsimbazaza and the Parc Zoologique d'Ivoloina (Feistner & Beattie, 1998). In July 1997 the Alaotran Gentle Lemur captive breeding program was formally accepted as a European Endangered Species Programme (EEP), and the first International Studbook was produced later that year (Feistner & Beattie, 1997). As of November 1998 there are 30 Alaotran Gentle Lemurs in captivity outside Madagascar, of which 15 are captive bred. There are six participating institutions in the EEP: JWPT, Bristol Zoological Gardens, Edinburgh Zoo (UK), London Zoo (UK), Rotterdam Zoo (The Netherlands), and Mulhouse Zoo (France).

Research in Captivity. The captive *Hapalemur griseus alaotrensis* have also been the focus of a research program. The initial study on infant development was continued as the lemurs bred, developing a data base involving six infants including two sets of twins (Steyn & Feistner, 1994; Taylor & Feistner, 1996; Courts, 1995). The frequency of twinning has been high—50% of ten surviving litters. Such a high twinning rate has not occurred in captive *H. g. griseus* (Haring & Davis, 1998) but at least 40% of wild *H. g. alaotrensis* litters were twins (Mutschler et al., in press). These studies on captive *H. g. alaotrensis* have shown that adult females park their infants initially, before carrying them. Most carrying is done by the mother, although infants will jump on and be carried by other group members when alarmed. The female stops carrying when infants reach 11–13 weeks of age. From a week of age infants mouth potential food items and start ingesting solid food at 5–6 weeks. They are weaned at 20–24 weeks (Beattie & Feistner, 1998).

The captive Alaotran Gentle Lemurs have also been the subjects of studies of diet and nutrition. In part these studies were a response to a health problem. In 1992 a vine *Polygonum baldschaunicum* was given to the Alaotran Gentle Lemurs. All individuals became sick and three subsequently died. Ill health was associated with elevated blood urea and creatinine levels (Fidgett, 1996) and oxalate crystals and kidney damage were found at post-mortem examination (Robert & Allchurch, 1995). This was unexpected as this plant is fed to other lemurs e.g., *Varecia variegata* and *Lemur catta*, with no ill effect, and in fact another member of the Polygonaceae, smartweed *Polygonum glabrum* is fed on by Alaotran Gentle Lemurs in the wild (Mutschler, this volume). Further studies of nutrition in captive *Hapalemur* are needed since four of five recent deaths in captive *H. g. griseus* at DUPC have also occurred through kidney disease (a combination of polycystic kidneys and glomerulo-nephritis) (Haring & Davis, 1998).

Folivores are challenging to maintain in captivity, since it is often difficult to provide sufficient quantities of appropriate forage. Dietary intake and composition of selected forage items have been analyzed. (Fidgett, 1995; Fidgett et al., 1996). In addition, a comparison of the nutrient composition of plant species grown locally on Jersey Island, but related to those consumed by the lemurs in Madagascar, has also been undertaken. As a result of the study the forage component of the diet was increased, the fruit component decreased, and the pellet feed was changed to Leaf Eater Diet (Mazuri). These changes resulted in a marked improvement in pelage and general condition.

Field Research. Research in the field has continued. A long-term study of behavioral ecology was undertaken (Mutschler, 1998) which has significantly increased our knowledge of this endangered primate. Alaotran Gentle Lemurs are found in groups of up to 9 individuals, but median group size is 3 (n = 78). There is no sexual dimorphism; adult females weigh 1251g (n = 27) and adult males 1228g (n = 31). Most groups contained one breeding female, but 35% of groups contained two, indicating that this taxon cannot be classified as being organized in monogamous family groups (Mutschler et al., in press). Details about diet and nutrition are given in Mutschler (this volume). In brief the lemur was shown to be exclusively folivorous with an extremely restricted diet of only 11 different plant species, with 97% feeding time spent on only four Graminaceae. Work on radio-collared individuals has clearly shown that these lemurs are cathemeral, the first time this has been clearly documented in a lemur other than *Eulemur* (Mutschler, 1998; see also Donati et al., this volume).

The extended field presence also provided the opportunity to train a national university student in field techniques and supervise his research project (Randrianarisoa, 1998). Positive signs of protection of wild Alaotran Gentle Lemurs, like reduced hunting pressure, in and around the study sites, have occurred recently. The tendency from substantial animal losses (20%) in one study population in late 1995, to an increasing study population (11%) in 1996 and early 1997 was encouraging and was probably linked to the permanent presence of researchers and their open communication with the local people (Mutschler, 1997).

The Research Link between Wild and Captive Lemurs. The research work carried out on the wild and captive lemurs is mutually beneficial. For example, understanding more about the diet and nutrition of *Hapalemur griseus alaotrensis* provides information on how the lemurs use their habitat and what plants are particularly important for their survival. These data can be used to guide conservation planning in Madagascar, and are also of relevance in guiding captive management, both from the nutritional and behavioral perspective (Feistner et al., 1998). Information acquired in captivity on infant development and social behavior assists in interpreting observations of the less easily visible wild lemurs.

Molecular genetic studies based on hair samples from both wild and captive-bred lemurs are also underway. Hair samples from 80 wild individuals sampled in 1996 and 1997 are being used to assess social structure and mating system, with hair from captive individuals of known paternity being used for comparison. In addition, samples of 38 individuals from six different area around Lac Alaotra are being used to estimate genetic variability in the whole population (Nievergelt, in prep.).

Program Development—Anthropogenic Factors

Alaotra is not only an area of biological importance, but is also an important economic center for fishing and rice production. The marshes form a natural barrier, protecting the lake from siltation from erosion and provide a breeding ground for fish.

Moreover the marshes are the source of a wide range of secondary products providing supplementary food, materials for houses, baskets, and mats, fodder for domestic animals, and medicinal plants. Since people and their needs are such a feature of this environment it was decided to commence socioeconomic studies to understand the importance of natural resources to the local people and to assess their attitudes towards conservation, with a view to promoting biodiversity conservation in the area (Razafindrakoto, 1997; Randrianarivo, 1998). These studies provide a basis from which to plan effective education and awareness activities.

Education and Community Awareness. Conservation action and sustainable resource use can only work in areas where state control is weak, if there is consensus about management decisions among the vast majority of the resource users. A grass-roots approach with dialogue and discussions with local people have been a key aspect of developing community activities around Alaotra. A description of these activities is given by Durbin (this volume). In brief a whole range of activities, aimed at involving all sectors of the community, from school children to village elders and government representatives, has been initiated. These include both formal education through schools and teacher training as well as more informal festivals. In all these activities the Alaotran Gentle Lemur has been a symbol of the wetland habitat. Another conservation education poster, emphasizing habitat protection and the link between ecosystem health and the health of human communities has been produced by JWPT in Madagascar.

DISCUSSION

Although focusing on one lemur may seem a somewhat narrow approach, this program has several features which may be more generally applicable and it could thus serve as a model for other conservation efforts. The program started as a species-focused effort and has expanded to encompass a wetland habitat. The Alaotran Gentle Lemur has become a flagship species representing the marsh ecosystem.

The program, which has been running since 1990, is characterized by a multidisciplinary approach in which the factors affecting the lemur's survival and the continued integrity of its habitat have gradually been addressed. A common theme throughout this program has been collaboration and dialogue, both between individuals and institutions. Contributors to the program, which is run by JWPT, include the Government of Madagascar, in particular the Ministère des Eaux et Forêts, the Ministère de l'Enseignement Secondaire et de l'Education de Base, and the Ministère de l'Enseignement Supérieur, several zoological institutions including two in Madagascar, universities, both within and outside Madagascar, and other NGOs. Some of these collaborations are formalized in protocols of collaboration and other accords. Many disciplines are involved in this conservation program: captive breeding (in country and *ex situ*), research on lemurs in the field and captivity, research on ecology, socioeconomic research, and environmental education.

The degradation in the fauna and flora of Lac Alaotra has been documented in several studies (Young & Smith, 1989; Feistner, 1993, 1994; Pidgeon, 1993, 1996). There is little doubt that the removal of vegetation around the lake, resulting in siltation with eroded lateritic soils, the introduction of exotic fish, the reduction in aquatic flora, uncontrolled pesticide use and human resource use are having a large negative impact on the lake's biodiversity (Pidgeon, 1993, 1996). Several species are already in severe decline or locally extinct (*e.g.*, birds: *Aythya innotata, Nettapus auritus, Tachybaptus rufolavatus, Platalea alba*; fish: *Rheocles alaotrensis, Ratsirakia legendrei*; plants: *Nymphaea stellata*).

The continued survival, of the Alaotran Gentle Lemur, which depends on the continued integrity of the Alaotran marshes, will thus have major spin-offs for biodiversity conservation in the region. In turn a healthy ecosystem will provide a sustainable living for the thousands of people of Alaotra.

REFERENCES

BEATTIE, J. C., AND A. T. C. FEISTNER. 1998. Husbandry and breeding of the Alaotran gentle lemur *Hapalemur griseus alaotrensis* at Jersey Wildlife Preservation Trust. International Zoo Yearbook, **36**: 11–19.

BLOXAM, Q. M. C., AND L. DURRELL. 1985. A note of the Trust's recent work in Madagascar. Dodo, Journal of the Jersey Wildlife Preservation Trust, **22**: 18–23.

COURTS, S. E. 1997. Development of the fourth captive-bred infant Alaotran gentle lemur *Hapalemur griseus alaotrensis* at Jersey Wildlife Preservation Trust. Unpublished report, Jersey Wildlife Preservation Trust.

DURRELL, L. 1991. Notes on the Durrell Expedition to Madagascar September—December 1990. Dodo, Journal of the Wildlife Preservation Trusts, **27**: 9–18.

FEISTNER, A. T. C. 1993. Preliminary field study of *Hapalemur griseus alaotrensis* at Lake Alaotra, Madagascar: A brief report. Unpublished report, Jersey Wildlife Preservation Trust, Jersey.

———. 1994. Dry Season studies of *Hapalemur griseus alaotrensis* at Lac Alaotra, Madagascar: A brief report. Unpubl. report, Jersey Wildlife Preservation Trust, Jersey.

FEISTNER, A. T. C., AND J. B. BEATTIE. 1997. International Studbook for the Alaotran gentle lemur *Hapalemur griseus alaotrensis* Number One 1985–1996. Jersey Wildlife Preservation Trust, Jersey.

———. 1998. International Studbook for the Alaotran gentle lemur *Hapalemur griseus alaotrensis* Number Two 1997. Jersey Wildlife Preservation Trust, Jersey.

FEISTNER, A. T. C., T. MUTSCHLER, AND A. L. FIDGETT. 1998. Nutrition of wild and captive Alaotran gentle lemurs *Hapalemur griseus alaotrensis*, p.6. *In* Diet, foraging behavior and time-budgets in nonhuman primates—How field studies may help in improving the welfare of captive primates. European Federation for Primatology Workshop, Paimpont, France.

FEISTNER, A. T. C., AND M. RAKOTOARINOSY. 1993. Conservation of gentle lemur *Hapalemur griseus alaotrensis* at Lac Alaotra Madagascar: Local knowledge. Dodo, Journal of the Wildlife Preservation Trusts, **29**: 54–65.

FIDGETT, A. L. 1995. Diet and nutrition of the Alaotran Gentle Lemur *Hapalemur griseus alaotrensis* at the Jersey Wildlife Preservation Trust. MSc thesis, University of Aberdeen.

———. 1996. Blood biochemistry values in Alaotran gentle lemurs *Hapalemur griseus alaotrensis* at the Jersey Wildlife Preservation Trust. Dodo, Journal of the Wildlife Preservation Trusts, **32**: 63–66.

———, A. T. C. FEISTNER, AND H. GALBRAITH. 1996. Dietary intake, food composition, and nutrient intake in captive Alaotran gentle lemurs *Hapalemur griseus alaotrensis*. Dodo, Journal of the Wildlife Preservation Trust, **32**: 44–62.

HARCOURT, C., AND J. THORNBACK. 1990. Lemurs of Madagascar and the Comoros. IUCN Red Data Book. IUCN, Gland and Cambridge.

HARING, D., AND K. DAVIS. 1998. Management of the grey gentle or Eastern lesser bamboo lemur *Hapalemur griseus griseus* at Duke University Primate Center, Durham. International Zoo Yearbook **36**: 20–34.

IUCN. 1996. 1996 IUCN Red List of Threatened Animals. IUCN, Gland.

MITTERMEIER, R. A., W. R. KONSTANT, M. E. NICOLL, AND O. LANGRAND. 1992. Lemurs of Madagascar: An Action Plan for their Conservation 1993–1999. IUCN, Gland.

MITTERMEIER, R. A., I. TATTERSALL, W. R. KONSTANT, D. M. MEYERS, AND R. B. MAST. 1994. Lemurs of Madagascar. Conservation International, Washington, D.C.

MUTSCHLER, T. 1997. Field studies of the Alaotran gentle lemur *Hapalemur griseus alaotrensis*: An update. *In* Feistner A. T. C., and J. C. Beattie, International Studbook for the Alaotran gentle lemur *Hapalemur griseus alaotrensis* Number One 1985–1996, Jersey Wildlife Preservation Trust, Jersey.

———. 1998. The Alaotran gentle lemur (*Hapalemur griseus alaotrensis*): A study in behavioural ecology. PhD thesis, University of Zürich, Zürich.

——, AND A. T. C. FEISTNER. 1995. Conservation status and distribution of the Alaotran gentle lemur *Hapalemur griseus alaotrensis*. Oryx, **29**: 267–274.
——, AND C. NIEVERGELT. 1998. Preliminary field data on group size, diet, and activity in the Alaotran gentle lemur *Hapalemur griseus alaotrensis*. Folia Primatologica, **69**: 325–330.
MUTSCHLER, T., C. NIEVERGELT, AND A. T. C. FEISTNER. 1994. Biology and conservation of *Hapalemur griseus alaotrensis*. Unpublished report. Jersey Wildlife Preservation Trust, Jersey.
——. 1995. Human-induced loss of habitat at Lac Alaotra and its effect on the Alaotran gentle lemur, pp. 35–36. *In* Patterson, B. D., S. M. Goodman, and J. L. Sedlock, eds., Environmental Change in Madagascar. The Field Museum, Chicago.
——. In press. The social organisation of the Alaotran gentle lemur (*Hapalemur griseus alaotrensis*). American Journal of Primatology.
NIEVERGELT, C., T. MUTSCHLER, AND A. T. C. FEISTNER. 1998a. Group encounters and agonistic interactions in *Hapalemur griseus alaotrensis*. Folia Primatologica, **68(S1)**: 412.
——. 1998b. Group encounters and territoriality in wild Alaotran gentle lemurs (*Hapalemur griseus alaotrensis*). American Journal of Primatology, **46**: 251–258.
PETTER, J.-J., AND A. PEYRIERAS. 1970. Observations éco-éthologique sur les lémuriens malgache du genre *Hapalemur*. La Terre et La Vie, **24**: 365–382.
PIDGEON, M. 1993. Progress report project Onjy—Further investigations of the Lake Alaotra region of Madagascar with emphasis on Madagascar Pochard *Aythya innotata*. Unpublished report, Missouri Botanical Gardens.
——. 1996. An ecological survey of Lake Alaotra and selected wetlands of Central and Eastern Madagascar in analysing the demise of Madagascar Pochard *Aythya innotata*. Missouri Botanical Gardens, Antananarivo.
POLLOCK, J. 1986. A note on the ecology and behaviour of *Hapalemur griseus*. Primate Conservation, **7**: 97–100.
PRIMACK, R. B. 1993. Essentials of Conservation Biology. Sinauer Associates Inc., Sunderland.
RANDRIANARISOA, A. J. 1998. Contribution à la quantification des aliments du *Hapalemur griseus alaotrensis* et à sa dynamique sociale. Mémoire de Fin d'Etudes, ESSA, Université d'Antananarivo.
RANDRIANARIVO, A. 1998. Etude socio-economique de la zone sud du Lac Alaotra. Unpublished Report, Jersey Wildlife Preservation Trust.
RAZAFINDRAKOTO, J. 1997. Rapport Final: Etude sur l'utilisation des ressources naturelles au Lac Alaotra. Unpublished Report, Jersey Wildlife Preservation Trust.
ROBERT, N., AND A. F. ALLCHURCH. 1995. Poisoning of Alaotran gentle lemur (*Hapalemur griseus alaotrensis*) by Russian vine (*Polygonum baldschuanicum*) at the Jersey Wildlife Preservation Trust. Verhandlungsberichte des Internationalen Symposiums über die Erkrankungen der Zootiere **37**: 207–210.
RYDER, O. A., AND A. T. C. FEISTNER. 1995. Research in zoos: A growth area in conservation. Biodiversity and Conservation, **4**: 671–677.
SOULÉ, M. E. 1985. What is conservation biology? BioScience, **35**: 727–734.
——. ed. 1986. Conservation Biology: The Science of Scarcity and Diversity. Sinauer Associates, Sunderland.
STEYN, H., AND A. T. C. FEISTNER. 1994. Development of a captive-bred infant Alaotran gentle lemur *Hapalemur griseus alaotrensis*. Dodo, Journal of the Wildlife Preservation Trusts, **30**: 47–57.
TAYLOR, T. D., AND A. T. C. FEISTNER. 1996. Infant rearing in captive *Hapalemur griseus alaotrensis*: singleton versus twins. Folia Primatologica, **67**: 44–51.
WHARTON, D., M. PEARL, AND F. KOONTZ. 1988. Minutes from a lemur conservation workshop. Primate Conservation, **9**: 41–48.
WILSON, E. O. 1992. The Diversity of Life. Harvard University Press, Cambridge.
YOUNG, H. G., AND J. G. SMITH. 1989. The search for the Madagascar pochard *Aythya innotata*: Survey of Lac Alaotra, Madagascar October-November, 1989. Dodo, Journal of the Jersey Wildlife Preservation Trusts, **26**: 17–34.

TEACHING PRIMATOLOGY AT THE UNIVERSITÉ DE MAHAJANGA (NW MADAGASCAR)

Experiences, Results, and Evaluation of a Pilot Project

Urs Thalmann,[1] and Alphonse Zaramody[2]

[1] Anthropological Institute and Museum
University of Zurich
Winterthurerstrasse 190
CH-8057 Zurich, Switzerland
email: uthal@aim.unizh.ch (corresponding author)
[2] Département des Sciences de la Terre
Université de Mahajanga
BP 652, Mahajanga (401), Madagascar

ABSTRACT

In 1995, the course "Introduction to Anthropology/Primatology" was organized at the Université de Mahajanga in collaboration with the Anthropological Institute of the University of Zurich as a pilot project. The defined goals were: (1) presentation of primatology as an interesting and important subject, which may be of special relevance for Madagascar; (2) initiation of a test case to see how it would operate in general; (3) provision of a starting point for future adjustments and potential extensions. Two questionnaires filled out by students before and after the course showed that they came from all over the island (such that knowledge would potentially be widely distributed later), that they were relatively old for the study level achieved, and that knowledge of the natural richness of Madagascar at the beginning of the lectures was lower than expected. Students judged the course to be interesting and worth following, feeling that it should be offered again; many believed that they had understood quite a lot. In fact, however, some of the messages of the lecture course did not transfer very well. As a pilot project, the course was successful in several realms but failed to catalyze longer term teaching and research activities for various reasons. An evaluation of the experiences made and of results of the questionnaires is presented.

RÉSUMÉ

Un projet pilote, sous la forme d'un cours intitulé "Introduction à l'Anthropologie/Primatologie" dispensé à l'Université de Mahajanga, s'est déroulé en 1995 en collaboration avec l'Institut d'Anthropologie de l'Université de Zurich. Les objectifs poursuivis furent la présentation de la primatologie en tant que sujet intéressant et important notamment à Madagascar, la mise en place d'une épreuve-test destinée à comprendre le fonctionnement global et l'identification d'une base sur laquelle des ajustements et d'éventuelles extensions ultérieures pourront reposer. Des questionnaires remplis par les étudiants avant et après le cours ont montré que leurs origines géographiques englobait l'ensemble de l'île (les connaissances acquises pourraient ainsi être largement disséminées), qu'ils étaient relativement âgés par rapport à leur niveau universitaire et que leurs connaissances sur les richesses naturelles de Madagascar au début du cours étaient moins importantes que celles attendues. Les étudiants ont estimé que le cours était intéressant et utile à suivre, estimant que l'expérience devait être renouvelée et nombreux ont été ceux qui ont dit avoir compris beaucoup de choses alors qu'en fait certains messages du cours n'ont pas été très bien transmis. En tant que projet pilote, le cours a remporté plusieurs succès mais n'a pas réussi, pour plusieurs raisons, à catalyser un enseignement et des activités de recherche à plus long terme. Une évaluation des expériences réalisées et les résultats des questionnaires sont ci-dessous présentés.

INTRODUCTION

Teaching an "Introduction to Anthropology/Primatology" at a university is usually not a very noteworthy event *per se*. It may simply be part of a particular education program. On the other hand, everybody would probably agree that teaching and education play a key role in any community or society and should be part of a conservation strategy.

Madagascar has been recognized as a biological "megadiversity country", exceptionally contributing to the earth's faunal and floral diversity. At the same time, it has also been recognized as a "megadiversity hotspot region" (WCMC, 1992; Mittermeier et al., 1994). Threats to the natural environment of the island are of major and immediate concern. Lemurs are an especially conspicuous group of Malagasy mammals and are completely endemic to the island (on the Comoros they were probably introduced; Harcourt & Thornback, 1990). The recognized 14 extant genera (33 species) constitute about 64% of all strepsirhine genera (species: 63%) and 16% of all primate genera (species: 12%) (based on Mittermeier et al., 1994; Mittermeier & Konstant, 1996/97; Zimmermann et al., 1998). Their attractive appearance, exclusive natural distribution, and predominantly forest-dwelling life-style predestine lemurs to be ideal flagship species that may serve as ambassadors for conservation issues outside and within Madagascar (see Durbin, this volume). As members of the Order Primata and as our close relatives, they may be particularly valuable for placing humans in a natural and evolutionary context, thereby pointing to their dependence upon—and increasing responsibility—for their very own habitat. Simply stated: lemur conservation equals forest conservation. Forest conservation is essential to prevent erosion, floods, droughts, etc., thus allowing sustainable and predictable production of food and other basic goods. "No forest" does not only mean no lemurs, it may also mean not enough food and not enough goods. Although this is all common sense and not new, it should always be kept in mind that the "big picture" should not be forgotten.

In 1991, an Accord de Coopération was initiated between the Anthropological Institute and Museum of the University of Zurich (Switzerland) and the Département des Sciences de la Terre de l'Université de Mahajanga. The accord was derived and adjusted from an existing document between the Université de Mahajanga and a French university, with which the Anthropological Institute already cooperated. The major points addressed in the skeleton agreement lasting for an initial period of five years were the promotion of common research and teaching activities, mutual information on projects, and the expressed intention to facilitate corresponding activities in order to enhance research on lemurs and support teaching activities through a cooperative approach. This would eventually allow for the development of local research capacity. The idea behind such a direct bilateral and limited agreement was to start cooperation on a small scale, appropriate to specific needs and possibilities, while assuring a high degree of flexibility, suitability, and influence by the directly concerned signatories. As a long term goal the establishment of a research station was envisioned. AZ was delegated by the Université de Mahajanga as the contact, and UT served the same role for the Anthropological Institute of Zurich. The Accord de Coopération was initiated with the Université de Mahajanga for several reasons: (1) Mahajanga is one of three universities in Madagascar that provides the possibility to study natural sciences. A basic (although insufficient) infrastructure is present, but the number of the permanent teaching staff is inadequate. (2) As the capital of the Mahajanga Province (150,023 km^2, 25.6% of Madagascar's surface), it is an important regional center. (3) The Mahajanga Province still has important forests of various types. (4) Some regions of the province have hardly been studied. (5) With the Station Forestière d'Ampijoroa and Réserve Naturelle Intégrale d'Ankarafantsika, a potential field site was close, and other projects were already established. (6) Initial successful activities could have had a catalyzing effect to promote additional or extended research and teaching activities in fields other than primatology.

Research activities on lemurs initiated by students of the Anthropological Institute of University of Zurich started in 1993 and focused mainly on behavior, ecology, and distribution of lemurs (species concerned: *Avahi occidentalis*, *Cheirogaleus medius*, *Eulemur mongoz*, *Lepilemur edwardsi*, and *Propithecus verreauxi coronatus*), with attention turning more recently also to phylogeny of lemurs. The first teaching course at the Université de Mahajanga was scheduled for the academic year 1994/95, when UT was in Madagascar for prolonged post-doctoral research on lemurs. The course "Introduction to Anthropology/Primatology" was designed as a pilot project with several goals: (1) Primatology should be introduced and presented as an interesting and important field—perhaps of special relevance for Madagascar—showing human beings in a natural and evolutionary context. (2) The course should serve as a test case to see how teaching would work in general. (3) To provide a starting-point for future adjustments and potential extensions. From the outset, it was not intended to offer a special, one-time course on lemurs alone (Fig. 1). The approach chosen was to teach the general part at the same level and as far as possible in the same way as at University of Zurich. This approach was decided to remove any criticism with respect

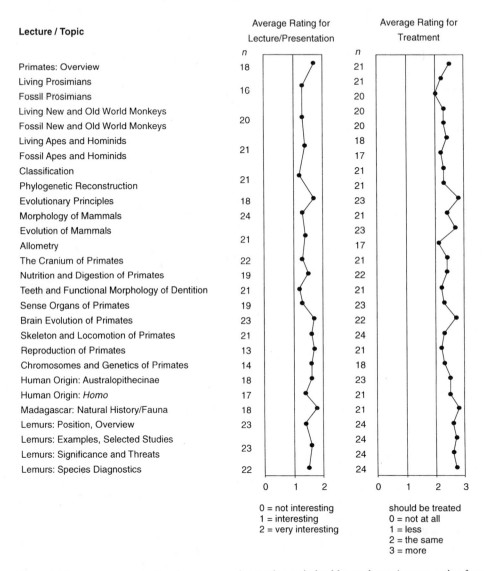

Figure 1. Topics presented during the course, and reception as judged by students. Average rating for presentation and chosen topics, and average rating for treatment of topics.

to patronizing prejudices. Following some preparation through the general introduction, lemurs were embedded within the framework presented; essentially, a very pragmatic approach was chosen concerning lemurs and Madagascar's wildlife. Lemurs were mostly presented as an important part of Madagascar's natural heritage, which contributes exceptionally to biodiversity worldwide.

We are reporting here experiences made during this pilot teaching project, as well as presenting results of questionnaires distributed at the beginning and at the end of the course (yielding information to build on for potential follow-up activities) and trying to assess the impact of this pilot project. Finally, we try to discuss it in the light of ongoing discussions about cooperation with developing countries. This limited teaching project illuminates on a small scale many aspects of the issue "Research Partnership with Developing Countries."

MATERIAL AND METHODS

Preparation

The course was proposed to the Université de Mahajanga in late 1993, and then prepared in early 1994. The course was scheduled for the academic year of 1994/95. An infrastructure was established and officially handed over in 1994 to the section of natural sciences in front of the assembled teaching staff: five overhead projectors; a slide projector with some 220 slides; 14 recent fundamental books (mostly in English); casts of primate skulls (including fossil hominids) and a cast of a complete human skeleton; and two folders with prepared handouts and documents for the lecture. The content of the course was shortly outlined. The material could serve for potential follow-up teaching activities or serve other teachers in their lectures. The technical equipment was overhauled, simple, and robust material that had proven its usefulness at the University of Zurich. A stock of spare parts and consumables was also provided. Handouts of different complexity were prepared, in French as far as possible. As a matching commitment, the Université de Mahajanga provided partial support for the transportation and importation of the teaching material. Teaching (in French) took place from 14 February to 2 March 1995, usually in blocks of three lectures.

The core of the course "Introduction to Anthropology/Primatology" was based on the same course offered to first-year biology students and non-biologists (e.g., teachers of natural sciences for secondary schools, students of sports, psychology, or prehistory) at the University of Zurich by Prof. Dr. R. D. Martin (Martin et al., 1997). UT was actively involved for several years in teaching these and other courses at the University of Zurich.

Planned were 30 hours of lecture covering 28 topics as listed in Figure 1. According to actual experience during teaching, the schedule, and content were slightly adjusted, mainly the extent to which topics were covered. About 17% of the course was dedicated to an overview of living and fossil primates, 8% to methodological aspects, 11% to the primate order as member of the mammal class, 33% to functional complexes using a comparative approach, 8% to human origins *sensu stricto*, and about 23% to Madagascar and its lemurs. The pilot course was originally proposed and planned for no more than 30–40 persons, including some voluntary first-year students of natural sciences and interested teachers. This approach was proposed to give (1) at least some students the possibility to continue primatological studies if extended teaching activities were subsequently established, (2) to present the course to teachers to judge its worth, (3) to discuss necessary adjustments with students and teachers, and if welcomed (4) support a teacher to take over the course as soon as possible for the future. This concept could not be realized under the difficult circumstances prevailing (see also participation in the results).

Questionnaires

Two questionnaires were prepared on site, one to be filled out by attending students before the course started (Q1: 33 questions), and one after it ended (Q2: 42 questions). The questionnaires were predominantly developed to provide a baseline of information to start with (Q1), and secondarily for limited comparison with this information after the course, and to provide information on emerging issues during the course (Q2). Both questionnaires were filled out anonymously. An introductory statement explained the goals of the questionnaires. Questions were designed to yield basic sociographic information providing some indication of the composition of the class (Q1: 13/Q2: 10), information on students' biological interests (2/2), general biological and specific knowledge on Madagascar's wildlife (4/2), and information on knowledge or appraisal of aspects related to conservation issues in a wider sense (13/13). In the second questionnaire, sociographic questions were asked again, as participation in the course fluctuated markedly. Also, some questions referring to biological knowledge and conservation (specifically concerning Madagascar) were repeated to give some idea about the impact of the course. Additional questions (n = 14) referred to the reception of the lecture by students and other points that could be of interest for follow-up developments. Both questionnaires ended with the question whether information given in the questionnaires could be analyzed and eventually be published or communicated in a conference.

Structurally, questions were of different types: simple questions concerning age, gender, origin, etc., and simple or compound multiple-choice questions. Two questions in the second questionnaire gave students the opportunity to express themselves freely. A multiple choice question was considered technically correctly answered, if—according to the specific question—a complete set of marks was given on the answer sheet. A biologically correct answer demanded in addition that these marks be put in the appropriate place. Hence, a biologically correct answer is also correctly answered in a technical sense, whereas a technically correct answer is not necessarily biologically correct. A subset of questions and answers is presented in the result section.

RESULTS

Participation

The number of attending students fluctuated as a result of several factors (Fig. 2). The course had to start during a period when teaching was postponed because of strikes at the Université de Mahajanga. Time constraints imposed by ongoing field research made it impossible to wait for the start of all teaching activities, otherwise the course would have had to be cancelled. After the first official notice, participation increased to over 120 students. Subsequently, the course was no longer announced for "third-year" students. Participation remained stable before the number of students dropped understandably when teaching was on Saturday at 0700, and fell to 15 students when official announcement of the course ceased. This happened following internal discussions at the university about the course. However, despite the lack of further announcement, the number of students increased again at the end of the course. Based on information from the second questionnaire, only eight students were able to follow the entire course. On average, students who answered the second questionnaire followed

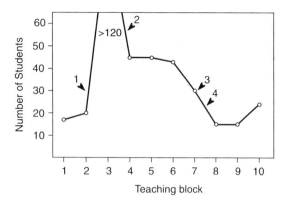

Figure 2. Participation. Number of students (n) attending the different teaching blocks and events influencing participation. 1: first official announcement. 2: exclusion of 3rd year students. 3: Saturday morning 0700. 4: no further official announcement.

about 80% of the course ($n = 24$; range: 46–100%). Apart from AZ who followed the entire course, no other member of the university staff ever attended a lecture.

Questionnaires

The first questionnaire (Q1) was filled out by 17 students, the second questionnaire (Q2) by 24 students. None of the students explicitly objected to an analysis and eventual publication of results (Q1: 13 agreed/0 disagreed/4 gave no answer. Q2: 24/0/0). For the analyzes of Q1, two questionnaires were partially excluded, as these students arrived late when questions were discussed in class. Of the remaining 15 questionnaires, one student (7%) answered all questions in a technically correct manner, four students (27%) tried to answer all questions, and 10 students (67%) did not answer at least one question. In Q2, three students (12.5%) answered all questions technically correctly, six (25%) tried to answer all questions, and 15 (63%) did not answer at least one question.

With the exception of two students in Q1, all students provided a technically correct answer at least once to a question of a particular structural type. In both questionnaires, technically incomplete or missing answers were concentrated on questions directly related with biological knowledge.

Sociographic Data

Data are partially presented for the two questionnaires separately (Table 1, Figs. 3, 4). As participation fluctuated heavily, it was impossible to identify persons individually based on the questionnaires. Q1 was filled out by three female and 14 male students (Q2: 9f/14m). Students came from almost all over Madagascar including very remote places (Fig. 3). They had to travel for considerable distances and time to follow their studies in Mahajanga (Table 1). On average they were about 27 to 28 years old, and had spent a long time at the university without achieving concluding degrees (Fig. 4).

Biological Knowledge

Asked about examples for species of 18 extant mammalian orders listed (Fig. 5), 26 examples were given in total, of which two (8%) were inappropriate. Most exam-

Table 1. Sociographic data from questionnaires 1 and 2. Age of students, distance, and travel time to reach Mahajanga, and years at the Université de Mahajanga

	Questionnaire 1					Questionnaire 2				
	n	mean	median	range	S. D.	n	mean	median	range	S. D.
Age*	17	28.4	28.0	22–33	3.0	23	27.9	27.0	23–34	2.9
Distance in km	17	721	650	4–1400	386	—	—	—	—	—
Travel time in days*	17	2.2	2.0	1–5	1.2	23	2.8	2.5	1–7	1.6
Years at university*	12	5.1	5.0	3–8	1.4	21	5.6	6.0	3–8	1.4

*No significant difference between questionnaires 1 and 2 (Mann-Whitney U; $0.220 \leq p \leq 0.581$).

ples were given for the primate order (n = 7). Individual students gave on average one example (n = 15, mean = 1.3, mode = 0, median = 0, range: 0–6). Three students gave technically correct answers regarding the natural presence or absence of eutherian orders during the last 3000 years in Madagascar or close to its coasts. On average, individual students made only seven explicit statements (n = 15, mean = 6.8, mode = 7, median = 7, range: 4–12) out of the total of 18 statements that were expected. Of the totally 270 expected statements for all students, only 123 were made of which 56 state-

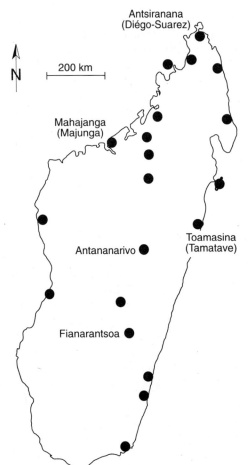

Figure 3. Origin of students. Dots indicate communities from which at least one student came.

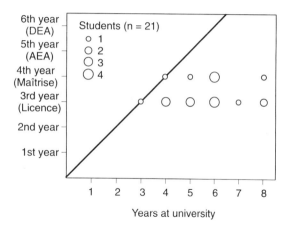

Figure 4. Study success. Years students are at the university (horizontal axis) and level reached (vertical axis). An ideal student would be on the line and have the corresponding certificate by the end of each study year.

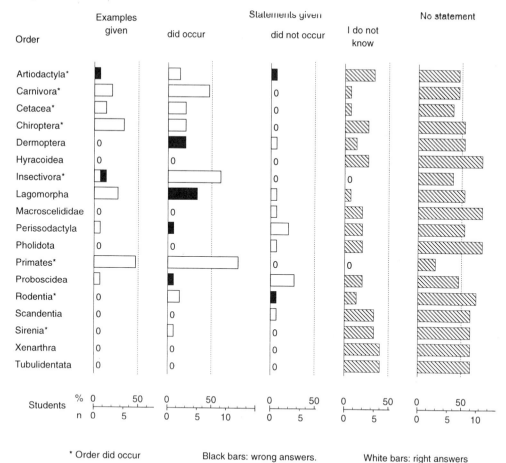

Figure 5. Knowledge about mammal orders before the course. Compiled answers to a compound multiple-choice question about extant eutherian orders. White bars: appropriate answers. Black bars: inappropriate answers. Hatched bars: "I do not know" and "no statements".

ments (46%) were correct and 13 (11%) were inappropriate. The optional statement "I do not know" was selected 54 times (44%).

The question about the worldwide occurrence of lemurs was asked in Q1 and Q2 (Fig. 6). In Q1, 13 students (87%) indicated lemurs for Madagascar, two students (13%) did not try to answer the question. Four students (27%) gave a technically correct answer, but none was biologically correct. Students indicated lemurs for Africa (n = 5), Asia (n = 3), South America (n = 3), but also for Australia (n = 3), Europe (n = 3), North America (n = 3), and Antarctica (n = 1). Of the eight choices expected per student, three were made on average. In Q2, 19 students (79%) indicated lemurs for Madagascar, five students (21%) did not answer the question. Five students (21%) gave a technically correct answer of which three (13% of all students) were also biologically correct. Six students still indicated lemurs for regions other than Madagascar (Africa = 6, Asia = 2, South America = 4, Australia = 3, Europe = 2, North America = 2), but none for Antarctica. The relation between correct and incorrect statements shifted significantly between the two questionnaires towards more correct answers (Wilcoxon Sign-ranked test; tied $p = 0.01$). However, most students made no statement regarding the occurrence of lemurs in other regions. Of the expected 192 statements for all students, 71 (37%) were made. This question did not allow for an optional statement "I do not know". Students had either to guess the right answer or to mark eventually a wrong answer. A comparison between the question with an "I do not know"-option (Fig. 5) and no such option (Fig. 6) revealed no significant differences regarding the technical correctness of answers given.

Conservation Issues

Students were asked several questions related to conservation. Three are presented here. Two questions were asked in Q1 and Q2, one only in Q2. When asked about the role of forests for ecological equilibrium (Fig. 7a) all students said in both questionnaires that forests are "important" or "very important". Asked if they benefit from forests (Fig. 7b), 12 (75%) said "yes", none said "no", and three (25%) gave no answer (Q2: 23 "yes"/0 "no"/1 no answer). Benefits specified in Q1 were basically utilitarian, such as "wood for construction" or "firewood" (n = 5), basically non-utilitarian such as "to protect wild animals" or "to do research" (n = 2), or a combination of both (n = 6) (Q2: 5/3/13). Chi square testing between Q1 and Q2 reveals no significant difference. Asked in Q2 about their impression of the degree of threat to Malagasy forests (Fig. 7c), 13 students (54%) indicated "very heavily threatened", nine (38%) "heavily threatened", and two (8%) "a little bit threatened". Nobody indicated "not threatened" or "I do not know".

Reception of the Course

The different lectures were judged by students to be on average "interesting" to "very interesting" (Fig. 1). Peaks occurred for the primate overview, "evolutionary principles", several functional complexes such as "brain evolution" or "skeleton and locomotion", and Madagascar's natural history. Judgments whether topics should be treated "more extensively", to "the same degree", "less" or "not at all" ranged on average between "to the same degree" and "more extensively". Conspicuous peaks occurred again for "evolutionary principles", also for "evolution of mammals", "brain evolution", and all topics dealing with Madagascar and lemurs. Twenty-three students

Teaching Primatology at the Université de Mahajanga (NW Madagascar)

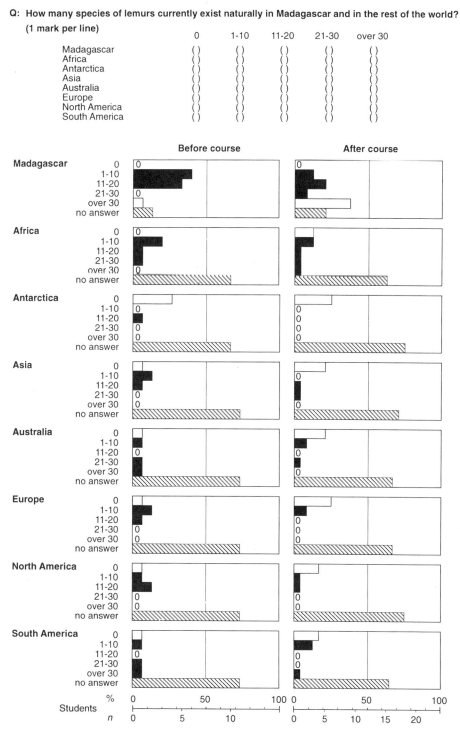

Figure 6. Knowledge about lemur distribution before and after the course. Compiled answers to a compound multiple-choice question about worldwide occurrence of lemurs. White bars: appropriate answers. Black bars: inappropriate answers. Hatched bars: no answer.

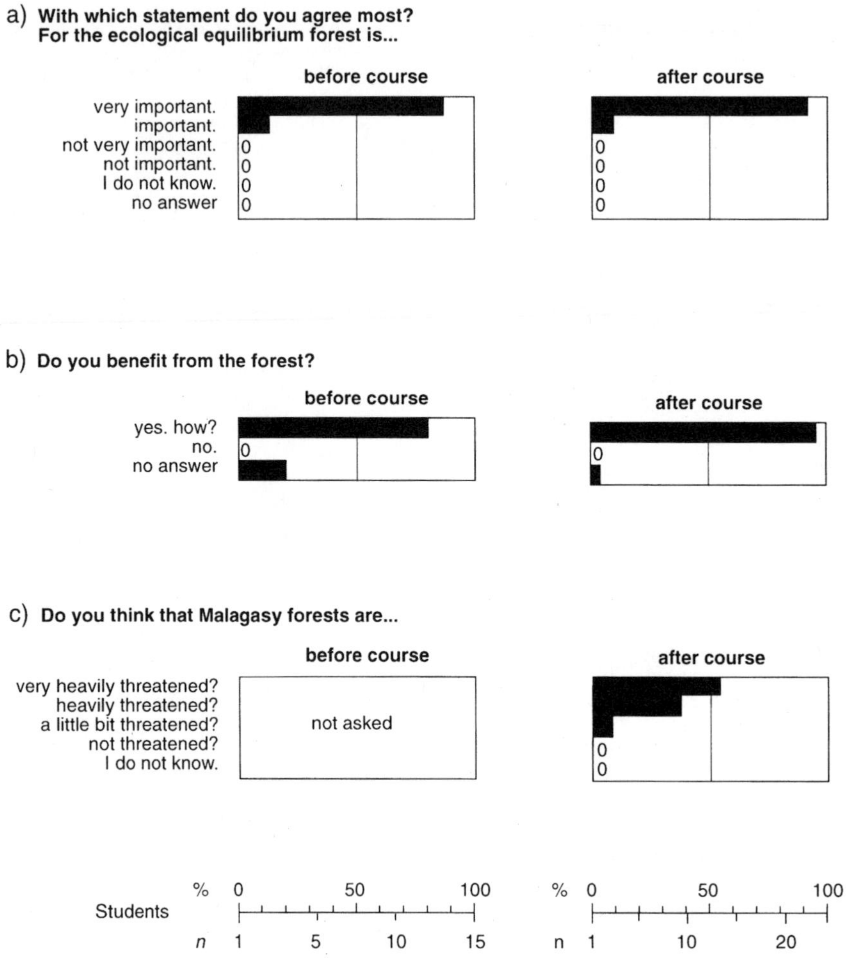

Figure 7. Simple multiple-choice questions and answers on appraisal of issues related to forests before and after the course.

(96%) said that during this course they have heard "very many" or "many new" things (Fig. 8a), and only one (4%) thought she had heard "not many new things". Twenty-two students (92%) would welcome the course being offered regularly (Fig. 8b), preferentially in the third or fourth year of their studies (Fig. 8c; n = 17 (71%)). Two (8%) proposed that the course should be split, with different parts offered in consecutive years.

Overall, 14 students (58%) thought that they have understood "very much" or "much" of the course, nine (38%) "not very much" and one (4%) "a little bit" (Fig. 8d). This self-assessment of understanding checked against the results of the question referring to worldwide lemur occurrence (Fig. 6) revealing that, of the 14 students claiming to have understood "very much" or "much", only two students gave a technically and biologically correct answer. Four students gave a technically incomplete answer, but marked the right range for the number of lemur species in Madagascar, and did not indicate lemurs for other regions than Madagascar. Eight students gave an answer that could not be accepted as appropriate.

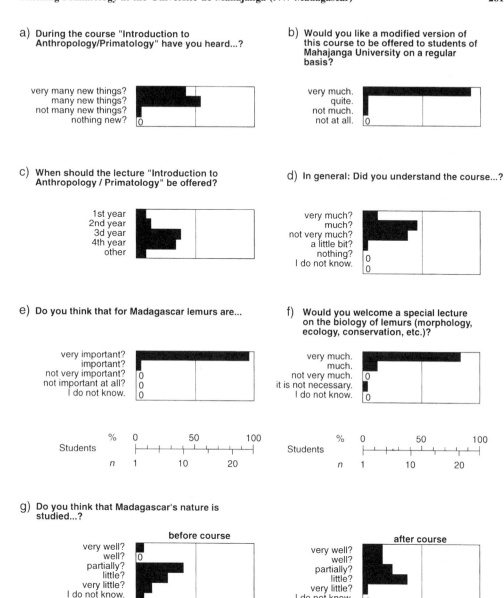

Figure 8. Different aspects related with the course. Answers to simple multiple-choice questions. a–f: Questions and answers from questionnaire 2 and to the same scale. g: Question and answer before and after the course with respective scale.

Asked in Q2 about how important lemurs are for Madagascar (Fig. 8e), 23 students (96%) indicated "very important", and one student (4%) "important". Nobody indicated "not very important" or "not important at all". A special course on lemur biology (Fig. 8f) would be welcomed "very much" or "much" by 23 students (96%).

Only one student (4%) thought that "it is not necessary". Asked about their impression as to how well studied Madagascar's wildlife is (Fig. 8g), in the first questionnaire one student (7%) believed "very well studied", none "well studied", six (40%) "partially studied", four (17%) "little", and three (20%) "very little studied". Corresponding values for Q2 were: "very well studied" = 4 (17%), "well studied" = 4 (17%), "partially studied" = 6 (25%), "little studied" = 9 (38%), "very little studied" = 1 (4%). Chi-square testing reveals no significant difference between Q1 and Q2. The question inviting students to give their impressions on the course ("What was good, bad or what should be changed?") were in the great majority and overall of a positive baseline. However, some criticized that either time was not sufficient (n = 5) or that teaching was too fast (n = 8). Five students criticized UT's French or accent in that language. Two students explicitly expressed their satisfaction about the handouts, and two criticized that handouts were sometimes not available in sufficient numbers. Three students complained about the irregular announcement of the course, and the irregular times for which it was scheduled.

Students' Motivation/Expectations/Miscellaneous

Students were asked, whether they would actively contribute to improving the situation/state of the Université de Mahajanga, if material was provided, and how much time they would invest (Fig. 9a). This question was asked because there were no curtains in the lecture room, and consequently slides could not be projected. Fourteen students (58%) answered that they would help "very willingly", nine (38%) "willingly", and one student (4%) "not very willingly". None excluded the possibility of helping. In terms of time (Fig. 9b), two students (8%) would invest "1–2 hrs/week", 13 (54%) "3–4 hrs/week (a morning)", four (17%) "5–6 hrs/week", and five (21%) ">7 hrs/week (a day)". If an English course were taught at the university (Fig. 9c), 20 students (83%) would participate "with certainty" and four (17%) "perhaps". No student said "I do not know" or excluded participation. Asked about their expectations, and wishes regarding their personal future, the future of the Université de Mahajanga and of Madagascar, the following picture emerged: relatively few students expressed concerns about their personal future, but many expressed their concern about the environmental and socio-economic situation in Madagascar. Regarding the Université de Mahajanga, students wished that research possibilities would be created or enhanced, that better and more material for teaching would be available. They also indicated that students were basically willing to take action of some kind, but various problems, such as lack of money, material, and support by the university staff, prevent that happening. The written criticisms concerning the course, expectations, and wishes for the future additionally showed that on average students had problems, sometimes quite serious problems, in expressing themselves correctly in French, the official teaching language at Malagasy universities.

DISCUSSION

The pilot teaching project "Introduction to Anthropology/Primatology" at the Université de Mahajanga met some of the predefined goals, but failed in achieving others. Primatology as a discipline was presented and found the interest of students in natural sciences. Although the course had to be offered and conducted following an

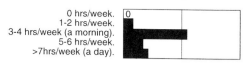

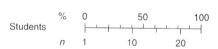

Figure 9. Questions and answers to simple multiple-choice questions after the course regarding motivational aspects.

irregular schedule, it attracted a considerable number of students (Fig. 2), and could potentially attract even more if included in a regular education plan. The second questionnaire clearly showed that students favored presentation as well as topics presented, and would welcome the course being offered on a regular basis, many topics even in extended form (Fig. 1). The origin of students from throughout all over Madagascar (Fig. 3) could lead to a wide distribution of gained understanding and knowledge. As students were aware of the fact that forests are important for ecological equilibrium (Fig. 7a), and that they benefit directly from forests (Fig. 7b), it was quite easy to point out the importance of lemurs. Indeed, after the course all students (Fig. 8e) thought that lemurs are very important or important for Madagascar. The conservation message regarding lemurs was obviously easily transmitted.

As a test case, the course was extremely revealing in positive as well as less positive realms. The importation of the teaching material caused some unexpected problems. Though documents testified that these items were for teaching, it took some time to bring the material through customs in the capital, Antananarivo. Due to technical communication problems between Zurich and Mahajanga, as well as Antananarivo and Mahajanga, it was impossible to prepare and synchronize the importation procedures. Another particularly unpleasant experience concerned the implementation of the course. As outlined above, preparation for this pilot teaching project started early and the university staff could have commented on the project, at latest when the material was officially handed over. It was therefore a surprise when arguments were advanced to stop the course once it was under way. None of the conflicts concerned the course as such directly or the lecturer directly. Problems were mostly of an administrative nature (new rules and procedures to implement courses) and were never directly revealed to the lecturer at the time. However, in view of the early preparation of the project and accompanying information, most of the problems could have been resolved at an earlier stage and outstanding difficulties could easily have been resolved through dialogue. It has to be said that these arguments were put forward by a clear minority of the teaching staff, and were obviously in disharmony with students' interests. As a compromise, teaching continued without official notice, and participation decreased as a consequence. The combination of unpredictable events such as strikes, unexpected problems in importing essential material for teaching, unexpected changes in rules and procedures, lack of transparency and communication reflecting internal problems, which obstruct carefully-prepared, time-, money-, and energy-consuming projects that had been explicitly welcomed by the university prevented a fruitful basis for cooperation. The positive reception by students (Figs. 1, 8), their expressed will to work and invest their time to improve the partially degraded state of the university (Fig. 9), as well as their strong wish to start or enhance research activities (see results section) is certainly encouraging. (How it would work in practice is a different issue.) However, the general knowledge and methodological foundation of students is not very promising for producing results in a short term. Many biologists may certainly have problems in naming an example for every extant eutherian order (Fig. 5). More surprising were the results from the question about worldwide occurrence of lemurs (Fig. 6). In this regard, better results were expected even before the course started. Especially interesting is a deeper analysis of the answers. In both questions (Fig. 5, 6), students avoided making clear statements, obviously when they were not sure about whether they were able to give a correct answer. Students usually showed what they knew, e.g. seven students gave an example for the order Primates, and 12 (80%) indicated that primates are naturally present in Madagascar (Fig. 5). A comparable picture also emerges in the question concerning lemurs (Fig. 6). Of course, most students knew that lemurs occur in Madagascar and they showed that; but obviously not that they occur naturally only in Madagascar. The structure of the question or possible unfamiliarity with questionnaires is not responsible for that picture, as in other contexts such questions were answered in a technically correct manner.

The questionnaires were a very successful instrument for obtaining information on a variety of issues. Sociographic data showed not only that gained knowledge would be potentially distributed widely but also that students are relatively old compared to the actual education level mastered. Obviously, a high percentage of the students do not successfully conclude study years with the corresponding certificate (Fig. 4). Reasons for this are complex and beyond the scope of this article, but it is certainly up to the university to deal with that problem. The actual biological knowledge was lower than expected for students in natural sciences, notably in a renowned megadiversity country. It was not evaluated in detail what students effectively understood and learned during the general part of the course. It was more important to get a general response, and a response regarding the pragmatic approach chosen for topics more relevant for Madagascar and conservation. Responses to the question about worldwide lemur occurrence in the second questionnaire were rather frustrating, as the subject was covered in detail and supported with many specially-prepared handouts. The message that lemurs are endemic to Madagascar obviously did not transfer satisfactorily. Several factors may be responsible for that: (1) the lecturer was unable to communicate messages: though UT's native language is German, and English is a much more familiar language than French, especially regarding science, his French was as good as that of most students. It proved impossible to find an up-to-date primatology book in French, and scientific terms therefore caused some problems. (2) Students could not follow and/or could not sufficiently understand the different lectures: Only eight students could follow the entire course, while in the second questionnaire, 42% of the students—less than expected—actually said they did not understand much of the course. At the same time, questionnaires (and experience) showed, that the students' ability to fully master the French language were sometimes excellent but also often very limited. (3) Information provided was too concentrated within the lecture sessions and for the entire time span of the course: Though it was firmly suspected that teaching was too fast and too intense, the pace was lowered but essentially maintained as long as no response came. For the topics dealing with Madagascar, the pace and intensity were definitely changed. The results of the questionnaire would better allow an assessment, as to how and to what extent such courses should be derived from corresponding courses at University of Zurich. Probably a combination of these three reasons was responsible for the partially unsatisfactory results. All three factors could be influenced and adjusted to generate better results (and certainly would be for future courses). The on site preparation of an extensive French glossary, a better spaced and more regular time schedule allowing for an immediate reaction and adjustment regarding the content and degree to which different topics should or could be treated, would certainly have helped. Constraints in operation at the time prohibited such immediate measures.

Nevertheless, the results from the second questionnaire regarding lemurs were distinctly better, though not satisfactory. The message that lemurs may be of a certain importance was obviously received by students, and a certain interest for lemurs arose as is shown by Figure 8f. Unfortunately, the first questionnaire did not ask how students view lemurs, and the second questionnaire did not ask on what basis students decided that lemurs are important. It is interesting that after the course students had the impression that Madagascar's wildlife is better studied than they thought before (Fig. 8g). Though the difference is statistically not significant, there is a tendency. The result of this question is tentatively interpreted to mean that students themselves thought that they know not very much about Madagascar's wildlife. After they heard "new things" (Fig. 8a), students discovered that much more is known than they expected and is accessible to them (at least on lemurs). The questionnaires also allowed for other perceptions. The distinct possibility to answer questions or make statements such as shown in Figures 5 and 6 seem to demonstrate, that students much prefer to show what they know or believe they know. They dislike to show what they do not know, or to make a statement that could be incorrect. In case of doubt, they prefer to make no statements at all. This typically human quality seemed to be especially developed, as none of the questionnaires had the character of an exam or was in the slightest way presented as such. In addition, questionnaires were filled out anonymously, and no repercussions in any form could be expected by individual students. It is suspected that a certain reserve hindered interactive exchange between students and the lecturer from the beginning, and a real response to the course was only possible through questionnaires and questions also allowing for cross-comparisons. Such across-comparison showed a striking difference between understanding of the course as judged by students on the one hand

and answers about worldwide lemur occurrence on the other hand. Of the 14 students claiming to have understood very much or much (Fig. 8f), 10 did not answer the question in a technically correct manner, eight did so in a definitely technically and biologically incorrect fashion. Critical and if possible continuous evaluation of ongoing teaching activities, including the lecturer, seem of the utmost importance to adjust activities immediately, especially when differences in the diffuse range of "cultural background" and "mentality" are suspected. It needs much empathy, patience, and a healthy capacity for self-criticism to bring education projects to a success.

Follow-up teaching activities were not catalyzed at the university through this pilot project, although research activities continued. The danger of unpredictable events such as strikes in combination with logistical and technical problems prevented a repetition of a modified version of the "Introduction to Anthropology/Primatology" in 1995/96 and 1996/97, not to speak of extended teaching activities. UT had to give priority to his research projects, which made it impossible to travel between the field site and the university and to modify and properly prepare the course.

The failure regarding induction of extended teaching and research activities can be attributed to several factors. (1) The Accord de Coopération was initiated under better and more optimistic conditions. It soon became apparent that it is easier for students from University of Zurich to initiate research projects than to promote long-term commitment and establish a suitable infrastructure for teaching and research capacity building at the Université de Mahajanga. This is much more expensive and time-consuming, and results do not show up within an immediate time span. The risk that such projects may fail due to unpredictable developments of all kinds proved to be high. (2) The position of AZ and UT were too weak either to facilitate working conditions in Madagascar or to invest enough time and energy into a joint long-term project, or open up ways that the Université de Mahajanga could develop its own activities. UT could be present only through post-doctoral research activities, which must produce valuable scientific results in a certain amount of time and need full attention. (3) Experiences made did not unambiguously show that future commitments on an extended level would necessarily be rewarding. The critical point for a definite breakthrough could not therefore be reached.

After having explored the pilot teaching project, minuscule compared to pertinent environmental and socio-economic problems in Madagascar, the opportunity should not be missed to evaluate it in the light of current developments of much wider importance: "In the course of the last few years three things have become increasingly clear. Firstly, without major contributions on the part of scientific research there will be no solutions to the growing worldwide problems which endanger life on earth. Secondly, research activities are needed in the South as much as in the North. Thirdly, however, in many countries in the South, the efforts undertaken so far to strengthen the research capacities have not resulted in establishing the research potential needed ... Moreover, the present global distribution of qualified researchers is inadequate, and their number insufficient. Thus the present situation does not allow for sustainable development, either of the South or of the North." (A. Freyvogel, Chairman of the "Swiss Commission for Research Partnership with Developing Countries", Maselli & Sottas, 1996, p. 13).

It may seem disproportionate to look at this teaching project in anthropology/primatology in such a large-scale context. This is particularly so because physical anthropology and primatology have the reputation of being of rather academic relevance, while problems of immediate concern in Madagascar are much more of a daily life nature than academic. Growing concern about the ongoing loss of biodiversity worldwide, the quality of Madagascar as a megadiversity country, and the exclusive distribution of lemurs perhaps allow for another perspective. Lemurs are perfectly suitable subjects for teaching and research training at any scientific level in Madagascar were they occur widely (as long as forests are present). Indeed, they are fancy research subjects in the north and favorite zoo animals that attract much attention, and may facilitate fund-raising for and by Malagasy researchers. Acquired skills in scientific methodology are applicable to a wide variety of topics. Though it was intended from the beginning with the Accord de Coopération to catalyze research activities by the Université de Mahajanga, the time was not yet ready for that. Meanwhile, documents have been elaborated at least in Switzerland which formulate not only strategies regarding research encouragement (DDC/ASSN, 1997) but also guidelines to initiate research partnerships (Freyvogel, 1998; KFPE, 1998), programs and funds to finance such partnerships, though limited (Freyvogel, 1998). One of the most important prerequisites to profit from such opportunities is that developing countries participate very actively in formulating research goals together with their partner, and are informed or can inform themselves about such developments and opportunities. It is clear that formulating research goals and the actual research process have to rely on a certain foundation to be promising. Presented results and experiences show that this is not yet the case at least at the Université de Mahajanga. Such foundations can only be built through adequate education to prepare and formulate projects, otherwise one or the other potential partner is excluded. A partner from the north could be excluded because goals are actually formulated from a northern perspective to meet, for example, maximal publication output to increase chances for fund raising. A partner from the south may be excluded through not having a suitable background to formulate and meet the criteria for such projects, even if he knows about opportunities. This problematic situation has only recently been fully recognized and partially formulated (Freyvogel, 1996, 1998). It is one of the institutionally bound reasons why the present attempt of co-operation has currently reached a state of hibernation, although the actual targets (students of the Université de Mahajanga) obviously welcomed the cooperation. The funds and support for this limited project were easily found. Fund-raising for sustainable teaching and research activities is of a completely different dimension and involves dealing with complex issues and at several levels. These are excellently and controversially presented in the compilation by Maselli and Sottas (1996).

Whether this limited pilot project remains a mere anecdote or helps at least to point realistically to the interwoven parts, colors, and nuances that make a big picture (that looks so different depending on the perspective), remains to be shown: "Do not let us indulge in illusions: neither the public, nor politicians, nor industrial leaders are waiting for scientists to tell them what to do. Scientists are not going to change the world. Yet lack of power is no reason to turn to fatalism; even small steps do influence events." (Freyvogel, 1996).

ACKNOWLEDGMENTS

First of all, we thank the students of the Université de Mahajanga who, by their attendance and with the questionnaires, provided so much valuable information, and César Rabeny who helped voluntarily as teaching assistant. UT is especially grateful to his co-author Alphonse Zaramody, who much protected and supported this project, and apologizes that for logistic reasons his views could not flow more into this article. Special thanks go to Maximilian Jaeger, who enthusiastically allocated technical material and financial support through the University of Zurich, and to the Zürcher Hochschulverein for additional financial support. The Anthropological Institute and Museum contributed teaching material. We thank Bob Martin and Peter Schmid for this help. Various support came from Elisabetha Joos-Müller and Gustl Anzenberger. The teaching was only possible because UT stayed for a prolonged time in Madagascar for research, funded by the Swiss National Fund for Science. Unaware that teaching is not allowed during such work, no repercussions followed, and permission was later kindly provided. UT thanks the Karons for the hospitality during his stay in Mahajanga. Alexandra Müller, Christophe Soligo, Bob Martin, and two reviewers are thanked for helpful comments on the manuscript.

REFERENCES

CARROLL, R. L. 1988. Vertebrate Paleontology and Evolution. W.H. Freeman and Co., New York.

DDC/ASSN Direction du développement et de la coopération/Académie suisse des sciences naturelle. 1997. Stratégie suisse pour l'encouragement de la recherche dans les pays en développement, 2nd edition. DDC/ASSN, Berne. (Available in G/F/E viawww.kfpe.unibe.ch).

FREYVOGEL, T. A. 1996. Scientific research partnership: north-south and south-south. Paper presented at Annual Conference of the Swiss Society for Tropical Medicine and Parasitology. Neuchâtel. (Available via www.kfpe.unibe.ch).

———. 1998. Forschungspartnerschaft mit Entwicklungsländern—die grosse Herausforderung unserer Zeit. Schweizerische Akademie der Naturwissenschaften SANW Info Spezial, **1/98**: 13–20. (Available in G/F/E via www.kfpe.unibe.ch).

HARCOURT, C., AND J. THORNBACK. 1990. Lemurs of Madagascar and the Comoros. The IUCN Red Data Book. IUCN, Gland & Cambridge.

KFPE Schweizerische Kommission für Forschungspartnerschaften mit Entwicklungsländern. 1998. Leitfaden für Forschungspartnerschaften mit Entwicklungsländern. KFPE, Berne. (Available in G/F/E via www.kfpe.unibe.ch).

MARTIN, R. D., G. ANZENBERGER, AND R. GREIF. 1997. Der Mensch im Rahmen der Primatenevolution. Companion text for lecture course. Anthropological Institute and Museum, University of Zurich.

MASELLI, D., AND B. SOTTAS (eds.). 1996. Research Partnerships for Common Concerns: Proceedings of the International Conference on Scientific Research Partnership for Sustainable Development north-south and south-south Dimensions. LIT Verlag, Hamburg.

MITTERMEIER, R. A., AND W. R. KONSTANT. 1996/97. Primate conservation: a retrospective and a look into the 21st century. Primate Conservation, **17**: 7–17.

MITTERMEIER, R. A., I. TATTERSALL, W. R. KONSTANT, D. M. MEYERS, AND R. B. MAST. 1994. Lemurs of Madagascar. Conservation International, Washington, D.C.

WCMC (World Conservation Monitoring Center). 1992. Global Biodiversity-Status of the Earth's Living Resources. Chapman & Hall, London.

ZIMMERMANN, E., S. CEPOK, N. RAKOTOARISON, V. ZIETEMANN, AND U. RADESPIEL. 1998. Sympatric mouse lemurs in north-west Madagascar: a new rufous mouse lemur species (*Microcebus ravelobensis*). Folia Primatologica, **69**: 106–111.

16

LEMURS AS FLAGSHIPS FOR CONSERVATION IN MADAGASCAR

Joanna C. Durbin[1]

[1] Jersey Wildlife Preservation Trust
BP 8511, Antananarivo (101), Madagascar
jwpt@dts.mg

ABSTRACT

Lemurs are the best known Malagasy wildlife, attracting international attention to Madagascar's unique biodiversity and to the ecological changes that threaten their existence. Lemurs have been the underlying impetus for the creation of some of Madagascar's protected areas, which aim to conserve entire forest ecosystems. The presence of rare lemurs has drawn researchers, funding, and conservation measures to certain areas. Lemurs are being used as indicator species for ecological monitoring. Lemurs are also prime attractions for visitors, thus making a major economic contribution to a growing tourist industry and, potentially, to the future maintenance of protected areas. However, international and national support is insufficient to ensure conservation as local people strive to maintain their livelihoods. The Black Lemur Forest Project, Parc Ivoloina's environmental education center, Parc Botanique et Zoologique de Tsimbazaza and Jersey Wildlife Preservation Trust's Project Alaotra, demonstrate that lemurs can provide an effective focus for education and awareness programs at the local level leading to community-based conservation initiatives.

RÉSUMÉ

De la nature malgache, on connaît surtout les lémuriens qui attirent l'attention internationale sur l'unique biodiversité de Madagascar et sur les changements écologiques qui menacent leur existence. A cause des lémuriens, certaines aires protégées de Madagascar ont été créées pour conserver des écosystèmes forestiers entiers. La présence de lémuriens rares dans certaines zones a eu pour effet d'amener des chercheurs dans ces zones et avec eux, des financements et des mesures de conservation. Les lémuriens sont pris comme des espèces indicatrices pour le suivi écologique. Ils constituent aussi l'un des plus grands attraits touristiques, apportant ainsi une contribution importante à l'industrie touristique naissante et à l'entretien des aires protégées à l'avenir. Toutefois, l'appui national et international ne suffisent pas pour assurer la conservation dans ces zones où les populations villageoises s'efforcent de maintenir leurs moyens d'existence. Le "Black Lemur Forest Project", le Centre d'Éducation Environnementale du Parc d'Ivoloina, le Parc Botanique et Zoologique de Tsimbazaza et le Projet Alaotra du Jersey Wildlife Preservation Trust démontrent que les lémuriens peuvent fournir un centre d'intérêt efficace pour les programmes d'éducation et de sensibilisation au niveau local, menant à des initiatives de conservation venant de la communauté.

INTRODUCTION

Madagascar is famous for its diverse and unique wildlife, of which the lemurs are the best known. They represent exceptional primate diversity, with 5 families, 14 genera, 33 species, and 51 taxa (including subspecies) occurring naturally nowhere else (see Table 1; Mittermeier et al., 1994; Zimmerman et al., 1998), making Madagascar third highest on the world list of primate species diversity after Brazil and Indonesia (Mittermeier et al., 1994).

Ecological changes that have occurred in Madagascar since the arrival of humans around 2000 years ago are resulting in the disappearance of forests at an alarming rate (MacPhee & Burney, 1991; Nelson & Horning, 1993). This, in turn, has caused environmental problems such as soil erosion, reduced soil fertility, flooding, and drying up of water supplies, in addition to loss of useful forest products. Around 70% of the human population of over 16.3 million live in rural areas (EIU, 1998a) and rely on the continuing productivity of natural resources for their livelihoods. With an annual population growth of 3.1% (EIU, 1998a), increasing demands will be made on the remaining natural resources. Madagascar, which has been ranked twelfth poorest nation in the world with a GNP per head of $ 230 US per year (EIU, 1997), provides an example where environmental degradation is one of the factors contributing to poverty in the poorest developing countries.

The loss of forests, in which nearly all Madagascar's endemic biota resides, also puts many species at risk of extinction. At least 8 genera and 15 species of lemurs, all larger than the remaining species, have become extinct in Madagascar in the last 2000 years (Godfrey et al., this volume). Of those remaining, 10 are considered critically endangered, 7 endangered, and 19 vulnerable (over 70% of taxa), making Madagascar the world's highest primate conservation priority (Mittermeier et al., 1994).

In recognition of the urgency and importance of conservation action in Madagascar, many measures have been taken aiming to halt environmental degradation and maintain biological diversity. This paper reviews the contribution of some conservation initiatives linked to lemurs and describes several current projects that show how lemurs are continuing to play an important role in helping to draw attention to conservation issues, at the local level as well as internationally.

A flagship species, often a charismatic large vertebrate, is one that can be used to promote a conservation campaign because it arouses public interest, such as the Mountain Gorilla *Gorilla gorilla berengei* of Rwanda, Democratic Republic of Congo, and Uganda (Butynski et al., 1990). Lemurs, although mostly smaller and more cryptic

Table 1. List of currently recognized extant lemur taxa[1]

Family	Species	Common Name
Family Cheirogaleidae	*Microcebus murinus*	Gray Mouse Lemur
	Microcebus myoxinus	Pygmy Mouse Lemur
	Microcebus ravelobensis	Golden Brown Mouse Lemur
	Microcebus rufus	Brown Mouse Lemur
	Microcebus spp.[2]	Mouse Lemur spp.
	Allocebus trichotis	Hairy-eared Dwarf Lemur
	Mirza coquereli	Coquerel's Dwarf Lemur
	Cheirogaleus major	Greater Dwarf Lemur
	Cheirogaleus medius	Fat-tailed Dwarf Lemur
	Phaner furcifer electromontis	Amber Mountain Fork-marked Lemur
	Phaner furcifer furcifer	Eastern Fork-marked Lemur
	Phaner furcifer pallescens	Pale Fork-marked Lemur
	Phaner furcifer parienti	Pariente's Fork-marked Lemur
Family Megaladapidae	*Lepilemur dorsalis*	Gray-backed Sportive Lemur
	Lepilemur edwardsi	Milne-Edwards' Sportive Lemur
	Lepilemur leucopus	White-footed Sportive Lemur
	Lepilemur microdon	Small-toothed Sportive Lemur
	Lepilemur mustelinus	Weasel Sportive Lemur
	Lepilemur ruficaudatus	Red-tailed Sportive Lemur
	Lepilemur septentrionalis	Northern Sportive Lemur
Family Lemuridae	*Hapalemur aureus*	Golden Bamboo Lemur
	Hapalemur griseus alaotrensis	Lac Alaotra Bamboo Lemur
	Hapalemur griseus griseus	Eastern Lesser Bamboo Lemur
	Hapalemur griseus occidentalis	Western Lesser Bamboo Lemur
	Hapalemur simus	Greater Bamboo Lemur
	Lemur catta	Ring-tailed Lemur
	Eulemur coronatus	Crowned Lemur
	Eulemur fulvus albifrons	White-fronted Brown Lemur
	Eulemur fulvus albocollaris	White-collared Brown Lemur
	Eulemur fulvus collaris	Collared Brown Lemur
	Eulemur fulvus fulvus	Common Brown Lemur
	Eulemur fulvus rufus	Red-fronted Brown Lemur
	Eulemur fulvus sanfordi	Sanford's Brown Lemur
	Eulemur macaco flavifrons	Sclater's Black Lemur
	Eulemur macaco macaco	Black Lemur
	Eulemur mongoz	Mongoose Lemur
	Eulemur rubriventer	Red-bellied Lemur
	Varecia variegata rubra	Red Ruffed Lemur
	Varecia variegata variegata	Black-and-white Ruffed Lemur
Family Indriidae	*Avahi laniger*	Eastern Woolly Lemur
	Avahi occidentalis	Western Woolly Lemur
	Propithecus diadema candidus	Silky Sifaka
	Propithecus diadema diadema	Diademed Sifaka
	Propithecus diadema edwardsi	Milne-Edwards' Sifaka
	Propithecus diadema perrieri	Perrier's Sifaka
	Propithecus tattersalli	Golden-crowned Sifaka
	Propithecus verreauxi coquereli	Coquerel's Sifaka
	Propithecus verreauxi coronatus	Crowned Sifaka
	Propithecus verreauxi deckeni	Decken's Sifaka
	Propithecus verreauxi verreauxi	Verreaux's Sifaka
	Indri indri	Indri
Family Daubentoniidae	*Daubentonia madagascariensis*	Aye-aye

[1] List based on Mittermeier et al. (1994) and Zimmerman et al. (1998).
[2] During the International Primatological Society Congress in Antananarivo in August 1998, Rodin Rasoloarison presented information on several undescribed species of *Microcebus* from western Madagascar. The descriptions of these new taxa are currently in preparation.

than many flagship species, are appealing, furry creatures. Primates, in general, hold a special fascination as our closest relatives in the animal kingdom, and other small primates such as the Lion Tamarins *Leontipithecus* spp. of Brazil have been shown to be effective as flagship species (Dietz et al., 1994). In addition to the educational and political support for conservation that such species can provide, a focused, single-species approach can also provide a means of protecting many additional plant and animal species and their habitats (Durrell & Mallinson, 1987). There has been criticism that conservation of flagship species can be very expensive, may not automatically ensure conservation of entire ecosystems, and may even hinder the conservation of other endangered species (Simberloff, 1998). The extent to which such considerations have a negative impact on more general conservation objectives clearly depends on the role of the flagship species in each case. The relative advantages of the flagship concept are considered for the examples given below.

The Contribution of Lemur Researchers

Lemurs have been the focus of numerous research projects which have helped to draw international attention to Madagascar's wildlife and to the conservation importance of particular study sites, not only for lemurs but for forest ecosystems. In some cases, these research sites already had formal protection. A network of ten strict nature reserves was created to protect representative ecosystems as early as 1927 (Nicoll & Langrand, 1989). These included, for example, the Réserve Naturelle Intégrale (RNI) d'Ankarafantsika (60,520 ha) adjacent to Station Forestière (SF) d'Ampijoroa, which has been the site of many lemur studies (Fig. 1). From the 1950s, 23 Réserves Spéciales (RS) were added, several of which aimed to protect habitats for particular endangered species. Examples include the RS de Périnet-Analamazaotra (Andasibe) which protects 810 ha of rainforest for *Indri indri*, and the RS de Nosy Mangabe, a 520 ha forested island, for *Daubentonia madagascariensis* (Nicoll & Langrand, 1989).

In other cases, the efforts of primate researchers, in partnership with Malagasy authorities, have resulted in formal protection for forests around research sites. The RS de Beza Mahafaly, which protects 580 ha of gallery forest and spiny bush in southwest Madagascar, was created in 1986 as a result of collaboration between the Département des Eaux et Forêts of the Ecole Supérieure de Sciences Agronomiques of the Université d'Antananarivo, and Yale and Washington universities (Nicoll & Langrand, 1989). The Parc National (PN) de Ranomafana was created in 1991, protecting 41,601 ha of rain forest in southeast central Madagascar, also as a result of promotion by researchers (Wright, 1992). Ranomafana, famous as the site where *Hapalemur aureus* was discovered in the 1980s (Meier et al., 1987), is a much frequented research base from which 15 doctorate research projects (eight on lemurs) and 59 masters-level studies (50 by Malagasy researchers) have been completed (ICTE, 1998).

Similar efforts continue and dossiers are in preparation for the classification of new sites as protected areas as a result of lemur-based research. These include Sahamalaza in the northwest, a critical site for the conservation of *Eulemur macaco flavifrons* (Meier et al., 1996), and Daraina in the northeast, for *Propithecus tattersalli* (S. Rajaobelina, pers. comm. 1998). Neither of these endangered taxa occurs in any existing protected area. Efforts are also underway to create a reserve at Anjozorobe (Fanamby, 1998; Rakotondravony & Goodman, 1998) with the collaboration of the Group d'Etude et de Recherche sur les Primates de Madagascar (GERP), where seven

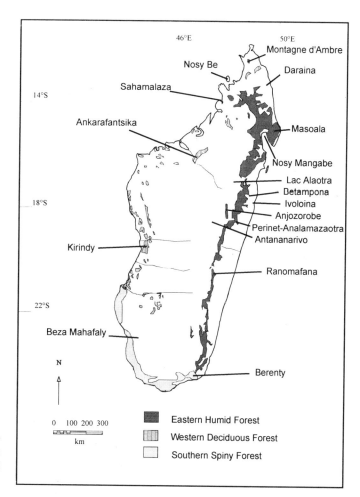

Figure 1. Map of Madagascar showing existing primary forest cover and localities mentioned in the text.

lemur species are known to occur, including *P. d. diadema*, and several others have been reported by local people, including *Indri indri* (Razanahoera, 1998).

At PN de Masoala, lemur research is providing information of direct relevance for park management. *Varecia variegata rubra* and *Eulemur fulvus albifrons* are being used as indicator species to monitor the evolution of human pressures on the park. Population density and group composition are recorded by repeated observations at a number of sites around the park each year during the lemurs most active season (mid-October to February) so that changes in ecology likely to be caused by human pressures can be monitored (Rakotondratsima, 1996).

The first lemur release into the wild was undertaken in 1997, using five *Varecia variegata variegata* from the USA to augment existing populations at the RNI de Betampona (Britt et al., 1998). This project, undertaken by the Madagascar Fauna Group and the Duke University Primate Center, in collaboration with the Association Nationale pour la Gestion des Aires Protégées (ANGAP) and the Ministère des Eaux et Forêts (MEF) is helping to draw international attention to lemurs, to Madagascar and to the Betampona Reserve. Publicity and funding were boosted when John Cleese (of Monty Python fame) organized donation of profits from the European premiere of the film "Fierce Creatures" (Chepisi & Young, 1997), which featured a Ring-tailed Lemur as one of the stars. John Cleese also made a much-publicized visit to the release site in April 1998, and helped to make a film that explained the goals of the project to an international audience (Kershaw & Cleese, 1998). Critics could argue that the funds spent on this relatively expensive species conservation project would have been better spent on ecosystem conservation. It certainly looks excessive, from the local Malagasy point of view, to spend so much money on five lemurs. In defense of the project, it is almost certain that most of these funds would not have been channeled to conservation or even to Madagascar, without the strong flagship species appeal. In addition to testing the viability of releasing captive-bred lemurs as a strategy to reinforce small, isolated populations, the project also aims to improve the level of protection of the reserve through the presence of project personnel and through an education and awareness program undertaken in conjunction with Parc Ivoloina (Britt et al., 1998). This flagship species project should contribute very tangibly to the conservation of the 2228 ha eastern rain forest reserve.

Extensive research on lemurs and other fauna and flora at the Kirindy forestry concession in western deciduous forest has contributed to knowledge of the impact of forestry exploitation and other forms of disturbance on the forest ecosystem (Ganzhorn & Sorg, 1996). In addition, the almost permanent presence of researchers, particularly since the Deutsches Primatenzentrum installed a field research base at Kirindy in 1993, has apparently reduced hunting pressure and thus assisted with conservation (J. Ganzhorn, in litt., 1998). Furthermore, the fees received by the concession managers, the Centre de Formation Professionelle Forestière (CFPF), from researchers, have provided a continuous income which has assisted them during a period of transfer to ecotourism management objectives (J. Randrianasolo, pers. comm., 1997) after it became clear that sustainable forest exploitation, using the techniques employed in Madagascar, was not economically viable, and CFPF was having little success in achieving its original aims of increasing professionalism and sustainability of forestry exploitation (Cuvelier, 1996).

Tourism and Conservation

Researchers are not the only ones attracted by Madagascar's lemurs. Tourists are arriving in Madagascar in ever-increasing numbers. Annual non-resident arrivals increased by 23% in 1997 from 1996, but still only numbered a modest 100,762 (Ministère du Tourisme, 1998). Natural areas and wildlife are a major attraction, and viewing lemurs is a principal target for visitors to national parks or reserves. As an indicator, visitors to SF d'Ampijoroa have risen over ten-fold from 203 in 1990 to 2154 in 1997 (Responsable d'Ecotourisme, Projet de Conservation et Développement Intégré d'Ankarafantsika, pers. comm., 1998). Figures recorded by ANGAP for all protected areas they manage show an eight-fold increase in six years from 5836 in 1992 to 46,872 in 1997 (Fig. 2; Cellule Ecotourisme de l'ANGAP, pers. comm., 1998). The number of visitors to State-owned protected areas (not including private protected areas like Berenty or Kirindy) is just under half that of non-resident arrivals in 1997, showing how important the protected area network is to tourism in Madagascar. Nearly all of these tourists visited four main sites, as these have some infrastructure, are of relatively easy access, and are already better known.

Tourism clearly has huge potential to boost the national and local economy, and

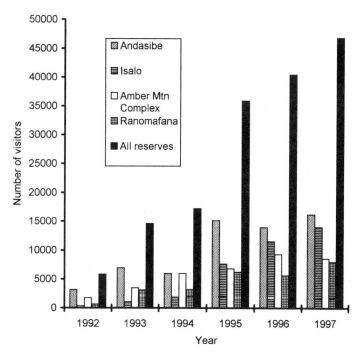

Figure 2. Graph showing evolution of visitor numbers to protected area in Madagascar based on Cellule Ecotourisme d'ANGAP (pers. comm., 1998). Amber Mountain Complex includes Parc National de Montagne d'Ambre, Réserve Spéciale d'Ankarana, Réserve Spéciale d'Analamerana, and Réserve Spéciale de Forêt d'Ambre.

already rivals major exports such as coffee and prawns as the primary foreign currency earner, even at the current relatively low level (EIU, 1998b). Eco-tourism can also provide financial incentives for conservation, both contributing to costs of protected area maintenance and providing local people with benefits directly linked to the protection of a forest area and its lemurs. For example, CFPF can now amply cover running costs of Kirindy from visitor revenues, especially since receiving a record 558 visitors and over $ 3000 US in August 1998 (J. Robson, pers. comm., 1998). This exceptional monthly figure was probably augmented by visitors associated with the International Primatological Society congress held in Madagascar in August 1998, giving another example of the contribution of researchers, in this case as specialized tourists. ANGAP have helped to reinforce local benefits by making half of visitor entrance fees to protected areas available to local communities for development projects (Durbin & Ratrimoarisaona, 1996). While tourism-related revenues can play a very important role as incentives for local people to protect forests and wildlife, they cannot yet be expected to cover the costs of maintaining the entire national protected area network of 44 parks and reserves. The total income from entrance fees was only approximately $ 100,000 US in 1997, and half of this sum is returned to local community development projects around the reserves visited by tourists, leaving only $ 50,000 US to contribute to administration and management expenses for the entire network (Ministère de l'Environnement, 1997). Many of the reserves are remote and rarely visited, or their status prohibits tourism, as is the case for Réserves Naturelles Intégrales, so cannot benefit from tourism.

LOCAL CHALLENGES TO EFFECTIVE CONSERVATION IN MADAGASCAR

Ensuring the support of local people is the key to success of conservation projects in Madagascar. Internationally and nationally supported initiatives, including declaration of protected areas, are important, but without the support of local people, forests will continue to be cleared. Forests are disappearing in Madagascar, mainly through slash-and-burn agriculture, creation and maintenance of cattle pasture, and uncontrolled fires, as a result of a complex combination of factors including:

1) the nature of the forests and climatic conditions;
2) demand for more agricultural land for the subsistence needs of a growing population;
3) lack of labor and cash for more intensive agricultural practices;
4) strong cultural traditions to continue slash-and-burn;
5) inconsistent support for conservation from government technical services and authorities leading to inefficient application of laws and policies;
6) and, perhaps most importantly, as a result of the "tragedy of the commons".

The "tragedy of the commons" refers to the scenario in which it is in each individual's interest to over-exploit natural resources for short term gain where there is no secure management structure to ensure sustainable use for long term gains in perpetuity (Ostrom, 1990). In the classic illustration, each herdsman gains more in the short-term by increasing the number of his cattle on common land, even beyond the carrying capacity of the land, leading to over-grazing. The costs of over-grazing are shared between the cattle owners, but the benefit of adding another cow goes directly to the individual herdsman (Hardin, 1968).

Recognizing the importance of winning local support, several protected areas became the focus of integrated conservation and development projects during the first phase of the National Environmental Action Plan (NEAP) (1991–1997). Development projects, implemented in peripheral zones, aimed to provide alternatives for local people to compensate for lack of access to resources within the protected area. However, in general, these projects were expensive and often were not as successful as had been anticipated (World Bank, 1996). Some of them suffered from top-down planning and inadequate communication with local communities, in particular neglecting the development of local management structures to implement negotiated resource use agreements, which could respond to the problem posed by the tragedy of the commons. They also concentrated efforts in a few areas, not really addressing the broader problems of conservation in Madagascar (Richard and O'Connor, 1997). The second phase of the NEAP, underway since 1997, has promoted new laws and policies giving local rights for sustainable resource management in contracts with the State which show great promise to address these local resource management issues on a more widespread scale (World Bank, 1996; ONE, 1997).

Using Lemurs as Flagships to Promote Local Support: The Black Lemur Forest Project

A few projects are using lemurs as flagships to implement communication and education projects with the aim of promoting local support for conservation. The Black Lemur Forest Project is a conservation education project based in Nosy Be, and is part of a larger research project concerning *Eulemur macaco* and its habitats in north west Madagascar. Active since 1988, the members of the small project team aim to increase public awareness (both in Madagascar and abroad) of the value and threats to Madagascar's unique biological heritage, to engender support for conservation efforts in Madagascar, to explain their work to local people and encourage their involvement, and to provide information to tourists to encourage responsible and sustainable tourism which can benefit local village communities. They work closely with local schools, giving talks and slide shows, and organizing classroom activities and forest walks. They have also enabled a local community group to open a visitor center, sell souvenirs, and work as guides, thus benefiting directly from the Black Lemurs in their adjacent forest (Andrews, 1998). Although on a very small scale, this locally-managed project provides direct benefits to the local people and provides a model for sustainable eco-tourism.

Local Zoos Contributing to Local Education and Awareness: Ivoloina and Tsimbazaza

The Parc Ivoloina Education Center concentrates on raising awareness of environmental issues and appreciation of wildlife among young people of the east coast Toamasina (Tamatave) area. School visits are organized to the zoo and its education center, which has imaginative, inter-active displays explaining different habitats found in Madagascar. There is also a teaching room, with displays illustrating the carbon cycle, the water cycle and other environmental concepts. During school visits, children and their teachers are given a short course on environmental inter-relationships, biodiversity in Madagascar, local conservation problems and suggestions for practical solutions, followed by a guided tour of the zoo. For three years running, special "Saturday" classes have been held, which aim to help children from surrounding primary schools to pass the national exam necessary to move up to secondary school. In addition, in collaboration with the local education service and the government Bureau Programme d'Education Environnementale (BPEE), eight schools around Ivoloina and Betampona have been involved in teacher training and school visits to encourage the integration of environmental aspects into the usual class work (Katz, 1998).

The captive lemur collection at Ivoloina and life-size panels of extinct lemurs give children an appreciation of lemurs that most would never otherwise have had. A surprisingly large proportion of the Malagasy population have never seen a lemur. Thus the captive collections and associated education programs, such as at Ivoloina and at Antananarivo's Parc Zoologique et Botanique de Tsimbazaza (PBZT), have a very important role to play in education.

PBZT is running a mobile zoo that takes some animals from the collection to towns surrounding the capital on their weekly market day. Four such outings have been undertaken so far in 1998. The visit to Mahitsy, north of Antananarivo, in May 1998 was extended to four days by popular demand, during which period over 5000 school children visited the exhibition in addition to the general public (P. Ravokatra, pers. comm., 1998).

A Lemur as a Flagship for Wetland Conservation: The Jersey Wildlife Preservation Trust Alaotra Project

Lac Alaotra is the largest lake in Madagascar, and with the surrounding marshes and rice fields constitutes a wetland area of 1000 km^2. Wetlands are among the most productive and diverse ecosystems and are often of great economic and biological importance. Lac Alaotra is no exception, harboring a particularly high density of aquatic birds, including an important population of an endemic Malagasy duck species, Meller's Duck *Anas melleri*, and three animals endemic to the area: the Lac Alaotra Bamboo Lemur *Hapalemur griseus alaotrensis*, the Madagascar Pochard *Aythya innotata*, and the Alaotra Grebe *Tachybaptus rufolavatus* (Nicoll & Langrand, 1989). Lac Alaotra is one of the most important rice production areas in Madagascar, rice being the most important food crop in the country (PLI, 1990). A significant proportion of the local population also relies on fishing and on marsh products such as construction and weaving materials.

The Alaotra wetlands sustain more than 450,000 people (Préfecture d'Ambatondrazaka, pers. comm., 1998), who rely on the continuing productivity of the ecosystem. The wetland area has already undergone dramatic negative ecological changes within the last century and unless measures are taken to limit future degradation, productivity will continue to decline. Deforestation on hills around the lake has caused serious soil erosion. Large areas of marshes have been transformed into agricultural land. Introduced fish and plant species have caused the decline of some native species through direct competition and by alteration of ecosystem equilibrium. The increasing human population contributes to pollution of the lake, as do fertilizers and pesticides used in nearby agricultural fields. As a result, there have been changes in the fauna inhabiting the wetland area. The endemic pochard and grebe have not been seen recently and may be extinct. Fewer and smaller fish are now caught by an increasing number of fishermen using more efficient techniques. Rice production has fallen as fields and canals become silted by erosion and choked by weeds (Pidgeon, 1996).

The Jersey Wildlife Preservation Trust (JWPT) has been running an education and awareness campaign around Lac Alaotra since December 1996, using the endangered *Hapalemur* as a flagship, as part of an integrated species conservation program including captive-breeding and research (Feistner, this volume). The aim of the education campaign is to increase local awareness of the ecological changes occurring in the area, and to catalyze local management initiatives that promote long-term productivity of the ecosystem and maintenance of biodiversity. Initially, the project worked mainly through schools, collaborating with the BPEE and with local education services. A teacher training program was combined with competitions which encouraged children to learn about their local environment and the changes that have occurred, and to represent these via theatrical sketches, poems, and drawings. The finals of these competitions were attended by parents and other villagers. By mid-1998, ten teacher training sessions have been held for 249 teachers of 31 primary schools and seven secondary schools around the lake.

In November 1997, the approach was expanded by holding environmental festivals (*Fetin'ny Zetra* or "Festival of the Marshes") celebrating the value of the marshes, and promoting the *bandro* (local name for the *Hapalemur*) as an emblem of wetland health in 20 villages to the east and west of the lake in the form of a competition. School teachers and children were joined by village leaders and youth groups to organize the event, and activities and displays were prepared by people throughout the community. Expression of the local *Sihanaka* culture was encouraged, which meant that people could express themselves freely using their local dialect with traditional forms of performance, and people with less formal education could participate. The elders appreciated the respect given to traditions and participated actively.

During these events it became clear that many people recognized that clearance of the marshes contributes to their problems. According to local people, the marshes provide a breeding ground for fish, a very important source of protein and livelihood in the region, and they also play a role in controlling water flow and so protect adjacent rice fields from flooding and drying out. Further, the marshes provide products for weaving (an important source of income for women), for houses, and for medicinal plants. These sentiments were not only reflected in countless speeches, songs, and poems, but people of several villages gave stronger demonstrations of their convictions. Most remarkably, eight villages had actually started planting marshes in denuded zones to promote marsh rehabilitation and the people of ten villages publicly made a traditional oath (*joro*), in which they requested benediction from the ancestors and made a commitment to respect wildlife and to protect a local marsh area. One village had created a pond for the school to breed endemic fish which are becoming rare in the lake. The finals of the festival competitions drew crowds of over 1000 people. They were attended by regional administrators, local mayors, representatives of central government environmental and educational ministries, and were covered by local and national radio, television, and newspapers.

These village initiatives for natural habitat restoration are really exceptional and demonstrate great commitment, showing that people are highly motivated to improve the management of their local wetland. For example, at one village, Vohitsara, 600 people worked on two separate days to replant an area of over a hectare. They would not have done this if they did not really believe strongly that marsh rehabilitation is in their interest.

At a regional level, there is now a federation of local non-governmental organizations, called Alaotra Ramsar Convention (ARC), promoting the designation of Lac Alaotra as a Ramsar site, which encourages sustainable use to maintain wetland functions and biodiversity. The Ramsar Convention, an international treaty which provides the framework for international cooperation for the conservation and wise use of wetlands (Frazier, 1996), was ratified by Madagascar in 1998. In addition, a recent regional fair at Alaotra was given the slogan *Zetran'Alaotra* ("marshes of Alaotra") and both the Préfet (the most important regional government administrator) and the Ministre d'Agriculture made speeches supporting the maintenance of areas of marshes in order to safeguard the future productivity of the region. The emblem of the fair, and of the ARC federation, and of many of the village festivals, was a *bandro* sitting in a clump of papyrus. The Lac Alaotra Bamboo Lemur has truly become a flagship for the conservation of an entire wetland ecosystem.

CONCLUSIONS

The projects and initiatives described above demonstrate that lemurs have played an important role in drawing international attention to Madagascar and to the conservation of its biological diversity. This has, in turn, encouraged the national government to take conservation measures such as the creation of protected areas. More recently, several projects have demonstrated that lemurs can also help to draw attention and support at a very local level, a type of support without which a conservation project cannot succeed in Madagascar. These projects demonstrate that the use of flagship species does not mean exclusive concentration on lemurs and their welfare at the cost of other members of the surrounding ecosystem. Lemurs can successfully be used as a means of catalyzing local initiatives, leading to more general conservation of biodiversity and of ecosystem functions that are of great benefit to local people.

ACKNOWLEDGMENTS

I would like to pay tribute to the motivation and efforts of my Malagasy colleagues, in particular Lala Jean Rakotoniaina, Jonah Randriamahefasoa, and Hasina Randriamanampisoa, which have led to the impressive impact of the education and awareness component of the JWPT Alaotra project. We are very grateful for the support given to this project by the British Ambassador to Madagascar, His Excellency Robert Dewar, and for funding provided by the British Government. I would also like to thank Lee Durrell, Frank Hawkins, Steve Goodman, Joel Ratsirarson, and Urs Thalmann for the helpful comments they made on an earlier draft of this paper. Frank Hawkins kindly assisted with preparation of the map and Hasina Randriamanampisoa translated the abstract.

REFERENCES

The unpublished reports cited in this chapter can be accessed via the Jersey Wildlife Preservation Trust, BP 8511, Antananarivo (101), Madagascar.
ANDREWS, J. 1998. Black Lemur Forest Project. BLFP, Nosy Be, Madagascar, 5 pp.
BRITT, A., C. R. WELCH, AND A. S. KATZ. 1998. First release of captive-bred lemurs into their natural habitat. Lemur News, 3: 8–11.
BUTYNSKI, T. M., S. E. WERIKHE, AND J. KALINA. 1990. Status, distribution and conservation of the Mountain Gorilla in the Gorilla Game Reserve, Uganda. Primate Conservation, 11: 31–41.
CHEPISI, F., AND R. YOUNG. 1997. Fierce Creatures. Sony Pictures Entertainment, USA/UK, film.
CUVELIER, A. 1996. Problems and ways of improving forest exploitation in Madagascar, pp. 133–148. In Ganzhorn, J. U., and J.-P. Sorg, eds., Ecology and Economy of a Tropical Dry Forest in Madagascar. Primate Report, 46–1: 1–382.
DIETZ, J. M., L. A. DIETZ, AND E. Y. NAGAGATA. 1994. The effective use of flagship species for conservation of biodiversity: the example of lion tamarins in Brazil. In Olney, P. J. S., G. M. Mace, and A. T. C. Feistner, eds., Creative Conservation: Interactive Management of Wild and Captive Animals. Chapman & Hall, London.
DURBIN, J. C., AND S.-N. RATRIMOARISAONA. 1996. Can eco-tourism make a major contribution to the conservation of protected areas in Madagascar? Biodiversity and Conservation, 5: 345–353.
DURRELL, L., AND J. MALLINSON. 1987. Reintroduction as a political and educational tool for conservation. Dodo, 24: 6–19.
EIU. 1997. Mauritius, Madagascar, Seychelles Country Report. 4th quarter 1997. The Economist Intelligence Unit, London.
———. 1998a. Madagascar Country Profile 1998–99. The Economist Intelligence Unit, London.
———. 1998b. Madagascar Country Profile 1997–98. The Economist Intelligence Unit, London, UK.
FANAMBY. 1998. Création d'une réserve forestière régionale pour la préservation de la biodiversité à Anjozorobe: Rapport de diagnostic. Association Fanamby, Antananarivo, Madagascar, 150 pp.
FRAZIER, S. 1996. An overview of the world's Ramsar sites. Wetlands International, Gloucester, UK.
GANZHORN, J. U., AND J.-P. SORG. (eds.). 1996. Ecology and Economy of a Tropical Dry Forest in Madagascar. Primate Report 46–1: 1–382.
HARDIN, G. 1968. The tragedy of the commons. Science, 162: 1243–1248.
ICTE. 1998. Publications, reports, and theses resulting from research at Ranomafana National Park, Madagascar (update October 26, 1998). Institute for the Conservation of Tropical Environments, Antananarivo, Madagascar, 25 pp.
KATZ, A. 1998. Parc Ivoloina Update. MFG Member News, **July 1998**: 2.

KERSHAW, J., AND J. CLEESE. 1998. Born to be Wild—Operation Lemur with John Cleese. WNET/BBC, UK, film.
MACPHEE, R. D. E., AND D. A. BURNEY. 1991. Dating of modified femora of extinct dwarf *Hippopotamus* from southern Madagascar: implications for constraining human colonisation and vertebrate extinction events. Journal of Archaeological Science, **18**: 695–706.
MEIER, B., R. ALBIGNAC, A. PEYRIERAS, Y. RUMPLER, AND P. C. WRIGHT. 1987. A new species of *Hapalemur* (Primates) from south east Madagascar. Folia Primatologica, **48**: 211–215.
MEIER, B., A. LONINA, AND T. HAHN. 1996. Expeditionsbericht Sommer 1995: Schaffung eines neuen Nationalparks in Madagaskar. Zeitschrift des Kölner Zoo, **39**: 61–72.
MINISTÈRE DE L'ENVIRONNEMENT. 1997. Convention sur la Diversité Biologique; Premier Rapport National. Ministère de l'Environnement, vi + 48 pp.
MINISTÈRE DU TOURISME. 1998. Depliant d'informations statistiques du tourisme 1997. Ministère du Tourisme, Antananarivo, Madagascar, 2 pp.
MITTERMEIER, R. A., I. TATTERSALL, W. R. KONSTANT, D. M. MEYERS, AND R. B. MAST. 1994. Lemurs of Madagascar. Conservation International, Washington, D.C.
NELSON, R., AND N. HORNING. 1993. AVHRR-LAC estimates of forest area in Madagascar, 1990. International Journal of Remote Sensing, **14**: 1463–1475.
NICOLL, M. E., AND O. LANGRAND. 1989. Madagascar: revue de la conservation et des aires protégées. WWF, Gland.
ONE. 1997. Ce qu'il faut savoir sur la GELOSE; Gestion Locale Sécurisée des Ressources Renouvelables. Office Nationale pour l'Environnement, Antananarivo, Madagascar, 26 pp.
OSTROM, E. 1990. Governing the commons: the evolution of institutions for collective action. Cambridge University Press, Cambridge.
PIDGEON, M. 1996. An ecological survey of Lake Alaotra and selected wetlands of central and eastern Madagascar in analysing the demise of Madagascar Pochard *Aythya innotata*. Wilmé, L., ed. Missouri Botanical Garden and World Wide Fund for Nature, Antananarivo, Madagascar.
PLI (Projet Lutte Intégrée). 1990. Protection intégrée en rizières à Madagascar: possibilités et limites. Ministère de la Recherche Scientifique et Technologique pour le Développement et Ministère de la Production Agricole et du Patrimoine Foncier, Antananarivo, Madagascar, 31 pp.
RAKOTONDRATSIMA, M. 1998. Méthodologie de suivi des lémuriens diurnes dans la peninsule de Masoala. Wildlife Conservation Society, Antananarivo, Madagascar, 3 pp.
RAKOTONDRAVONY, D., AND S. M. GOODMAN. 1998. Inventaire biologique, forêt d'Andranomay, Anjozorobe. Recherches pour le Développement, Série Sciences Biologiques, **13**: 1–110.
RAZANAHOERA, M. R. 1998. Les lémuriens, pp. 94–96. *In* Rakotondravony, D., and S. M. Goodman, eds., Inventaire biologique, forêt d'Andranomay, Anjozorobe. Recherches pour le Développement, Série Sciences Biologiques, **13**: 1–110.
RICHARD, A. F., AND S. O'CONNOR. 1997. Degradation, transformation and conservation: the past, present, and possible future of Madagascar's Environment, pp. 406–418. *In* Goodman, S. M., and B. D. Patterson, eds., Natural Change and Human Impact in Madagascar. Smithsonian Institution Press, Washington, D.C.
SIMBERLOFF, D. 1998. Flagships, umbrellas and keystones: is single-species management passé in the landscape a era? Biological Conservation, **83**: 247–257.
WORLD BANK. 1996. Staff Appraisal Report: Madagascar Second Environment Program. World Bank, Washington, D.C.
WRIGHT, P. C. 1992. Primate ecology, rainforest conservation, and economic development: Building a national park in Madagascar. Evolutionary Anthropology, **1**: 25–33.
ZIMMERMANN, E., S. CEPOK, N. RAKOTOARISON, V. ZIETEMANN, AND U. RADESPIEL. 1998. Sympatric mouse lemurs in north-west Madagascar: A new rufous mouse lemur species (*Microcebus ravelobensis*). Folia Primatologica, **69**: 106–114.

ABOUT THE EDITORS

Madame Berthe Rakotosamimanana is Professor in the Département de Paléontologie et d'Anthropologie Biologique at the Université d'Antananarivo, Faculté des Sciences. She served as the Secretary General of the XVII International Primatology Society (IPS) Congress held in Antananarivo between 10 and 14 August 1998. For several decades she has been one of the pivotal primatologists on Madagascar and has helped untold numbers of national and foreign researchers working on the island. Her research interests include paleontology and functional morphology of primates.

Hantanirina Rasamimanana is Professor at the Ecole Normale Supérieure of the Université d'Antananarivo. She was responsible for the scientific program of the IPS held in Antananarivo. For over a decade she has been studying the social behavior of the Ring-tailed Lemur in the Berenty Reserve in southern Madagascar.

Jörg U. Ganzhorn is Professor at the University of Hamburg, Germany, where he is a member of the Zoological Institute and Museum. He has been working on the community ecology of lemurs for nearly 15 years and has published extensively on this subject. He is co-editor of another Plenum Press book on lemurs entitled "Lemur social systems and their ecological basis" published in 1993.

Steven M. Goodman is Field Biologist at the Field Museum of Natural History, Chicago, and Coordinator of the Ecology Training Program of World Wide Fund for Nature, Antananarivo. Over the course of the past ten years he has been studying aspects of the natural history of the island. He also serves on the faculty of the Université d'Antananarivo and resides most of the year in Madagascar.

INDEX

Abutilon
  pseudocleistoganum, 218
Acacia, 218
  delbeata, 215
  rovumae, 179, 218
Académie Malgache, 21
Acalypha, 217
Acanthaceae, 214, 216
Accipiter, 133
  henstii, 78
ACCTRAN, 58
activity budgets, 98, 225
Adansonia, 197
  rubrostipa, 132
ADF content, 231
adipose tissue, 100
aDNA, 4
Aerva
  madagassica, 216
Aframomum
  angustifolium, 216
Agavaceae, 216
Agave
  sisalana, 179, 180
age, 164
Ageratum
  conyzioides, 204
aggression
  targeted, 110, 111
Ailuropoda
  melanoleuca, 231
Albizzia, 196
  chinensis, 215
  gummifera, 215
  polyphylla, 218
Alectroenas
  madagascariensis, 191
alkaloids, 187, 208, 209, 210, 212
Allocebus, 21
  trichotis, 24, 56, 271
Allophyllus
  cobbe, 216
Alluaudia

Alluaudia (cont.)
  ascendens, 217
  procera, 217
Aloe
  capitata, 218
  vahombe, 180, 218
Alouatta, 233, 235
  fusca, 235
  palliata, 120, 161, 235
  pigra, 233
  seniculus, 192
Amaranthaceae, 216
Ambararata, 32
Ambatondrazaka, 278
Ambatovaky, 24, 34
Ambohitantely, 24, 31, 34, 35, 45, 47
Ambolisatra, 30, 34, 46
amino acid
  content, 232
  racimization, 4
Amparihingidro, 40
Ampasambazimba, 22, 24, 27, 28, 29, 30, 31, 33, 34, 35, 37, 38, 39, 41, 42, 46, 47, 48, 66
Ampijoroa, 141, 251, 272, 274
Ampoza, 30, 31, 32, 41, 45, 46, 48
Anacardiaceae, 196, 214, 216
Anacardium, 132
Analamazaotra, 24, 34, 272
Analamera, 24, 34, 35
Analavelona, 29, 48
Anas
  melleri, 278
Anavoha, 32, 40
Andasibe, 209, 272
Andetobe, 28
Andohahela, 23, 24, 32, 34
Andrafiabe, 26, 28
Andrahomana, 24, 41, 46
Andranomena, 24, 34
Andreba, 223
Andringitra, 24, 29, 34
Anjamena, 133
Anjanaharibe-Sud, 24, 34

Anjohibe, 22, 24, 28, 29, 34, 38, 40, 41, 42, 45, 46, 48
Anjohikely, 22
Anjohin'olona, 28
Anjozorobe, 272
Ankarafantsika, 29, 34, 35, 44, 251, 272, 274
Ankarana, 20, 21, 22, 24, 26, 28, 29, 33, 34, 35, 38, 41, 42, 48
Ankaratra, 46
Ankazoabo, 11
   Cave, 24, 32, 33
Ankilitelo, 21, 22, 24, 26, 28, 32, 33, 34, 35, 38, 39, 41, 42, 46
Ankoatra, 28
Annona, 216
Annonaceae, 195, 214, 216
annual cycle, 99
Antananarivo, 31
Antenoaka, 28
Anthocleista, 215
   madagascariensis, 215
   rhizophoroides, 215
antipredator behavior, 133
Antsirabe, 37, 46
Antsiroandoha, 26, 28
Antsohihy, 29
Aotus
   trivirgatus, 120, 122, 204, 208, 235
Aphloia
   theiformis, 195, 215
Apocynaceae, 196, 214, 216
Aquifolicaeae, 214
Araliaceae, 196, 214
Archaeoindris, 2, 3, 6, 26
   fontoynontii, 38
Archaeolemur, 2, 3, 6, 10, 38, 41, 46, 197
   edwardsi, 38, 41, 46
   majori, 38, 41, 46
Arecaceae, 195
Argyreia
   vahibora, 229, 230, 232
Aristolochia, 216
Aristolochiaceae, 216
Artabotrys, 214
Artiodactyla, 257
Asclepiadaceae, 216, 217
ash, 231
Association Nationale pour le Gestion des Aires Protégées (ANGAP), 273, 274, 275
Asteraceae, 210, 214, 217, 229
Ateles
   geoffroyi, 161
   paniscus, 204
Avahi, 25, 28, 44, 134, 140, 222, 233, 235
   laniger, 26, 27, 35, 44, 45, 70, 231, 233, 234, 235, 271
   occidentalis, 44, 234, 251, 271
Aythya
   innotata, 243, 247, 278

Azadirachta
   indica, 179, 180, 218
Azima
   tetracantha, 179, 180, 218

Babakotia, 3, 6
   radofilai, 38, 41, 46
banana, 180
bandro, 278, 279
basal metabolic rate (BMR), 121, 132, 222, 233
Bauhinia, 217
   grandidieri, 217
Bay of Antongil, 29, 45
behavioral threshold, 203
Beloha, 39
Belo-sur-Mer, 32, 41, 46
Bemaraha, 44
Berenty, 24, 25, 34, 35, 175, 187, 209, 274
Betafo, 46
Betampona, 24, 34, 273
Betsiboka River, 29
Bevoha, 11
Beza Mahafaly, 24, 34, 35
Bignoniaceae, 217
biological clock, 98, 102
bipedalism, 73
Black Lemur Project, 276
Blotia, 215
Boerhavia
   diffusa, 180
Bombaceae, 193, 197
Boraginaceae, 217
Boscia
   longifolia, 217
Bosqueia, 215
Bougainvillaea
   spectabilis, 218
Brachyteles
   arachnoides, 235
Breonia, 195
   perrieri, 132, 133
Bridelia
   tulasneana, 215
Bristol Zoological Gardens, 244
Brunfelsia
   hopeana, 180
Buddleia, 215
Burasaia, 195
Burseraceae, 195, 214, 217
Buteo
   brachypterus, 133
Byttneria, 218

Cabada
   virgata, 217
Cactaceae, 217
cactus, 180

Index

Caesalpinia
  bonduc, 217
  pulcherima, 179, 180
Caesalpiniaceae, 217
caffein, 208
calcium, 231
Callimico
  goeldii, 206, 208
Callithrix
  argentata, 208
  geoffroyi, 208
  jacchus, 204, 206, 208
Calophyllum, 195
Camellia
  thaeiformes, 216
Campylospermum
  anceps, 215
  lanceolatum, 215
Canarium
  madagascariense, 195
Canthium, 216
Capparidaceae, 217
Capparis, 217
  seppiaria, 180, 217
captive management, 245
Carex, 214
Carissa
  edulis, 214
Carnivora, 257
Cassia, 217
  siamea, 217
Cassipourea, 216
  gummiflua, 196
Casuarina, 180
Catharanthus
  roseus, 216
cathemeral activity, 119–134, 223, 233, 235
cattle pasture, 276
Cave of the Lone Barefoot Stranger, 28
Cebuella
  pygmaea, 161, 204, 208
Cebus
  apella, 133, 204
Cedrolopsis
  grevei, 218
Ceiba
  pentandra, 121, 193
Celastraceae, 217
Celtis
  bifidis, 179, 180, 219
  philippensis, 179, 180, 219
Centre de Formation Professionelle Forestière (CFPF), 71, 162, 274
Ceratogymna, 192
Cercopithecus
  ascanius, 161
  nictitans, 204
  pygerythrus, 204

Ceropegia, 217
Cervus
  elaphus, 161
Cetacea, 257
Cheirogaleidae, 20, 161, 271
Cheirogaleus, 9, 11, 12, 13, 58, 61, 62, 63, 64, 65, 85, 89, 139, 140
  major, 20, 27, 44, 48, 57, 58, 62, 140, 204, 271
  medius, 27, 46, 57, 58, 62, 139–154, 159–170, 190, 204, 208, 251, 271
chemical
  scores, 231
Chiroptera, 257
Chrysobalanaceae, 196
Chrysophyllum
  bovinianum, 195
  sambiranensis, 195
cinchona bark, 208
circadian activity, 121
Cissampelos, 180
Cissus
  quadrangularis, 219
Citrullus
  vulgaris, 180
Clerodendron, 216, 219
Clidemia
  hirta, 215
climbing, 73
clinging, 73
Clitoria
  heterophylla, 218
Clusiaceae, 195
Coffea, 216
Colobus, 233, 235
  guereza, 231, 235
  polykomos, 231, 234
  satanas, 231, 234, 235
Colubrina, 218
combat, 167
Combretaceae, 193, 195, 217
Combretum, 217
Commelina, 217
Commelinaceae, 217
Commicarpus
  commersonii, 180, 218
Commiphora, 217
convergence, 21
Convolvulaceae, 217, 229
coprophagy, 223
Cordia, 217
  rothii, 179, 180, 217
  varo, 179, 180
Corydala
  madagascariensis, 195
Crassulaceae, 217
Crateva, 217
  excelsa, 175, 179, 180
  greveana, 217

Crossandra
 poissonii, 216
Croton, 217
 mongue, 215
Cryptocarya, 195
Cryptoprocta
 ferox, 78, 133
Cucurbitaceae, 217
Cuscuta
 sinensis, 229, 230, 232
Cussonia, 214
Cyclosorus
 gongylodes, 229, 230
Cynanchum, 217
Cyperaceae, 227, 229
Cyperus, 223, 227, 228, 229, 230
 madagascariensis, 227, 232
cytochrome *b*, 9, 11

daily path length, 123
Dalbergia, 132, 196
 monticola, 216
Dama
 dama, 161
Danais, 216
Daraina, 272
Daubentonia, 6, 9, 11, 12, 13, 37, 133
 madagascariensis, 3, 26, 27, 28, 44, 56, 271, 272
 robusta, 3, 37, 38, 39, 46
Daubentoniidae, 20, 271
day-length, 99
Delonix
 regia, 217
demographic structure, 164
Dennstaedtiaceae, 216
Dermoptera, 257
Dianella, 215
 ensifolia, 215
Dichaetanthera, 215
 heteromorpha, 196
 oblongifolia, 215
Dichrostachys
 tenuifolia, 215
Dicranopteris
 linearis, 216
Dicrostachys, 218
Didiereaceae, 209, 217
diet, 127
digestive specialization, 222
Dioscorea, 214
 fandra, 217
 nako, 217
Dioscoreaceae, 214, 217
Diospyros
 clusiifolia, 195
 gracipes, 196
disappearence rate, 165
dispersal, 151

distance effect, 36, 44
Dja Reserve, Cameroon, 192
DNA
 contamination, 4, 6
 polymerase, 4
 preservation, 4
 primer sequences, 4
Dombeya, 216
doubly labeled water, 84
Dracaena, 195
 reflexa, 195
Duke University Primate Center, 22, 243, 273
Dypsis
 ampasindavae, 195
 madagascariensis, 195
 nossibensis, 195

Ebenaceae, 195, 196, 214
Echinochloa, 231
 crusgalli, 227, 229, 230, 232
Edinburgh Zoo, 244
Eichhornia, 231
 crassipes, 229, 230, 232
Eidolon
 dupreanum, 191
Emila, 214
 humifusa, 214
energetic costs, 77, 223
energy gross, 234
Enterospermum, 216, 218
 pruinosum, 180
Eocene, 6
Ericaceae, 214
Erythroxylaceae, 214
Erythroxylum
 nitidulum, 214
estrous cycling, 111
Ethulia, 231
 conyzoides, 229, 230, 232
Eucalyptus, 121, 122, 180, 215, 218
Eugenia, 215
 goviala, 215
Eulemur, 9, 11, 55, 56, 59, 61, 62, 63, 64, 65, 66, 110, 112, 121, 122, 132, 133, 235, 245
 coronatus, 26, 27, 28, 56, 57, 59–60, 62, 110, 271
 fulvus, 25, 26, 27, 28, 29, 30, 35, 36, 56, 57, 59–60, 62, 65, 72–78, 94, 110, 121, 131, 187, 190, 197, 204, 210, 212
  albifrons, 120, 271, 273
  albocollaris, 271
  collaris, 12, 13, 271
  fulvus, 44, 271
  mayottensis, 235
  rufus, 26, 44, 69–70, 71, 93–94, 96, 98, 119–120, 121, 122–134, 154, 271
  sanfordi, 26, 271
 macaco, 24, 56, 57, 59–60, 62, 65, 110, 121, 129, 131, 132, 161, 189, 190, 204, 227

Eulemur (*cont.*)
  macaco (*cont.*)
    flavifrons, 12, 13, 271, 272
    macaco, 12, 13, 190–197, 271
    mongoz, 12, 13, 27, 29, 30, 45, 48, 56, 57, 59–60, 62, 110, 121, 122, 129, 131, 132, 133, 204, 251, 271
    rubriventer, 12, 13, 24, 56, 57, 59–60, 62, 78, 121, 129, 132, 190, 235, 271
Euphorbia, 217
Euphorbiaceae, 195, 196, 215, 217
European Endangered Species Programme, 244
Evonymiopsis
  longipes, 217
exotic fish, 246

Fabaceae, 192, 196
fat, 109
  deposition, 113
fatness, 101, 113
female
  aggressiveness, 175
  dominance, 95, 96, 109, 110, 113, 174, 186
  need, 95
  strategies, 186
Fenerive-Est, 197
Fernandoa
  madagascariensis, 217
fertilizers, 278
fetal growth, 96
fiber
  content, 231, 234
  intake, 131
Ficus, 132, 133, 215, 218
  assimilis, 192, 195
  cocculifolia, 218
  grevei, 218
  megapoda, 218
  trichopoda, 218
Filicium
  decipiens, 216
fires, 276
fishing, 245
Flacourtia
  indica, 218
Flacourtiaceae, 195, 196, 215, 218
floaters, 152, 161, 162, 166
flower production, 107
Flueggea
  obovata, 180
food
  competition, 174
  intake, 103, 179
  quality, 222
  resources, 185
forest structure, 73
Fox's rule, 47
fructose, 203

Gaertnera
  macrostipula, 216
Galago, 12, 13
  crassicaudatus, 153
  senegalensis, 204, 208
  zanzibaricus, 141, 151, 153
Galerie des Gours Secs, 28
gallotannin, 206
Gambeya
  boiviniana, 216
Garcinia, 195
  chapeleri, 215
geophagy, 210
Gleicheniaceae, 216
glucose, 203
Gluta
  tourtour, 196
Gorge d'Andavakoera, 23
Gorilla, 233
  gorilla, 161, 235
    berengei, 231, 270
    gorilla, 234
Graminaceae, 245
Grangeria
  porosa, 196
Grewia, 132, 195, 218
  cyclea, 132, 133
  grevei, 218
Grotte de la Cassure des Arcades, 28
Grotte de la Forêt Isolée, 28
Grotte Nord de la Cassure de Milaintety, 28
Grotte de Matsaborimanga, 28
growth rate, 100, 105
guanidine thiocyanate, 8
Guttiferae, 215

Hadropithecus, 2, 3, 6
  stenognathus, 38, 40, 46
hair growth, 101
Hapalemur, 12, 13, 55, 56, 59, 61, 62, 63, 64, 65, 66, 110, 132, 133, 222, 223, 231, 233, 236, 244
  aureus, 24, 56, 58, 231, 271, 272
  griseus, 22, 26, 27, 28, 44, 56, 57, 59, 62, 65, 121, 133, 236, 244
    alaotrensis, 161, 221–236, 241–247, 271, 278
    griseus, 231, 234, 235, 271
    occidentalis, 271
  simus, 21, 25, 26, 27, 28, 29, 35, 45, 46, 48, 56, 57, 59–60, 62, 65, 121, 204, 231, 271
Harungana
  madagascariensis, 215
Hazunta
  modesta, 216
Hedychium
  coronarium, 216
Hibiscus, 218

Hildebrandtia
  promontorii, 217
  valo, 217
hindgut fermentation, 236
Hippocratea, 180, 218
Hippocrateaceae, 218
Holocene, 9, 20, 21
Homalium
  micranthum, 196
home-range size, 123, 145
Homo sapiens, 204
hormonal change, 109
hormone levels, 103
hormones, 113
Hylobates
  klossii, 161
Hypericaceae, 215
Hypoestes, 216
Hypsipetes
  madagascariensis, 191
Hyracoidea, 257

IGF-1, 100, 107
Ilex
  mitis, 214
Indri, 30, 31, 48, 110
  indri, 26, 27, 28, 29, 30, 31, 35, 45, 46, 48, 231, 234, 271, 272, 273
Indriidae, 3, 20, 113, 235, 271
infanticide, 112
Insectivora, 257
insulin-like growth factor, 1, 98, 100
International Primatological Society, 275
interspecific competition, 77
Ipomoea
  cairica, 217
Isalo, 24, 34, 35
Isolona, 195

Jaccard's index, 25–26
Jersey Wildlife Preservation Trust (JWPT), 242, 277–279
Jussiaea, 215

Kalanchoe
  beauverdii, 217
  beharensis, 217
Ketanest, 86
Kianjavato, 29
Kibale Forest, Uganda, 192, 197
Kirindy Forest, 34, 45, 71, 78, 122, 162, 274, 275
Kjeldahl method, 225

Laboratoire de Paléontologie des Vertébrés, 21, 22, 37
lactation, 113
Lagomorpha, 257

Lake Alaotra, 222, 223, 242, 277
Lake Mahery, 46
Lake Tritrivakely, 46
Lamboharana, 40
Lantana
  camara, 216
Lauraceae, 195, 215
Lautembergia, 196, 215
  coriacea, 196
leaf production, 107
leaping, 73
Leersia, 231
  hexandra, 227, 229, 230, 232
Lemur, 12, 13, 55, 56, 59, 61, 62, 64, 66, 110, 112
  catta, 26, 27, 56, 57, 59–60, 62, 64, 93–94, 97, 98, 110, 173–187, 210, 212, 244, 271
Lemuridae, 3, 20, 56–57, 65, 66, 93–114, 271
Lemuroidea, 56
Lemurs
  extinct, 2
  extinction, 197
  subfossil, 2, 20
Leontipithecus, 272
  chrysomelas, 208
  rosalia, 208
Lepilemur, 1, 2, 3, 9, 11, 12, 13, 14, 25, 28, 33, 46, 48, 58, 61, 63, 65, 89, 233, 235
  dorsalis, 27, 271
  edwardsi, 22, 27, 33, 70, 146, 234, 251, 271
  globiceps, 34
  leucopus, 20, 27, 33, 34, 46, 57, 58, 231, 235, 271
  microdon, 25, 271
  mustelinus, 25, 33, 57, 58, 234, 271
  ruficaudatus, 27, 33, 34, 83–84, 85–89, 233, 271
  septentrionalis, 26, 27, 271
Lepilemuridae, 20
Leptadenia, 216
  madagascariensis, 216
Leucaena
  glauca, 218
Liliaceae, 195, 215, 218
locomotion, 70
Loesneriellia, 218
Loganiaceae, 195, 215, 218
Lokobe Forest, 190
Loky River, 32
London Zoo, 244
Loris, 12, 13
  tardigradus, 204, 208
Lorisids, 96
Loxodonta
  africana, 192
Ludia
  scolopioides, 195
lunar eclipse, 131
Lygopodium
  lanceolatum, 216

Macaca
  fascicularis, 206
  fuscata, 161
  mulatta, 208
Macaranga, 196, 215
  alnifolia, 215
  ankafinensis, 215
  obovata, 215
Macarisia
  lanceolata, 196
MacClade, 58
Macphersonia
  madagascariensis, 195
  radlkoferi, 195
Macroscelididae, 257
Madagascar Faunal Group, 273
Maerua
  filiformis, 180, 185, 217
Mahabobaka, 29
Mahajanga Plateau, 20
Mahavavy River, 29
Maillard reaction, 5
male subordination, 109
Malipighiaceae, 218
Malvaceae, 215, 218
Mammea, 215
Manambato River, 32
Manamby Plateau, 20
Mangoky River, 23, 33
Manombo (Farafangana), 24, 34
Manombo (Toliara), 28, 30, 45, 46, 48
Manongarivo, 24, 34
Mantady, 24, 34
Marojejy, 24, 34
Marsdenia, 217
Masinandraina, 37
Masoala Peninsula, 45, 273
mate-guarding, 168
mating tactics, 161
Medinilla, 215
  occidentalis, 215
megafauna, 21
Megaladapidae, 271
Megaladapis, 1, 2, 3, 6, 7, 10, 11, 12, 13, 14, 21
  grandidieri, 38, 41
  madagascariensis, 21, 38, 41
Megistostegium
  nodulosum, 218
Melastomataceae, 196, 215
Melia
  azedarach, 179, 180, 218
Meliaceae, 218
Melospiza
  melodia, 161
Menispermaceae, 195, 218
Merremia
  tridentata, 214
Mesopropithecus, 3, 6

Mesopropithecus (cont.)
  dolichobrachion, 38, 41
  globiceps, 38, 41, 46
  pithecoides, 38, 41, 46
metabolic
  chambers, 84
  demands, 185
  rate, 84, 113, 233
  strategy, 108, 112
Microcebus, 12, 13, 24, 25, 26, 28, 35, 85, 89, 190, 271
  murinus, 45, 46, 153, 168, 169, 170, 201, 202, 204, 206, 207, 208, 271
  myoxinus, 45, 56, 271
  ravelobensis, 56, 271
  rufus, 35, 45, 271
Mikoboka Plateau, 22
Mimosaceae, 215, 218
Ministère des Eaux et Forêt (MEF), 243, 273
Mirounga
  angustirostris, 161
Mirza, 12, 13, 21
  coquereli, 24, 204, 271
molecular studies, 245
monellin, 204
Monimiaceae, 195, 215
monogamy, 141, 152
monophyletic assemblages, 3
Montagne d'Ambre, 24, 34, 35, 45
Montagne des Français, 22, 23, 29, 34, 35, 38, 41, 42, 48
Monty Python, 274
Moraceae, 192, 195, 215, 218
Morondava, 29, 33, 46, 71, 209
morphological data, 3
mortality, 165
Mozambique Channel, 3
Mulhouse Zoo, 244
Mundulea
  scoparia, 218
Mussaenda, 196
Myrica
  spathulata, 215
Myricaceae, 215
Myrsinaceae, 215
Myrtaceae, 195, 215, 218

Namoroka, 24, 34, 44
National Environmental Action Plan (NEAP), 276
Neodypsis
  decaryii, 180
Neotina
  isoneura, 179, 180, 218
nest hole, 163
nestedness, 21
Nettapus
  auritus, 247
nightly path length, 123

non-convergence, 21
Noronhia, 132, 195, 216
Nosy Be, 190, 276
Nosy Mangabe, 272
Nourages, French Guiana, 192
nutrient
  partioning, 107
  storage, 109
Nyctaginaceae, 218
Nycticebus, 12, 13, 85
  coucang, 204, 208
Nymphaceae, 229
Nymphaea
  lotus, 229, 230, 232
  stellata, 247

Ochnaceae, 215
Ocotea, 195, 215
  similis, 215
Oenotheraceae, 215
Oleaceae, 195, 216
Oncostemum, 215
Onilahy River, 33, 46
Operculicarya
  decaryi, 216
Opuntia, 180
  vulgaris, 180, 217
orange, 180
Otolemur, 96
over-grazing, 276

Pachylemur, 3, 6, 38, 41, 56, 64, 66
  insignis, 41, 56
  jullyi, 41, 56, 57
Pachytrophe
  dimepate, 215
Palaeopropithecus, 1, 2, 3, 6, 7, 10, 11, 12, 13, 14, 38, 40
  ingens, 21, 26, 38
  maximus, 38
palynology, 45
Pan
  troglodytes, 192, 208
Pandanaceae, 195, 216
Pandanus, 216
  androcephalanthus, 195
papaya, 180
Papilionaceae, 216, 218
Papio
  anubis, 234
  cynocephalus, 161
Parc Zoologique d'Ivoloina, 244, 274, 277
Parc Botanique et Zoologique de Tsimbazaza, 176, 244, 277
Parkia
  madagascariensis, 192, 193, 195
Paropsia
  obscura, 196

parsimony analysis, 12, 13, 62
Parus
  atricapillus, 161
Passiflora
  foetida, 216
  incarnata, 216
Passifloraceae, 196, 216
paternal investment, 109
PAUP, 6, 58
Peloriadapis, 41
Pennisetum
  purpureum, 180
Peponidium
  horridum, 195
Périnet-Analamazaotra, 272
Perissodactyla, 257
Pervillea
  decaryi, 217
pesticide, 246, 278
Phaner, 21
  furcifer, 24, 45, 141, 204
    electromontis, 271
    furcifer, 271
    pallescens, 271
    parienti, 271
phenology, 107, 113
phenols, 187, 210
Philippia, 214
Pholidota, 257
photoperiod, 98, 99, 111
Phragmites, 223, 227, 231
  communis, 227, 229, 230, 232
  mauritianus, 218
Phyllanthus
  casticum, 217
phylogenetic
  analysis, 3
  relationships, 3
phylogeny
  Lemuridae, 55–66
Physena
  sessiliflora, 218
Pithecellobium
  dulce, 218
pituitary growth hormone, 100
Plagioscyphus, 195
plants
  introduced, 187
  secondary compounds, 209
Platalea
  alba, 247
Plectaneia, 216
Pleistocene, 20
Plesiorycteropus
  madagascariensis, 37
Poaceae, 218, 227, 229
Polyalthia
  richardiana, 195

Polyboroides
  radiatus, 78, 133
polygamous, 141
Polygonaceae, 229
Polygonum, 228
  baldschaunicum, 244
  glabrum, 229, 230, 232, 244
Polymerase Chain Reaction (PCR), 3, 4, 10, 11, 66
  amplifications, 10
  inhibitors, 5
Polypodiaceae, 229
Polyscias, 214
  nossibensis, 196
Pontederiaceae, 229
population density, 165
Poupartia
  caffra, 216
  minor, 216
  sylvatica, 132
Presbytis, 233, 234
  aygula, 235
  entellus, 161
  pileata, 235
  senex, 231
Primatology
  teaching, 249–267
Proboscidea, 257
Propithecus, 11, 12, 13, 30, 31, 33, 58, 61, 63, 97, 110
  diadema, 27, 28, 31, 32, 45, 48, 57, 58, 78
    candidus, 271
    diadema, 32, 112, 271, 273
    edwardsi, 78, 271
    perrieri, 26, 23, 271
  tattersalli, 12, 13, 26, 27, 28, 32, 48, 56, 57, 133, 271, 272
  verreauxi, 27, 28, 31, 32, 35, 45, 46, 48, 58, 72–78, 190, 201, 202, 204, 210, 211, 235, 271
    coquereli, 12, 13, 271
    coronatus, 251, 271
    deckeni, 271
    verreauxi, 69–70, 71, 112, 122, 133, 271
    verreauxoides, 32
protein
  highest crude, 231, 234
protein-to-fibre ratio, 231, 234
Protium
  madagascariense, 195
Protorhus
  ditimena, 214
  thouvenotii, 214
provisioning, 99, 103
Psiadia
  altissima, 214
Psidium
  cattleyanum, 215
  guayava, 215
Psorospermum
  androsaemifolium, 215

Pteridium
  aquilinum, 216
Pteropus
  rufus, 191
Pyrostria, 216

quadrupedalism, 73
quinine, 208
Quisivianthe
  papinae, 179, 180, 218

radio tracking, 87, 163
radio-collared animals, 142
radio-telemetry, 87
Ramsar, 279
Ranomafana, 24, 29, 34, 78, 272
Ratsirakia
  legendrei, 247
Ravenala madagascariensis, 196, 216
Ravensara, 215
  crassifolia, 215
  ovalifolia, 215
Red Data Book, 242
regeneration, 197
reproductive stress, 96
Rhamnaceae, 218
Rheocles
  alaotrensis, 247
rheostasis, 105
Rhizophoraceae, 196, 216
Rhodolaena
  bakeriana, 214
rice production, 245, 278
Rinorea
  greveana, 179, 180, 210, 212, 219
Rodentia, 257
Rompun, 86
Rosaceae, 216
Rothmannia, 216
Rotterdam Zoo, 244
Rubiaceae, 195, 196, 216, 218
Rubus
  roridus, 216
  rosaefolius, 216
Rutaceae, 196, 218

Sabicea
  diversifolia, 216
Saguinus
  fuscicollis, 161, 204
  midas, 204, 208
  oedipus, 161, 208
Sahamalaza, 272
Saimiri
  sciureus, 133, 204, 208
Salvadora
  angustifolea, 179, 180, 218
Salvadoraceae, 218

Sambirano, 20, 34, 44, 45
Sanseveria, 216
Sapindaceae, 195, 196, 216, 218
Sapotaceae, 195, 216
Scandentia, 257
Schefflera, 214
Schizaeaceae, 216
Sclerocroton
  melanostictus, 217
Scolopia
  hazomby, 195
sea level change, 3
seasonal physiology, 185
seasonality, 107
Secamone, 217
Securinega
  capuronii, 217
seed dispersal, 190, 197
Senna
  lactia, 196
Setcreasea, 180, 185
sex ratio, 164
Sida
  rhombifolia, 215
Sihanaka, 278
siltation, 246
Sirenia, 257
skeletal morphology, 57
slash-and-burn agriculture, 276
sleeping
  associations, 141, 147
  hole, 142
  sites, 146
Smilacaceae, 216
Smilax
  kraussiana, 216
social organization, 175, 176–177
sodium chloride, 205
soil
  phyllitous, 205
Solanaceae, 216, 218
Solanum, 216
  auriculatum, 216
  batoides, 218
  croatii, 218
speleothem, 45
St. Catherine's Workshop, 242
Stachytarpheta
  jamaicensis, 216
Stephanostegia
  megalocarpa, 196
Sterculia
  perrieri, 196
Sterculiaceae, 196, 216, 218
Sticherus
  flagellaris, 216
Streblus
  dimepate, 195

Streblus (cont.)
  mauritianus, 195
Strelitziacae, 196, 216
strepsirrhine, 11, 250
Strepsirrhini, 56
stress, 104
Strobilanthes, 214
Strychnopsis
  thouarsii, 195
Strychnos, 195, 218
subcutaneous fat, 98, 100
substrate use, 70
sucrose, 203
sugar, 203
  discrimination, 203
support
  feeding, 75
  locomotion, 74
Suregada
  laurina, 215
survivability, 152
Symphonia, 195
  louvelii, 215
  tanalensis, 215

T4, 107
Tachybaptus
  rufolavatus, 247, 278
Tamarindus
  indica, 132, 180, 185, 217
Tamatave, 277
Tambourissa, 195
  purpurea, 215
  trichophylla, 215
Tampolo, 197
tannins, 205, 206, 209, 212
Taq, 4, 5, 6
Tarenna, 132
Tarsius, 141
taste
  bud sensory cells, 203
  discrimination, 201–212
  perception, 203
  physiology, 203
  sampling, 205
  selectivity, 208
  signals, 203
  thresholds, 206
telemetry, 142
Terminalia, 217
  boivinii, 132, 133
  calophylla, 195
  catappa, 193
testis volume, 162
Tetrapterocarpon
  geayi, 217
thaumatin, 203
Thaumatococcus

Thaumatococcus (*cont.*)
　danielii, 204
Theaceae, 216
thermoregulation, 108
thyroxine, 98, 100
Tiliaceae, 195, 218
Toamasina, 277
Tolagnaro, 45
Toliara, 29
torpor, 140
tourism, 274
Treculia
　africana, 195
Trema
　orientalis, 216, 219
Treron
　australis, 191
Tristellateia, 218
Tristemma
　mauritianum, 215
Trophis
　montana, 195
Tsimanampetsotsa, 24, 34, 35
Tsirave, 22, 23, 26, 28, 32, 33, 34, 35, 38, 39, 40, 41, 42, 46, 48, 66
Tubulidentata, 257
Turraea, 218
Twisselman method, 225

Uapaca, 215
　densifolia, 215
　louveli, 195
Ulmaceae, 216, 219
Université de Mahajanga, 251, 253, 254, 262, 267
University of Zurich, 251, 253
Urena
　lobata, 215

Vaccinaceae, 216, 219
Vaccinium, 216
van Soest method, 225
van Soest-Wine method, 225
Varecia, 3, 12, 30, 55, 56, 59, 62, 63, 64, 66, 110
　variegata, 27, 28, 29, 45, 46, 48, 56, 57, 59–60, 61, 62, 78, 97, 190, 244
　　rubra, 12, 13, 271, 273
　　variegata, 12, 13, 271, 273
Verbenaceae, 216
Verezanantsoro, 24, 34
Vernonia, 214, 217
　amygdalina, 209
　pectoralis, 210, 211, 217
Violaceae, 219
Vitaceae, 219
Vohitsara, 279
Vondrozo, 29

Weende method, 225
Weinmannia
　bojeriana, 214
　rutenbergii, 214
World Wide Fund for Nature (WWF), 243

Xenartha, 257
Xerosicyos
　decaryi, 217
Xylopia, 195, 214

Zahamena, 24, 25, 34
Zanthoxylum, 196
Zehneria, 217
Zingiberaceae, 216
Zizyphus, 218
Zombitse, 29, 34, 48
Zonotrichia
　capensis, 161